NELSON

FIT FOR LIFE

HEALTH AND PHYSICAL EDUCATION FOR THE AUSTRALIAN CURRICULUM

LEVELS 9+10

2ND ED

ROB MALPELI
AMANDA TELFORD
CLAIRE STONEHOUSE
LEE ANTON-HEM
MICHAEL SPITTLE
SAM WATKINS
EMMÉ WILD
DAVID BAKKER

Nelson Fit For Life Health and Physical Education for The Australian Curriculum
Levels 9 and 10
2nd Edition
Rob Malpeli
Amanda Telford
Claire Stonehouse
Lee Anton-Hem
Michael Spittle
Sam Watkins
Emmé Wild
David Bakker

Product manager: Sarah Craig/Cathy Beswick-Davison
Content developer: Rachael Pictor
Project editor: Sutha Surenddar
Editor: Anne Mulvaney/MPS Limited
Proofreader: MPS Limited
Production controller: Karen Young/Sutha Surenddar
Permissions/Photo researcher: Liz McShane
Cover designer: Mariana Maccarini
Text designer: Mariana Maccarini
Project designer : Mariana Maccarini
Typeset by: MPS Limited
Cover: Shutterstock.com/Nicetoseeya
MPS Limited

Any URLs contained in this publication were checked for currency during the production process. Note, however, that the publisher cannot vouch for the ongoing currency of URLs.

Acknowledgements
We respectfully refer to Aboriginal and Torres Strait Islander Peoples as First Nations Peoples throughout the book.

ACKNOWLEDGEMENT OF COUNTRY
Nelson acknowledges the Traditional Owners and Custodians of the lands of all First Nations Peoples of Australia. We pay respect to their Elders past and present. We recognise the continuing connection of First Nations Peoples to the land, air and waters, and thank them for protecting these lands, waters and ecosystems since time immemorial.

WARNING:
First Nations Peoples are advised that this book and associated learning materials may contain images, videos or voices of deceased persons.

National Library of Australia Cataloguing-in-Publication Data
A catalogue record for this book is available from the National Library of Australia.
9780170463102

Cengage Learning Australia
Level 5, 80 Dorcas Street
Southbank VIC 3006 Australia

Cengage Learning New Zealand
Unit 4B Rosedale Office Park
331 Rosedale Road, Albany, North Shore 0632, NZ

For learning solutions, visit cengage.com.au

Printed in China by 1010 Printing International Limited.
3 4 5 6 7 26 25 24

CONTENTS

ABOUT THIS TITLE

This book has been written for the Australian Curriculum for Health and Physical Education Years 9 & 10. Each chapter explores the focus areas, with a range of activities and investigations to guide your learning.

Investigations

Investigations scaffold research topics and experiments, and allow you to conduct your own investigation into a topic. Practise your data analysis skills and how to present results with these.

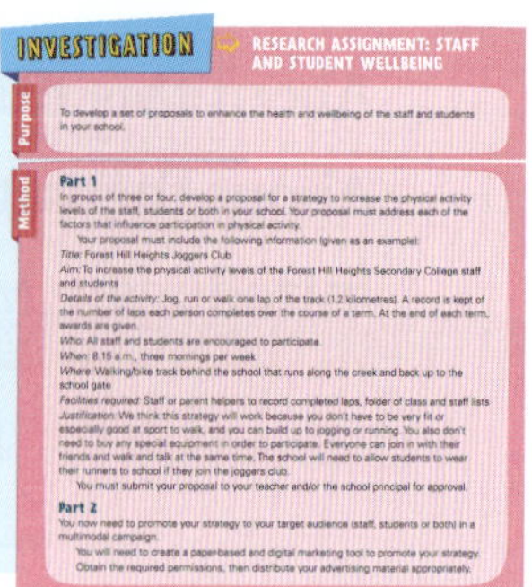

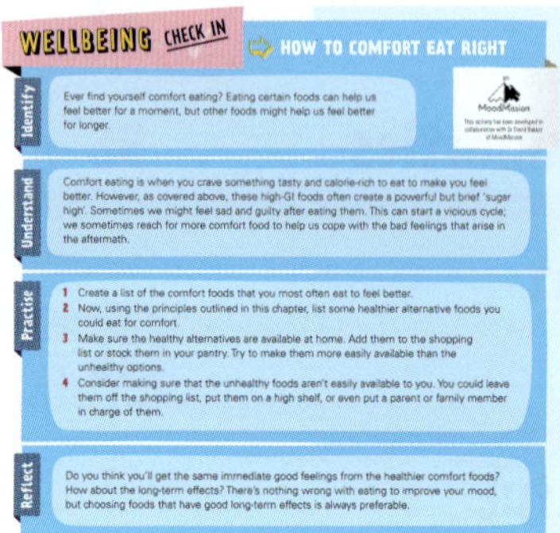

Wellbeing Check-in

In collaboration with Dr David Bakker of MoodMission, Wellbeing check-in activities introduce you to a specific strategy, based on scientific research, that is designed to improve your wellbeing. Practise and use these to discover which strategies work best for you!

Case studies

New and updated case studies further your learning by showing you how Health and PE knowledge can be applied in the real world. Learn how you can apply your Health and PE knowledge in the real world.

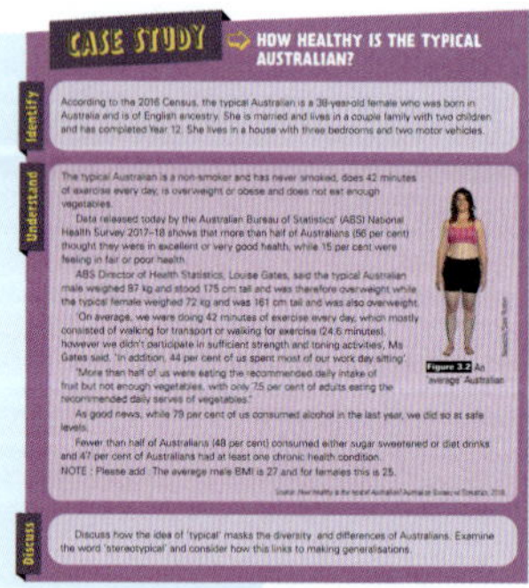

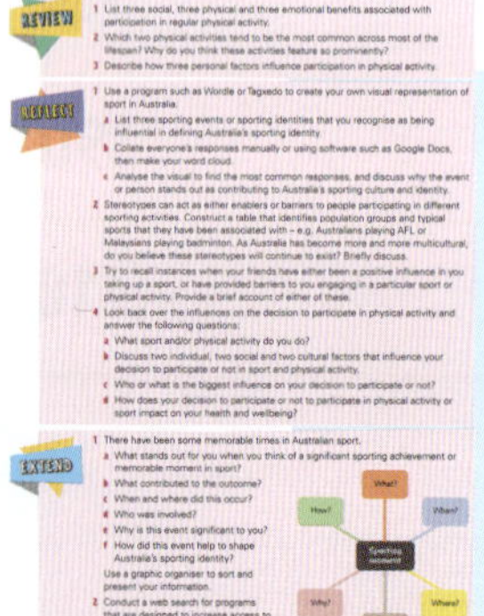

Review, reflect and extend

New review, reflect and extend activities end each section of a chapter. These provide differentiated summary questions to review your knowledge and explore further on each section. Cognitive verbs are highlighted throughout to help familarise you with these task words. Definitions for each are provided online in Nelson MindTap.

GET HANDS ON!

You can use this textbook alongside your print Workbook or the digital versions online in Nelson MindTap – look for the Workbook icon WB to complete the activities.

Face to face

These popular activities from our last edition are here to stay. Designed to complete in pairs, or a group, these provide opportunities for debate and collaboration and improve your ability to work in a team.

Up and moving

Health and Physical Education is a practical subject and these up and moving activities are designed to get you on your feet and moving! Explore different movement skills and practise games and sports with these short activities.

Fast facts

Grow your knowledge base with these weird or interesting facts! These short bites of information will help you stay awake with their exciting revelations!

A CUSTOMISABLE DIGITAL SOLUTION

Fit for Life for the Australian Curriculum 2e is designed so that it can be customised to suit any Health and PE classroom. The variety of digital resources will help you ensure the success of all learners. Everything is hosted in your online learning space, Nelson MindTap, with LMS integration options.

Videos

A range of videos are available in Nelson MindTap to assist with your learning. From animations detailing important or difficult concepts and guided meditations, to interviews and case studies to analyse, the videos and inquiry questions will enhance your learning in a visual and engaging way.

Nelson MindTap StudyHub

Accompanying the print book is access to the online eText in Nelson MindTap. Directly linking to weblinks, videos, worksheets and quizzes throughout, you can easily access all the information you need immediately. In addition, you can highlight or bookmark important content or take notes using the StudyHub.

Quizzes and Gradebook

Nelson MindTap provides both pre-chapter and chapter quizzes so you can track your own learning throughout the course.

Teachers can also use the Gradebook in Nelson MindTap to get an understanding of their students' knowledge allowing them to better manage and track class progress.

For teachers

Using Nelson MindTap, gain access to all of the student digital resources, access to the eText as well as access to additional teacher resources: explore the project-based learning outlines, the curriculum grids and assessment guide. Nelson MindTap's course customisation tools also offer teachers complete flexibility with adding links, resequencing, or hiding any content based on your school's approach and students' needs.

ABOUT THE AUTHORS

Rob Malpeli is regarded as a pre-eminent Physical Education leader and educator who has had over 40 published books in multiple Australian States as well as New Zealand. He has been a leading light in senior Physical Education and Years 7–10 Health Education for over 30 years who is recognised as developing contemporary and engaging resources for both teachers and students alike. Rob further supports teachers by conducting professional learning events throughout Australia and regularly runs student seminar/webinars that make learning fun, relevant and linked to real-world settings and examples.

Amanda Telford is the Professor of Educational Leadership at Australian Catholic University. Over the past 30 years Amanda has worked within both the secondary and tertiary sectors in educational leadership roles. Amanda has been an advisor for state and federal governments within health and physical education for children and youth. She has a strong background in learning and teaching and known for her vision to provide leadership to the teaching profession for both pre- and in-service teachers in Australia. Amanda has co-authored over 45 books and over 100 teaching and learning publications used throughout Australia and New Zealand.

Claire Stonehouse lectures at Deakin University in Health Education, Student Wellbeing and Sexuality Education. She is currently studying to gain her PhD. Claire has worked in many sectors of the community and has experience writing curriculum and educating children, young people and professionals across the board. Her areas of interest include parents as sexuality educators of their children, capacity building in sexuality education, respectful relationships education in schools and elite sports and helping young people open up conversations about mental health.

Lee Anton-Hem is a passionate educator with over 30 years of experience specialising in health, physical education and well-being. Lee currently lectures at universities and teaches in schools. When not teaching, she manages a health and physical education consultancy. Lee has written many educational resources and co-authored award-winning books. In 2008, Lee was awarded an Australian Learning and Teaching Council Citation for: Outstanding Contribution to Student Learning and in 2007 was the recipient of the RMIT University Teaching Award – Early Career Academic.

Michael Spittle is Professor of Motor Learning and Physical Education in the discipline of Sport and Exercise at Victoria University, where he teaches in areas of motor learning, skill acquisition, and applied movement science and researches skill acquisition in physical education and sport. Michael has published extensively in the area of games and sports, including prominent textbooks on game sense and skill acquisition in physical education and sport.

Sam Watkins has been a teacher of Physical, Health and Outdoor Education in both Victoria and Western Australia since 2003. He has worked in both private and public schools, teaching Years 7–12, and in various school leadership roles. Sam is currently Assistant Principal at Manor Lakes P-12 College.

Emmé Wild has taught Health and Physical Education for more than 20 years in England and Australia. She has actively been involved in curriculum writing in her capacity as Head of Department and educator, both at a school level and with the School Curriculum and Standards Authority of Western Australia (SCSA). Emmé was a board member for ACHPER WA and during her time in Western Australia, also chaired the 7–10 Health & Physical Education Course Advisory Committee for SCSA. After 18 years teaching and leading departments in Perth, she has since relocated to Far North Queensland. Emmé continues to teach Health and Physical Education with passion and purpose.

THANK YOU!

Nelson Cengage and our author team would like to thank the co-authors of the previous edition for their original contributions: Rachael Whittle, Kim Vandervelde, and Jonathan Fender. The publisher and authors would also like to thank Dr Mark Lock (Ngiyampaa) for First Nations sensitivity reading.

INTRODUCING MOODMISSION

Who is MoodMission?

Dr David Bakker is the Founding Director of MoodMission. He is a Clinical Psychologist working in private practice in Hobart, and an Adjunct Research Associate at Monash University. He has experience in rural mental health outreach, youth mental health, and disability support. David is particularly interested in how technology can be used to improve mental health. With many mental health apps focusing on low-tech interventions like diaries, David is working to utilise technology to enhance mental health. He has been involved in creating MoodPrism, a mood tracking app, as well as MoodMission.

Dr David Bakker

What does MoodMission do?

MoodMission is an app available for both iPhone and Android, that helps deal with stress, low mood and anxiety. It uses evidence-based strategies to overcome feelings of depression or anxiety by helping you find new ways of coping. MoodMission currently focuses on reducing feelings of depression, anxiety and stress, and is looking at expanding into positive psychology and maintaining good wellbeing as well as improving low mood.

MoodMission's vision is to be the best provider of easily accessible, evidence-based mental health and wellbeing support by:

- showing people new and better ways of dealing with mood and anxiety problems
- educating and enlightening people about their own psychology and the importance of practising self-care
- linking people in with professional or clinical supports.

MoodMission has conducted several peer review studies that support the use of MoodMission to improving mental wellbeing.

9780170463102

HOW TO USE MOODMISSION

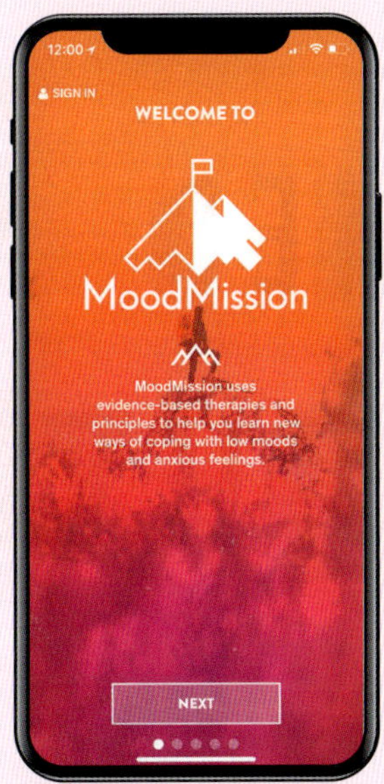

1. You can purchase the MoodMission app from the App Store or Google Play store.
2. On your first use of MoodMission, you will need to fill in a couple of surveys to get an idea of your initial mood.
3. When you want to use MoodMission, choose how you are currently feeling. Use the scales to indicate how distressing your current feelings are. You can also describe how you're feeling in more detail.
4. Five quick and easy to do strategies will appear on screen. Click on one to learn about why this strategy works.
5. If you don't like any of these strategies, or don't have the ability to complete them currently, you can choose the 'Show me 5 more' button.
6. Click 'Accept Mission' and complete the mission.
7. After completion, you can rate how successful the strategy was. The app then learns what strategies work best for you in specific situations.

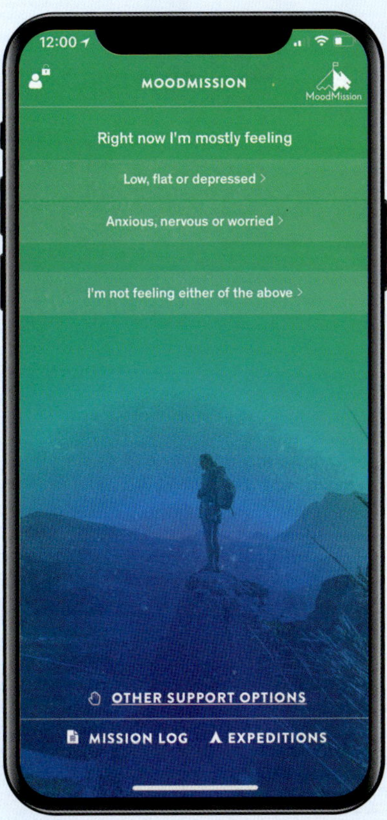

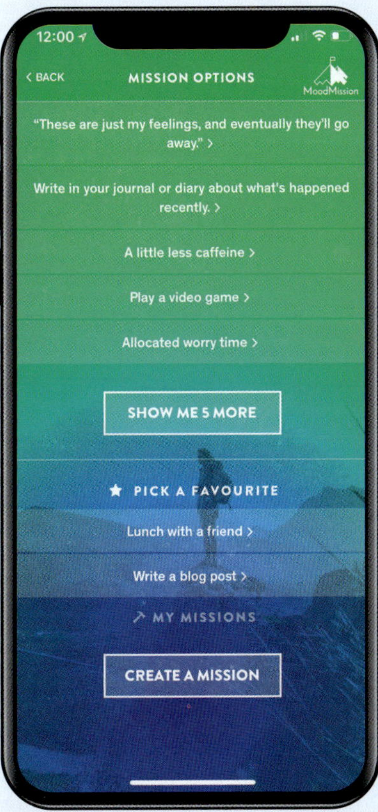

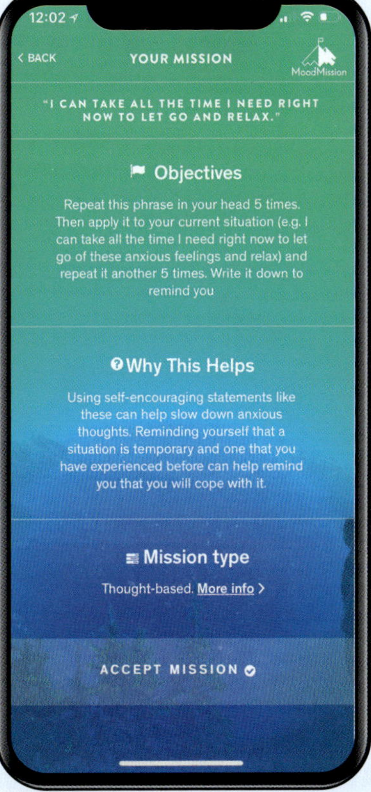

MAKING INFORMED DECISIONS ABOUT ALCOHOL AND OTHER DRUGS

1

IN THIS CHAPTER

Learn about a range of drugs, and the effect that these drugs can have on the wellbeing of you, your friends and your community. Explore how substance use can have an impact on your sense of self and shape the choices you make. Practise assertive behaviours and help seeking strategies, to appropriately respond to a range of drug-related situations.

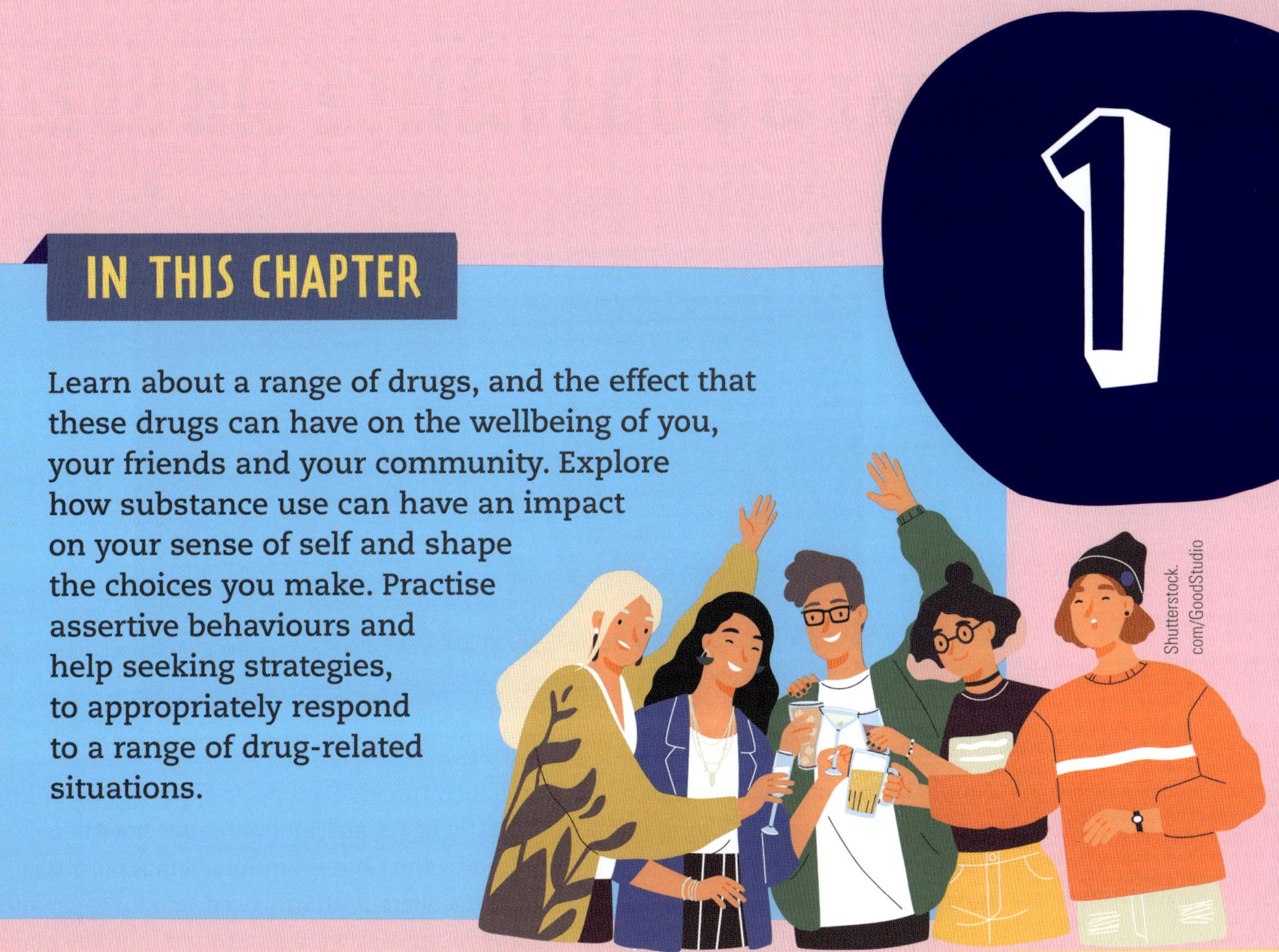
Shutterstock.com/GoodStudio

By the end of the chapter, you should be able to:

- evaluate how societal norms, stereotypes and expectations influence how young people view themselves and how to deal with these influences
- practise skills to deal with challenging or unsafe situations, such as refusal skills, communicating choices, expressing opinions and initiating contingency plans
- plan and practise responses to situations where you may be required to administer first aid
- evaluate options for managing drug-related situations where your own or others' health and safety may be at risk
- evaluate health information from a range of sources to make informed health decisions
- investigate community action initiatives young people have instigated that have had a positive influence on the health and wellbeing of their communities
- examine social, cultural and economic factors that influence choices about drug use.

WHAT IS SUBSTANCE ABUSE?

Quiz
Pre-chapter

Before you start, take the pre-chapter quiz to find out how much you already know.

Worksheet
1.1

Getty Images/Flashpop

Figure 1.1 What do you think of when you see the words 'substance abuse'?

euphoric a feeling of being on top of the world

narcotics drugs that can affect the brain and alter mood or behaviour

People have long used drugs to relax, stimulate or provide **euphoric** experiences. Wine consumption dates back to the time of the early Egyptians, and **narcotics** have been used since 4000 BCE. Marijuana was first used by the Chinese for medicinal purposes nearly 3000 years ago. In ancient societies, drugs were primarily used to help cope with aches, pains and other health issues. Healing properties were also found in many plant and animal materials.

psychoactive substances drugs that can alter the senses, awareness and mood

World Health Organization a specialised agency of the United Nations concerned with international health

Substance abuse refers to the harmful or dangerous use of **psychoactive substances**, including alcohol and other drugs. According to the **World Health Organization**, approximately 3.3 million people die every year from the harmful use of alcohol, representing 5.3 per cent of all deaths. In addition, approximately 13 million people worldwide suffer from drug use–related disorders.

The manufacture of drugs is becoming increasingly sophisticated, with drug users expecting to experience bigger and better 'highs'. However, there are never any 'winners' with substance abuse; the only people who benefit are the manufacturers and dealers.

FAST FACT

In 2020, an examination of sewer water found that Australians spent approximately $11.3 billion on methamphetamine (ICE), cocaine, MDMA and heroin. It is also estimated that Australians consume 11.5 tonnes of ICE annually. NSW was found to have the highest consumption of ICE, Cocaine and MDMA than anywhere else in Australia. Victoria was found to have the highest consumption of heroin.

Source: Drug use trends in Australia 2020, Drug-Safe Australia, https://drugsafe.com.au/drug-use-trends-australia-2020/

The Australian Government spends more than $1.5 billion a year on drug law enforcement, treatment, education and harm-reduction strategies and campaigns. Many people believe that more should be done to address the use of illegal drugs in

Australia. Understanding what these substances can do to the human body allows people to make informed choices about drug use. Informed choices promote long, happy and healthy lives.

WELLBEING CHECK IN

FINDING WAYS TO COPE

This activity has been developed in collaboration with Dr David Bakker of MoodMission

Identify

Coping with a problem means finding a way of dealing with it. Addiction to drugs or alcohol can form when we start to overuse them in order to cope.

Understand

When we feel down, anxious or experience any other negative emotion, we will naturally look for a way to cope. Some coping strategies are helpful, while others can be harmful. We might deal with the problem directly or, if this is not possible, find a way to deal with the unpleasant feelings caused by the problem.

People use drugs or alcohol for a wide range of reasons, but problems can emerge if they are relied upon for coping. Although drugs and alcohol can provide short-term relief from problems, they cause more problems in the long-term. If we have a range of coping strategies at our disposal, we can avoid using drugs and alcohol to cope.

Practise

1. Think back to the last time you felt down, anxious or upset. It doesn't have to be a big emotion; it might have been in the last day or week. How did you cope with it? What did you do to feel better?
2. Now think about other times you've coped with your feelings in the past. Did you use different ways of coping then?
3. Make a list of the coping options you have used in the past. It will be a varied list, as different coping strategies are suited to different problems. You might include things like 'talk to a friend' or 'go for a run'.
4. What are some strategies you haven't used much in the past, but may like to try in the future? These may be things you've seen friends or family do, have learned from this book or have found on apps like MoodMission.
5. Pair up with a classmate and share strategies you've listed. Did your partner list any strategies you think you could try too? If so, add them to your list.
6. The next time you feel like you need a boost, take a look at your list and try the option you think would suit best. If it's not as helpful as you hoped, maybe try another strategy. Things will get better!

Reflect

Did you find it hard to think of things that help you cope? Sometimes it can be tricky to remember what we did in the moment. The next time you face a problem, take a step back and notice how you deal with it. You might learn something about yourself.

PRESCRIPTION MEDICATION

Worksheet 1.2

Over-the-counter (OTC) and prescription medications have made a positive contribution to the health and wellbeing of Australians. Healthcare professionals and pharmacists prescribe and supply medicines to treat a variety of health conditions. The majority of Australians take these medications responsibly. However, the misuse of prescription and OTC medication is increasing, primarily because of their increased availability. Can you remember the last time you went to the doctor for a prescription or to your local pharmacy for medication? Can you remember what you purchased or were prescribed?

The most commonly abused pharmaceutical drugs are benzodiazepines, analgesics and stimulants.

Benzodiazepines

benzodiazepines psychoactive drugs usually prescribed to treat anxiety or sleep problems

depressant a drug that slows down the activity of the brain and central nervous system

Benzodiazepines are used for the treatment of anxiety. They are classed as **depressants**, as they slow down the activity in the central nervous system. They are commonly prescribed by healthcare professionals to reduce stress and anxiety, as well as to relax muscles and promote sleep. Benzodiazepines are generally taken orally in the form of a tablet or capsule, but they can also be injected. Drugs that are in the benzodiazepine family include diazepam (Valium), oxazepam, nitrazepam, temazepam, flunitrazepam and bromazepam.

Analgesics

simple analgesics non-opioids used to relieve pain and inflammation

opioid a psychoactive drug that helps to relieve pain

What did you do the last time you had a headache? Did you take any tablets? If you did, you probably took a medication known as an analgesic. Analgesics are grouped into **simple analgesics** and **opioid** analgesics. Opioid analgesics are the most commonly abused type. These include drugs such as morphine, codeine, methadone and oxycodone. Like benzodiazepines, they are classed as depressants because they slow down the activity in the brain and central nervous system. Simple analgesics or 'painkillers', such as paracetamol, aspirin and ibuprofen, are widely available.

Analgesics are usually taken orally. They come in many forms, including tablets, capsules, liquids and dissolvable powders. Sporting injuries, period pain and headaches are common reasons why young people take analgesics. In the short term, both opioid and simple analgesics help to relieve pain. Some may also relieve other symptoms, such as fever, inflammation or swelling.

If you take too many analgesics or use them too frequently, the following side effects may occur:

- nausea
- vomiting
- heartburn
- reduced appetite
- drowsiness
- dizziness
- kidney damage
- abdominal pain
- constipation
- diarrhoea
- headaches
- stomach problems
- skin rash.

Stimulants

stimulants drugs that increase the activity of the central nervous system

Stimulants can be prescribed or purchased over the counter. Prescription stimulants include amphetamines and methylphenidate. They are used to treat both psychological and physical disorders, including attention deficit hyperactivity disorder (ADHD)

and **narcolepsy**. Over-the-counter stimulants include caffeine medicines (such as 'No-Doz'), energy drinks, diet pills, pain relievers, motion-sickness tablets, cough and cold medications, pseudoephedrine (present in some cold and flu tablets) and some herbal remedies.

narcolepsy a sleep disorder where the person has disturbed sleep at night and often falls asleep during the day

Stimulants are often abused to get the user 'high', or to improve alertness, focus and attention. Higher doses of stimulants can lead to cardiovascular issues, which may lead to **stroke**. Stimulants can also reduce feelings of hunger, so they are often abused to promote weight loss. Side effects of stimulants include:

stroke the loss of brain function caused by a blocked or burst artery in the brain

- increased heart rate
- increased blood pressure
- increased body temperature
- loss of appetite
- lack of sleep.

Some people think prescription drugs are safer than illegal drugs, such as heroin or cocaine, but this is incorrect. Prescription drugs, like illegal drugs, can have some very powerful effects on the brain and body. Opioid analgesics act in a similar way to heroin, targeting the same area of the brain. The effects of taking prescribed stimulants are similar to taking cocaine and could cause, in the long term, paranoia and possible heart problems. Is it worth it?

Alamy Stock Photo/WENN Rights Ltd

Figure 1.2 Prince was a famous US musician who died due to drug abuse.

Abusing prescription drugs can have devastating consequences.

- Prince was an American musician. On 21 April 2016, aged 57, he was found unconscious and unresponsive alone in an elevator. An autopsy revealed that he had died from an accidental **overdose** of fentanyl. Fentanyl is a **synthetic** opioid, 50 times more powerful than heroin, and is used to relieve severe pain.

overdose to take more of a substance than is safe, resulting in serious health consequences and/or death

synthetic a substance that has been made artificially, by chemical reaction, usually to mimic a substance found in nature

FAST FACT

The COVID-19 pandemic has resulted in a shift in recreational drug use. MDMA, LSD and cocaine were less frequently consumed due to the closure of social and recreational venues. Lockdowns, increased emotional and financial stress and boredom resulted in a rise in the use cannabis, as well as in the non-medical prescription drugs such as benzodiazepines.

Tips for using medicines safely

1 Always follow the advice of your doctor or pharmacist.
2 Never take more than the recommended amount.
3 Read the instructions, and note any possible side effects.
4 Only use medication prescribed to you. Using someone else's prescription medication or letting a friend use yours is illegal and dangerous.
5 If taking prescribed medication, check with your doctor or pharmacist to ensure that any other medication you may be taking will not adversely interact with the medicine prescribed.
6 Check whether you are safe to drive after taking medication, as some types may make you feel drowsy or dizzy and/or affect your vision.
7 If in doubt, always ask your doctor or pharmacist!

Quiz
How can I use medicine safely?

1 Explain what each type of prescription drug is used to treat:
   a benzodiazepines
   b analgesics
   c stimulants.
2 Investigate why prescription drug abuse is on the rise.

1 Compare the side effects of abusing analgesics with those of stimulants.
2 Are prescription drugs as dangerous as illegal drugs such as heroin or cocaine? Discuss.

1 Research the Alcohol and Drug Foundation.
2 Using the information provided under 'Drug Facts', create an informative poster, pamphlet or brochure about the commonly used prescription and OTC drugs.

Weblink
Alcohol and Drug Foundation

WHAT ALTERNATIVES TO MEDICINE CAN I USE?

ALTERNATIVE HEALTH CARE

alternative medicine medical therapies, treatments and practices that are not used in conventional medicine

complementary medicine combining the use of alternative medicine with conventional medicine

chronic persistent or long lasting

In recent years there has been growing interest in **alternative** and **complementary** medicine and health care. Besides seeing a healthcare professional such as a doctor, many people seek alternative treatments to improve their health. A healthcare professional provides healthcare treatment and advice based on their qualifications and experience. A doctor, for example, can help to mend a broken bone. However, for a **chronic** health concern, such as arthritis, alternative therapy may be an additional option for treatment.

There are several alternative therapies that claim to promote health and wellness, including homeopathy, naturopathy, chiropractic, herbal medicine and aromatherapy. Many of these therapies are seen as alternatives to conventional medicine.

Homeopathy

Homeopathy focuses on stimulating your body's own healing powers to alleviate symptoms quickly and effectively using homeopathic medicines.

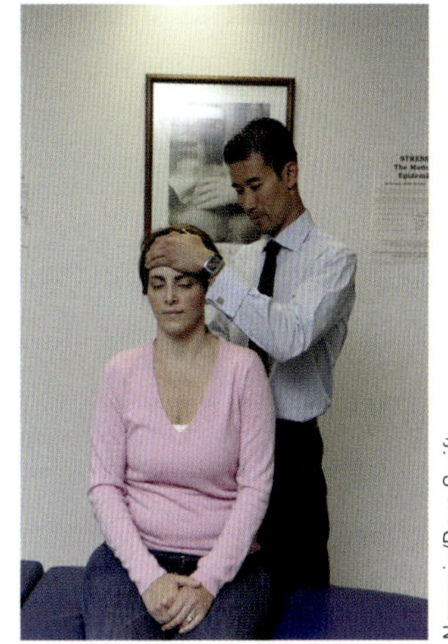

Newspix/Dave Swift

Figure 1.3 Chiropractors focus on the spine, so they often treat people suffering back or neck pain.

Naturopathy

Naturopathy focuses on the holistic treatment of an individual by addressing the symptoms of illness as well as seeking to resolve any underlying issues that may be causing the problem. Practitioners may use multiple treatments, including herbal medicine, dietary adjustments, massage or homeopathy.

Chiropractic

Chiropractic treatment focuses on the relationship between the spine, the **musculoskeletal system** and the nervous system. Health concerns such as back pain, shoulder pain or headaches may be alleviated using chiropractic treatment.

musculoskeletal system the system comprising the muscles, tendons, ligaments, bones, joints and tissues that make the body move

Herbal medicine

Herbal medicine uses herbs and plant extracts to treat health disorders and enhance wellbeing, aiming to return the body to a state of natural balance. Herbs have been used for medicinal purposes around the world for thousands of years. Indian Ayurvedic medicine and traditional Chinese medicine both use herbs to prepare specific treatments.

Herbal medicine is used to treat a variety of conditions, including arthritis, allergies, stress and depression, hormonal imbalances, colds and flus, headaches, sore muscles, skin complaints and upset stomachs. The common herb garlic is used to fight colds as well as reduce blood pressure.

Alamy Stock Photo/Tim Whitby

Figure 1.4 Herbal medicine has been used for thousands of years across the world in countries such as India, China and Vietnam.

Aromatherapy

Aromatherapy is the **therapeutic** use of aromatic and essential oils made from plants and flowers for mental and physical wellbeing. These are used to balance, harmonise and promote the health of the body, mind and spirit. Oils are applied through massage and inhalation, and to body surfaces such as the hair.

therapeutic healing

Massage

Massage is a form of relaxation or healing that focuses on different parts of the body. A massage therapist will use their hands to apply pressure to various muscles or joints in order to ease pain and tension in those areas. There are a number of different types of massage, including Swedish, Balinese, hot stone, aromatherapy and deep tissue.

Hypnosis

Also known as hypnotherapy, hypnosis uses guided relaxation techniques to induce intense concentration, often known as a 'trance'. An individual's attention will become so focused that all other distractions are temporarily blocked out. Hypnosis is used to treat many issues, including phobias, low confidence and low self-esteem.

INVESTIGATION ⇨ COMPLEMENTARY AND ALTERNATIVE HEALTH CARE

Purpose

- To investigate why more Australians are using complementary and alternative health methods instead of conventional medications.
- To examine five categories of complementary and alternative health therapies.
- To research the popularity of these methods and to share the findings.

Method

1. Conduct online research to define the word 'alternative'.
2. Add the word 'medicine' to your search, and brainstorm different kinds of alternative medicine.
3. In pairs, identify and describe **five** types of complementary and alternative health methods and discuss the health benefits of each method you have identified.
4. Investigate how many Australians currently use each of these forms of medicine.
5. Consider the phenomenon known as the 'placebo effect'. Discuss what is meant by this term.
6. What impact has the placebo effect had on studies involving alternative medicines and why?

Discussion

1. Share your research findings with your classmates.
2. Do you feel the benefits of alternative medicines are worth the money spent?

Newspix/Chris Mangan

Figure 1.5 First Nations Peoples have successfully used traditional healers for thousands of years.

Traditional medicine in Australia

First Nations Peoples have successfully used traditional healers and bush medicine for thousands of years. Indigenous traditional medicine is a complex system linked closely to their cultural beliefs, awareness of their land as well as the many plants and animals living on the land. A traditional healer within Indigenous culture is well respected with many community members believing they can cure many mental, spiritual and physical illnesses. While contemporary (western) medicine focuses on diagnosing and treatment, traditional medicine considers a meaningful explanation for the illness and the personal, family and community issues that may surround the illness. Traditional practices include the use of touch, breath and bush medicines to heal a person's spirit.

Many traditional healers work alongside doctors and other contemporary medical staff to ensure Indigenous patients receive a holistic treatment encompassing cultural tradition.

1. Determine the difference between homeopathy and naturopathy.
2. Detail how hypnosis can be used to improve health.

1. Create a mind map showing all the different types of alternative health care available.
2. Do you think alternative health care methods can provide the same effect as traditional medicines? Discuss your answer.

1. Use the Internet to investigate the traditional medicine practices of First Nations Peoples.

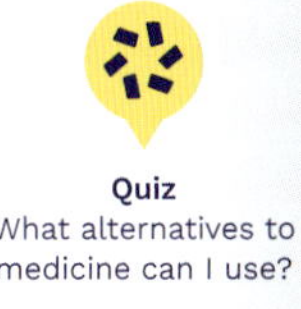

Quiz
What alternatives to medicine can I use?

HOW DO DRUGS AFFECT THE BODY?

FAST FACT
Science has shown that some chemicals and compounds interact adversely when mixed. The wrong 'cocktail' of drugs can kill.

Worksheet 1.3

All drugs, including caffeine and energy drinks, can affect the body. Drugs that alter brain function and affect the way people may think or feel are known as psychoactive drugs. Psychoactive drugs contain chemical substances that can affect the functioning of the central nervous system. Adverse effects of psychoactive drugs include confusion, changes in mood and behaviour, and altered levels of consciousness.

In order to understand why and how people use drugs, it is important to learn about the five main categories of drug use: experimental, recreational, situational, intensive and use leading to dependence.

Experimental use

People choose to experiment with drugs for many reasons. These include curiosity, to have fun, to fit in with friends, to improve sporting performance and to help ease a problem such as stress or depression. A first-time drug user may want to experiment with

iStock.com/sturti

Figure 1.6 Experimental use is when people try out drugs, often for the first time.

different types of drugs to gauge their reactions, and may be involved in the social use of drugs, which often involves experimentation.

Recreational use

Recreational use refers to the social or casual use of a drug, maybe on a weekend or as part of the person's social life. Often, drugs are taken to enhance a person's experience of a particular environment, such as a music concert or a nightclub.

FACE TO FACE Discussion

Most people who use drugs recreationally see this habit as being 'normal'. What do you think? In small groups, discuss whether you agree or disagree with this statement and provide reasons to support your decision.

Situational use

Situational use refers to the use of a drug to cope with the demands of a particular situation. For example, some Year 12 students may take caffeine tablets to help them stay awake while they are studying for examinations. Benzodiazepines may be prescribed by a doctor to help someone who may be dealing with a stressful situation, such as the death of a relative or friend.

Intensive use

Intensive use refers to the excessive use of drugs over a short period of time, or continuously over a few days or even weeks. **Binge drinking** is an example of intensive drug use.

binge drinking the excessive consumption of alcohol in a short amount of time, which can result in serious health and social concerns

Figure 1.7 Binge drinking is a form of intensive drug use, when a person drinks an excessive amount in a short time.

Shutterstock.com/itor

Dependence

Dependence refers to a person needing to use a drug to make them feel normal. This type of use is commonly referred to as a drug addiction. Drug addiction refers to the compulsive use of drugs, whereas dependence focuses on the impact of drugs on the body, both physical and psychological. Support is essential for recovery from dependence. This can come from friends, family, healthcare providers and counsellors.

Poisons helpline

Call the poisons helpline if you, a friend or a family member come in contact with a poisonous or toxic product and experience health difficulties. Remember, if the situation is life-threatening, **call 000**.

POISONS INFORMATION CENTRE
13 11 26
Call from anywhere in Australia – 24 hours a day

Figure 1.8 The Poisons Information Centre is open 24 hours a day.

CLASSIFICATION OF DRUGS

Drugs that affect the central nervous system, both legal and illegal, can be classified into three main groups: stimulants, depressants and hallucinogens.

Stimulants

Stimulants speed up the activity of the central nervous system. They increase alertness and reduce fatigue. Examples of stimulants include ecstasy, cocaine and methamphetamines (ice, speed, base).

Depressants

Depressants slow down the activity of the central nervous system. They increase fatigue and confusion, as well as impair coordination, memory and speech. Examples of depressants include:

- alcohol
- cannabis
- shisha
- GHB (gamma hydroxybutyrate)
- opiates and opioids (heroin, morphine, codeine)
- inhalants (solvents, aerosols, gases, nitrites).

Hallucinogens

Hallucinogens, also known as **psychedelics**, interfere with the central nervous system by altering the way reality is perceived. They can cause hallucinations that make you see or hear things that don't actually exist. Hallucinogens are either manufactured in laboratories or occur naturally in trees, vines, seeds, fungi and leaves. Examples of hallucinogens include:

- LSD (lysergic acid diethylamide)
- magic mushrooms
- cannabis
- ketamine.

psychedelics mind-altering psychoactive drugs that produce hallucinations; LSD and magic mushrooms are typical examples

INDIVIDUAL DRUGS

Cocaine

Drug classification: stimulant

Cocaine, commonly a white crystalline powder, is taken to speed up brain activity and make a person stay awake and alert. This bitter-tasting powder also has pain-relieving properties. Cocaine can be smoked, snorted, injected or rubbed onto the gums. It can also be manufactured into small rock-like crystals known as 'crack' and smoked.

Once cocaine or crack has been consumed, the immediate effects do not last long. Effects may peak after two to five minutes, and only last between 10 minutes and two hours.

FAST FACT

Only 2 per cent of 12–17-year-olds report having used cocaine before – 98 per cent have NEVER used this drug!

FAST FACT

Other names for cocaine include:

- Coke
- Charlie
- Star dust
- White lady
- Rocks
- Crack

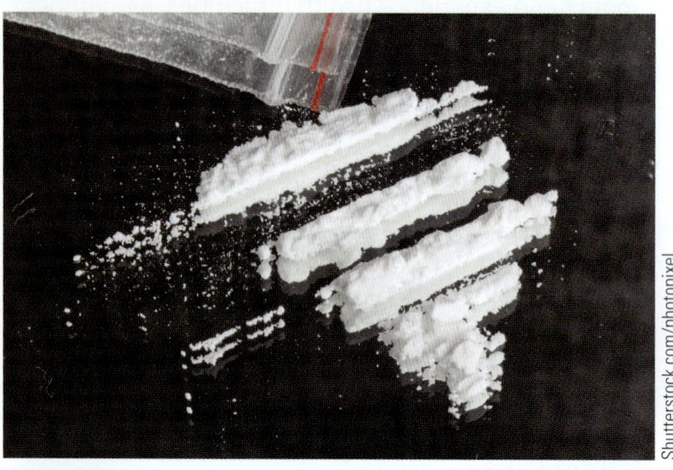

Shutterstock.com/photopixel

Figure 1.9 Cocaine is a stimulant which speeds up brain activity.

Effects of cocaine and crack

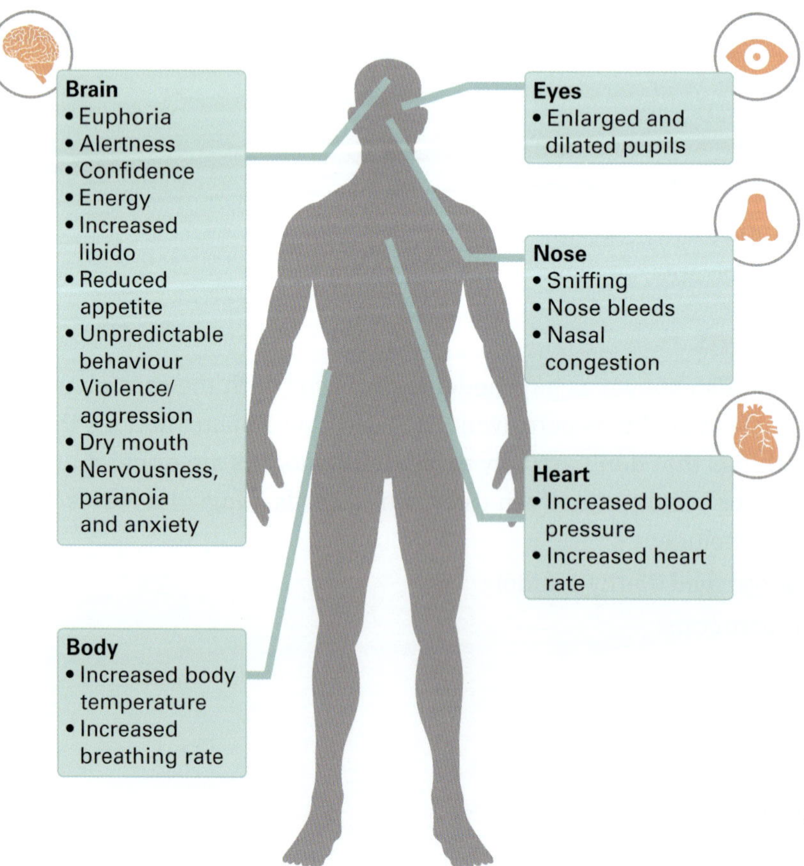

Figure 1.10 Short-term effects of cocaine and crack.

9780170463102

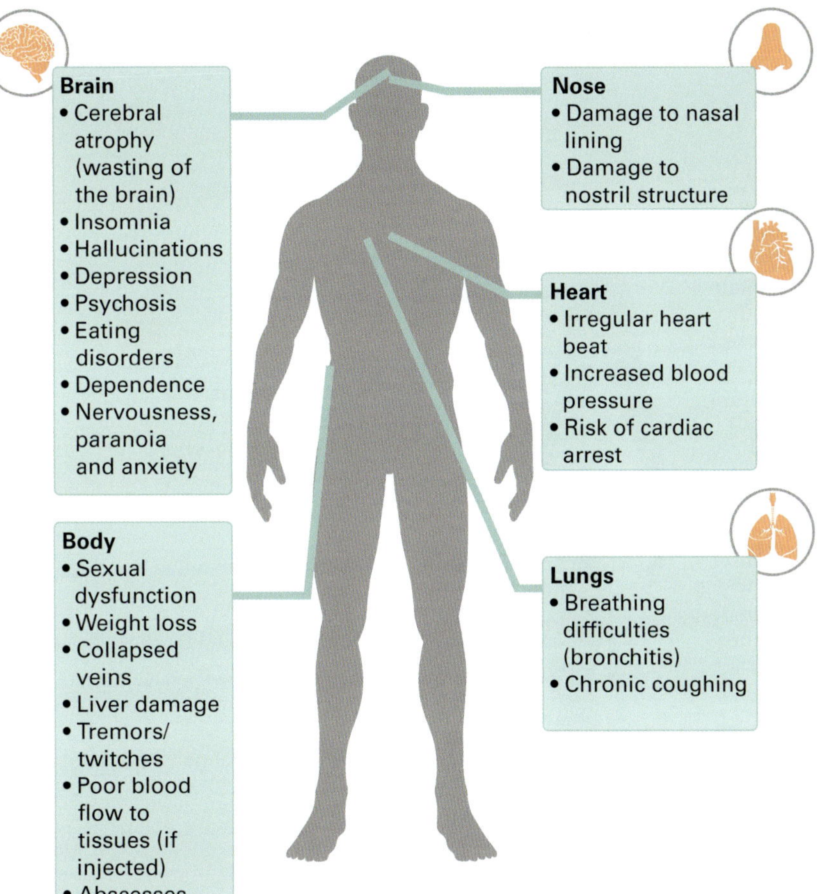

Figure 1.11 Long-term effects of cocaine and crack.

Methamphetamine

Drug classification: stimulant

Methamphetamine is a synthetic psychoactive substance manufactured from readily available chemicals. It is derived from amphetamine and was first manufactured in the 1950s and prescribed to treat a variety of conditions including depression, obesity and alcoholism. One major difference between methamphetamine and amphetamine is that the effects of 'meth' are more immediate and much stronger. Unlike amphetamine, methamphetamine is now considered too dangerous to be prescribed for use by healthcare professionals, as it is highly addictive. The drug comes in three forms, which differ in purity:

- speed
- ice
- base.

Methamphetamines can be swallowed, snorted, injected or smoked. They are often produced in illegal backyard laboratories, which are harmful to the environment.

FAST FACT

The 'purity' of a drug can vary from place to place. Drugs are often mixed with other substances, including caffeine, paracetamol, vitamin C, talcum powder and, in some cases, rat poison.

Remember, there is no such thing as a 'safe' drug. Manufacturers and dealers of illicit drugs are only out to make money; they are not interested in your health or wellbeing.

Effects of methamphetamine

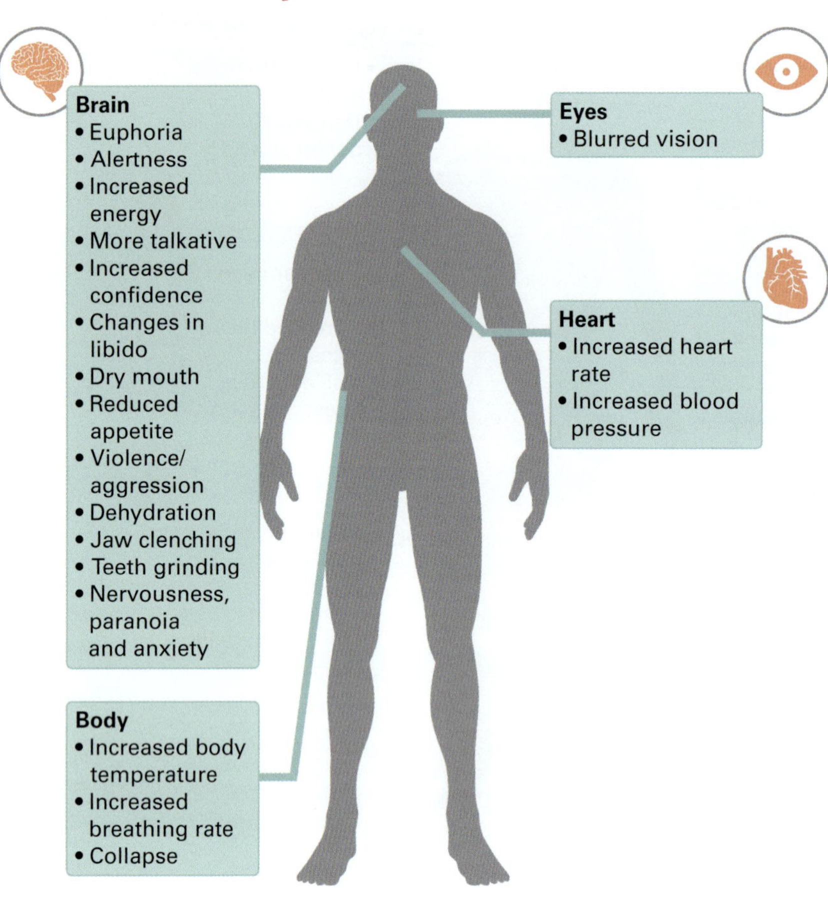

Figure 1.12 Short-term effects of methamphetamine.

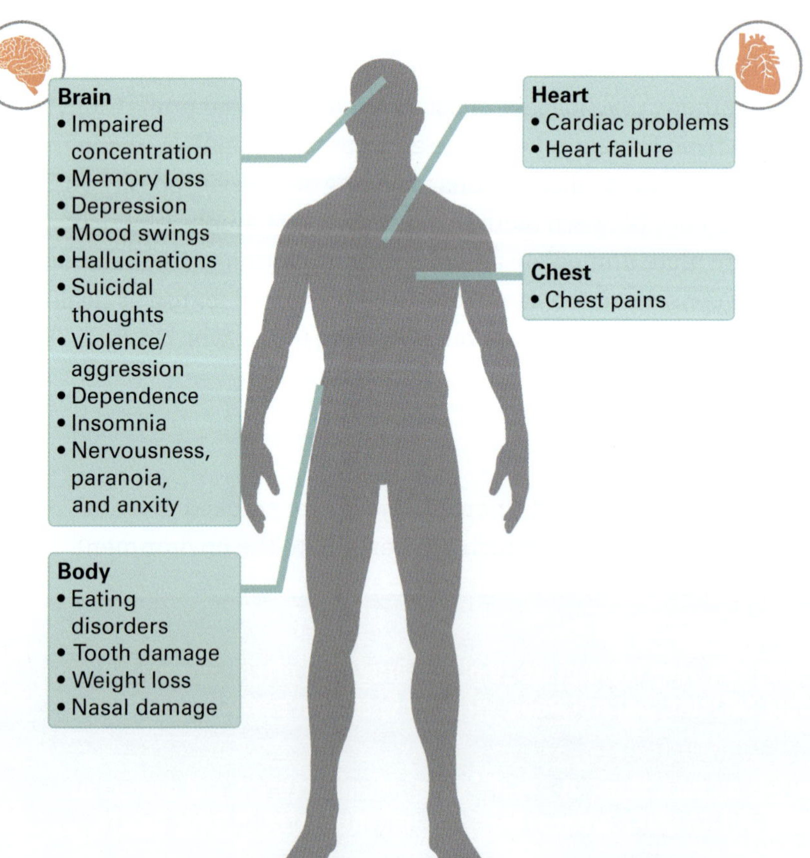

Figure 1.13 Long-term effects of methamphetamine.

Table 1.1 Three forms of methamphetamine

Methamphetamine (other names)	Form	Purity	Duration of effect
Speed (go-ee, whizz, uppers, revs)	Powder	10–20%	2–3 hours
Ice (crystal, meth, crystal meth, shabu, tina, glass)	White crystal-like powder	80%, approximately	12–24 hours
Base (pure, paste, wax, point)	Oily or pasty powder	20%, approximately	3–5 hours

CASE STUDY

IMPACT OF ICE

Identify

Young people in Darwin who have lost friends to ICE addiction and decide to take matters into their own hands:

- Young performers are collaborating with former ice users from a rehabilitation service
- Their performance 'Secrets' explores teenage ice use in Darwin
- Most of the cast members are local high school students

Understand

Arthur has lost three friends to drug abuse. The 25-year-old said ice use in particular was worse now among young people in Darwin than ever before.

"I moved to Darwin when I was 14 years old," Arthur said. "If I look back and compare now, it's worse. "It's just a lot of young kids doing this kind of stuff."

He's part of a group of young performers, mostly high school students, who have collaborated with recovering ice users from local rehabilitation service Banyan House.

Their dance performance 'Secrets' explores how ice, or crystal methamphetamine, use can destroy lives, friendships and families. "That motivates me, that this piece should be out there," Arthur said. The story begins with the downward spiral of a fictitious Darwin teenager, whose ice use starts at a house party where someone older tempts him to give it a try.

'Secrets' was first developed in 2013 and has been shown to a Darwin audience with different student cast members several times since then.

"I have friends that have been through this kind of thing," said Arthur, who was part of the original cast.

Francis said he started using ice as a teenager and recently spent 18 months in prison. He said working with the young cast members had given him further motivation to continue with recovery. "This really is a bad lifestyle, it really does end up very badly," he said.

Young performer Lilian, 16, said the cast members had watched the project change lives, including their own. "I know we've definitely changed lots of lives," Lilian said.

"Myself and the others would probably agree that we've saved many people from something very bad occurring later on in life for them.

Discuss

1 Discuss the motivation behind Arthur's involvement in this community initiative.
2 Reflect on why this initiative, created by high school students, is having a positive impact on the health and wellbeing of young ICE users in Darwin.
3 In pairs, propose a community-based initiative that could have a positive impact on the health and wellbeing of young people in your community. Share your ideas with your class.

Weblink
Alcohol: Why is alcohol so popular when it has so many negative side effects?
Do you plan to drink alcohol when you reach the legal drinking age? Why/why not? Watch the video and start the discussion!

Alcohol

Drug classification: depressant

Alcohol is widely consumed by teenagers in Australia and its use is a considerable problem. Alcohol is a depressant, which means that it slows down the activity of the central nervous system. When consumed in small quantities, alcohol acts as a relaxant, making you feel less anxious. In larger quantities, alcohol causes changes in the brain, resulting in **intoxication** or drunken behaviour.

intoxication the state that occurs when normal behaviour is changed or altered by substances such as alcohol or other drugs

Worksheet 1.5

FACE TO FACE Community action initiatives

Read the following scenarios. In pairs, discuss how you would respond and provide three strategies for each scenario.

1 You are at a beach party and everyone is sitting around a fire chatting and drinking. Some boys have started to act foolishly by throwing people's belongings, including thongs and phones, over the fire to each other. Naturally, people are getting annoyed.
2 Jason watches as someone suggests going for a swim in the ocean. He knows everyone has been drinking heavily and thinks this could be really dangerous. People are getting excited and start to take their clothes off as they run towards the water.

ALCOHOL FACTS

Approximately 1 in every 20 deaths globally is the result of an alcohol-related illness, accident or injury.

Source: 'Alcohol-related death', Alcohol Rehab Guide, https://www.alcoholrehabguide.org/resources/medical-conditions/alcohol-related-death/

FAST FACT

Australians spend more than $14.9 billion on alcohol each year.

Binge drinking

Although people generally consume alcohol in sensible quantities, teenage drinking, along with alcohol-fuelled violence and binge drinking are prevalent issues in today's society. These arise from the increased availability and vast selection of alcoholic products.

There are four main reasons that young people and adults drink:

- effects – to feel happy, cope with stress, change behaviour
- socialising – to fit in (peer acceptance), make friends
- curiosity – to try something new
- escape – to forget problems and life pressures.

The ability to 'hold your drink' among young people is often seen as a positive quality. Everyone is different, and the lives of young people are very different from those of adults.

Heading out for the night doesn't necessarily mean that you need to drink to get drunk. There is a huge difference between having a quiet beer or a glass of wine with mates and binge drinking! Binge drinking involves either excessive drinking (more than four standard drinks) over a short period of time, or drinking continuously over a number of days or weeks.

Shutterstock.com/Monkey Business Images

Figure 1.14 Binge drinking often happens at nightclubs or bars where people drink a lot of alcohol in one sitting.

Alcohol can seriously affect the decisions that people make. It is one of the primary causes of injury and death among young people. Ask yourself these questions: Would you get into a car with a drunk driver? Would you consider having unprotected sex? Usually, you would probably answer no to these questions. However, when drunk, you are more likely to put yourself in a risky situation than when you are sober. Binge drinking is a major problem in Australia. It can seriously affect your health, not only in the short term, but in the longer term too. Remember that alcohol does not only affect you; your actions under the influence of alcohol also affect everyone around you.

Alamy Stock Photo/PA Images

Figure 1.15 Musician Justin Bieber abused alcohol and drugs.

Canadian singer, songwriter and actor Justin Bieber was just 15 years old when he became a global pop sensation and teenage heart-throb. Within a couple of years, Bieber had accumulated huge fame and fortune. At the age of 19, the pressures of stardom started to take a toll on Bieber, causing him to make some very poor choices. Bieber admitted to heavily abusing alcohol and drugs, disrespecting women and breaking the law.

Pre-drinking

As alcohol prices continue to rise in bars and clubs, it has become increasingly common for partygoers to 'pre-drink' before they go out. Pre-drinking (or pre-loading) is a process whereby young people consume large amounts of alcohol before going to pubs and clubs, primarily to avoid paying the high alcohol prices at venues. Currently, more than 64 per cent of young Australians will pre-drink before heading out for the evening, with some consuming up to 25 standard drinks!

> **FAST FACT**
>
> Pre-drinking is also known as:
>
> - prinking
> - pre-funking
> - pre-gaming
> - pre-loading.

While pre-drinking has resulted in a significant increase in alcohol related crime and injuries including accidents, assaults, road trauma and alcohol poisoning, recent studies have shown that, in fact, young people are actually drinking less than previous generations. During COVID-19 lockdowns, additional research undertaken indicated that alcohol consumption among young people fell even further.

Young people today are more focused on academic success and more conscious about leading a healthy and active lifestyle. Parents are also playing a key role in helping their children make informed decisions about alcohol consumption.

Alcohol and energy drinks

In Years 7 and 8 you learnt about the dangers associated with caffeinated energy drinks. These dangers increase dramatically when the drinks are mixed with alcohol.

Scout Kozakiewicz

Figure 1.16 Mixing a depressant (alcohol) with a stimulant (caffeine in energy drinks) results in a toxic mix.

Mixing alcohol and energy drinks has become increasingly popular among young people. According to health experts, this is extremely dangerous because alcohol is a depressant and the caffeine in energy drinks is a stimulant. Mixing depressants with stimulants masks the effects of alcohol, making you feel more alert as you get more drunk. This means you may increase your risk of alcohol-related harm, including engaging in the following behaviours:

- drink driving
- violence
- unsafe sexual activity.

CASE STUDY

PARIS KAMPER

Identify

The tragedy of the 15-year-old Sydney girl who died after drinking a fatal cocktail of alcohol and energy drinks should serve as a warning to other young people, her mother said at her funeral.

Understand

In a heartfelt eulogy in front of more than 400 mourners at the service in northwest Sydney, Sandy Kamper said she hoped her daughter's death will save other lives.

Paris Kamper – described as a 'bright, talented light taken too soon' by her friends – was found unconscious at a home at Kenthurst, in Sydney's northwest on Friday, June 8 …

The much-loved teenager had reportedly been drinking alone and returned a blood alcohol reading of 0.4, which is eight times over the adult legal driving limit.

Energy drinks and sugary lollies were found near her on the property, while it's believed she had looked up an online recipe for making caffeine-laced alcoholic cocktails, NSW Police say. She was rushed to The Children's Hospital at Westmead but died there two days later …

The Hills police area commander, Superintendent Rob Critchlow says he did not realise how popular recipes for alcoholic cocktails were until he did his own online search.

'Five seconds on Google tells you it's a big problem,' Supt Critchlow told Network Seven last week. He warned about the risks of underage drinking, and sourcing information online relating to alcohol consumption.

'Alcohol poisoning victim's mum hopes her daughter's death will save lives' by Ben Graham, news.com.au, 20 June 2018.

Discuss

1. Paris Kamper was only 15 years old when she died. How much alcohol was reportedly found in her blood, and what was the reading compared to?
2. Police found energy drinks and sugary lollies near Paris. What do police believe Paris did that resulted in her being rushed to hospital?
3. What problem did Superintendent Rob Critchlow find, and what was his warning to others?
4. In pairs, investigate the dangers associated with mixing energy drinks and alcohol. Provide reasons for your answers.

FAST FACTS

1. Around 20 per cent of beverages sold in Australian convenience stores are energy drinks.
2. Denmark, Turkey, Finland, Uruguay, Iceland and Norway have resorted to outright bans on energy drinks due health risks linked to excessive caffeine consumption.

Safe drinking tips

Once you know the dangers of mixing energy drinks with alcohol, you may decide that you never want to drink them again! If you do choose to drink energy drinks, here are a few safety tips to consider:

- Caffeine (like alcohol) affects every person differently, so the amount one person can drink safely may not be the same as another person.
- Symptoms of caffeine **toxicity** include headache, upset stomach, shaking, feeling as if your heart is 'racing', difficulty breathing and not being able to sleep.
- Severe caffeine toxicity can lead to seizures, **psychosis**, irregular heartbeat and, very rarely, death.
- Seek medical attention immediately if you are having chest pain or severe reactions to caffeine, or if you are concerned about any symptoms you are experiencing.

Source: 'Energy drinks pack more punch than you might expect', Jenny Pogson, 9 May 2012, www.abc.net.au/health/thepulse/stories/2012/05/09/3471672.htm

toxicity the extent to which something is poisonous

psychosis an abnormal state of mind resulting in a loss of contact with reality

Drink spiking

Drink spiking is when alcohol or other substances are added to your drink without your knowledge. This practice is illegal in Australia and penalties can include fines

drink spiking placing a substance into a drink without the drinker's knowledge in order to harm them

or even imprisonment. People spike drinks for many reasons, including for a laugh at someone else's expense or to make it easier to commit a crime such as robbery or sexual assault. The majority of incidences, however, are not linked to criminal activity; these are commonly known as 'prank spiking'.

Unfortunately, there is no way to know if your drink has been spiked, but there are some warning signs, including:

- feeling faint or dizzy
- feeling sleepy
- feeling sick
- feeling drunk, even when you know you have had very little to drink
- doing things that you wouldn't usually do
- passing out
- waking up feeling confused or disoriented, or with headaches and blanks in your memory about the night before.

If you think you or your friend's drink may have been spiked:

- ask your friends or family to help move you or your friend to a safe place
- seek medical advice immediately. Your doctor may be able to test for drugs in your system if they are advised within 24 hours of a reaction to a potentially spiked drink.
- call 000 if the situation is life-threatening!

FAST FACT

Topping up your friend's glass with alcohol without them knowing is considered drink spiking. This is a crime.

© Commonwealth of Australia 2018

Figure 1.17 Always phone 000 if the situation is life-threatening.

GHB (date rape drug)

Drug classification: depressant

GHB (gamma hydroxybutyrate) is a substance that is found naturally in the human brain. The drug was originally developed as an anaesthetic, but is no longer used in Australia because of unwanted side effects. GHB is a colourless liquid commonly sold in small bottles. It can sometimes be bright blue – hence the nickname 'blue nitro'. GHB can either be swallowed or injected. On rare occasions, it may be consumed in tablet form. It is known as a '**date rape**' drug because it is often used to spike drinks, as it has no smell or taste, which makes it hard to detect.

date rape when a person is sexually assaulted by someone they know

Scout Kozakiewicz

Figure 1.18 GHB can be bright blue or colourless.

Effects of GHB

Currently, little is known about the long-term effects of GHB. However, it is considered to be highly addictive, and a high dose can kill.

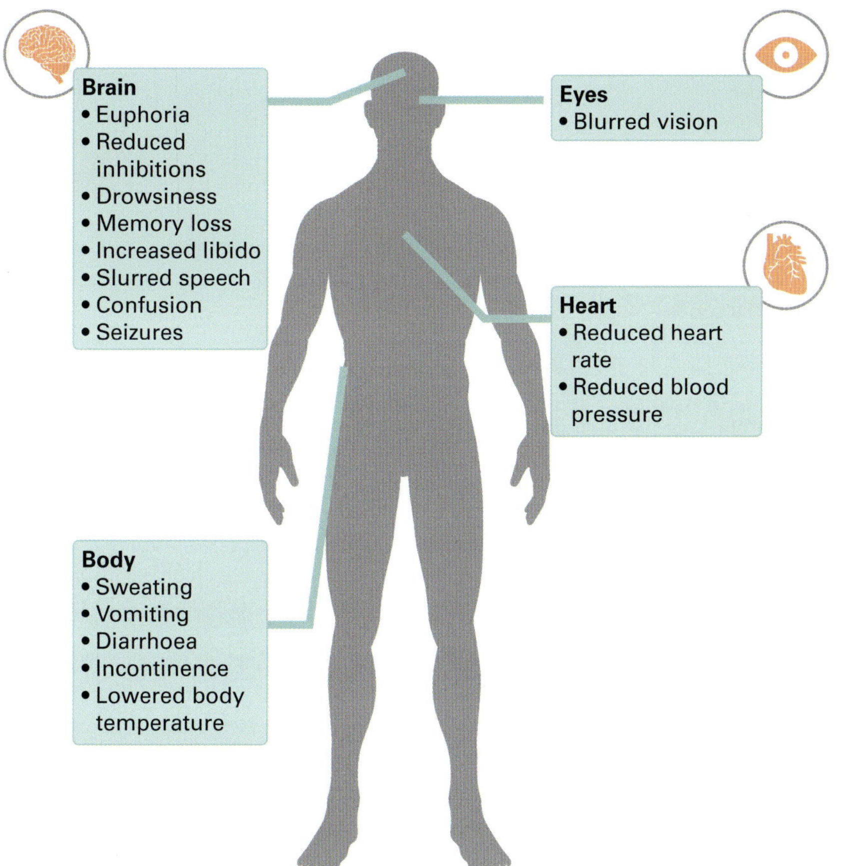

Figure 1.19 Short-term effects of GHB.

FAST FACTS

Other names for GHB include:

- G (most common)
- Blue nitro
- Grievous bodily harm (GBH)
- Fantasy
- Liquid E

Rohypnol (date-rape drug)

Drug classification: depressant

Originally prescribed to treat insomnia, Rohypnol or flunitrazepam is an extremely powerful sedative used to suppress the central nervous system in order to create a relaxed and sleepy state. Because of its potent effects, it is often associated with sexual assault and date rape. Rohypnol tablets are white. When dissolved in a fluid they are tasteless and have no distinct smell. Rohypnol is considered dangerous as it can stay in the body for up to eight hours, with the initial effects being felt within 30 minutes. It is illegal to use or possess Rohypnol in Australia unless you have been prescribed the drug by a healthcare professional.

FAST FACT

Other names for Rohypnol include:

- Roofies
- Date-rape drug
- Wolfies
- Forget-me pill

Effects of Rohypnol

Most long-term effects of Rohypnol are still unknown. Rohypnol is a highly addictive drug; regular use can lead to physical and psychological dependence. When mixed with alcohol or other drugs, Rohypnol is potentially lethal. It can slow down the heart rate and breathing, which can lead to unconsciousness, coma and, ultimately, death.

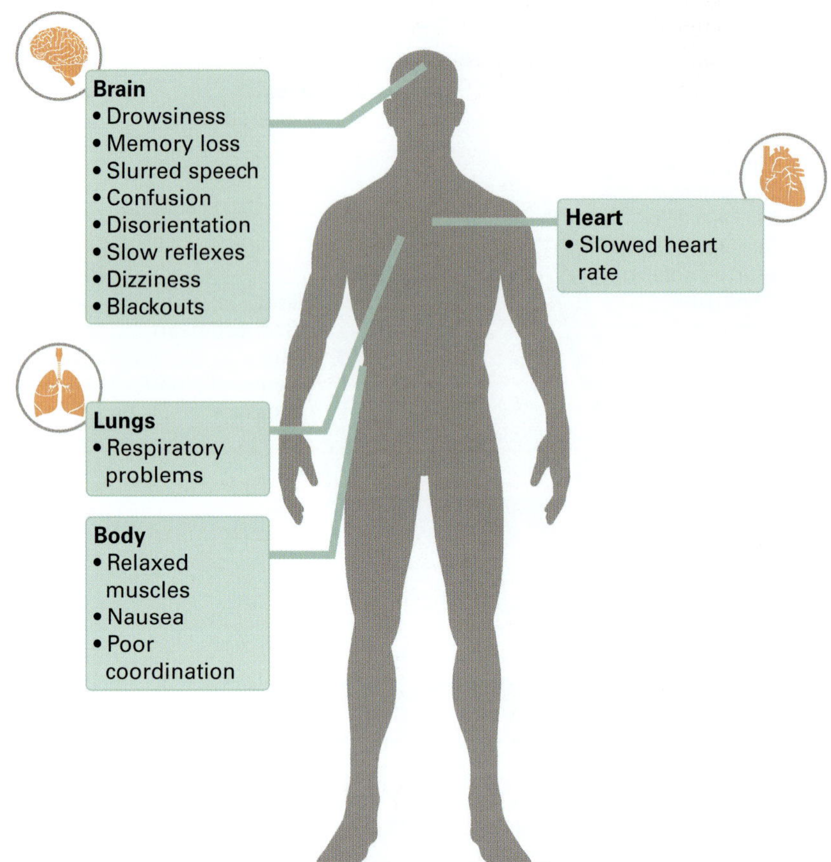

Figure 1.20
Short-term effects of Rohypnol.

Heroin

Drug classification: depressant

Heroin is an illegal, highly addictive drug made from the opium poppy. It belongs to the opioid family, which includes morphine, oxycodone, codeine and methadone. Heroin is a depressant that slows the activity of the central nervous system. This drug is usually a fine powder that can vary in colour from white to brown. Heroin is generally injected but can be snorted or smoked.

FAST FACT

Other names for heroin include:

- Smack
- Gear
- H
- Hammer
- Harry
- Horse
- Junk

Effects of heroin

One of the most damaging long-term effects of heroin use is addiction. This is characterised by compulsive drug seeking and the need to feel the 'rush' just to experience normality. The more you take heroin, the greater the risk of overdosing. Long-term abuse can result in dependence, depression, mental dysfunction, infertility, organ failure and death.

Shisha (flavoured tobacco)

Drug classification: depressant

Shisha is a flavoured tobacco that is smoked through a hookah or water pipe. It was originally developed in India and the Middle East. Common tobacco flavours include apple, grape and rose. The shisha is heated by

FAST FACT

Other names for shisha include:

- Hubbly bubbly
- Hookah

9780170463102

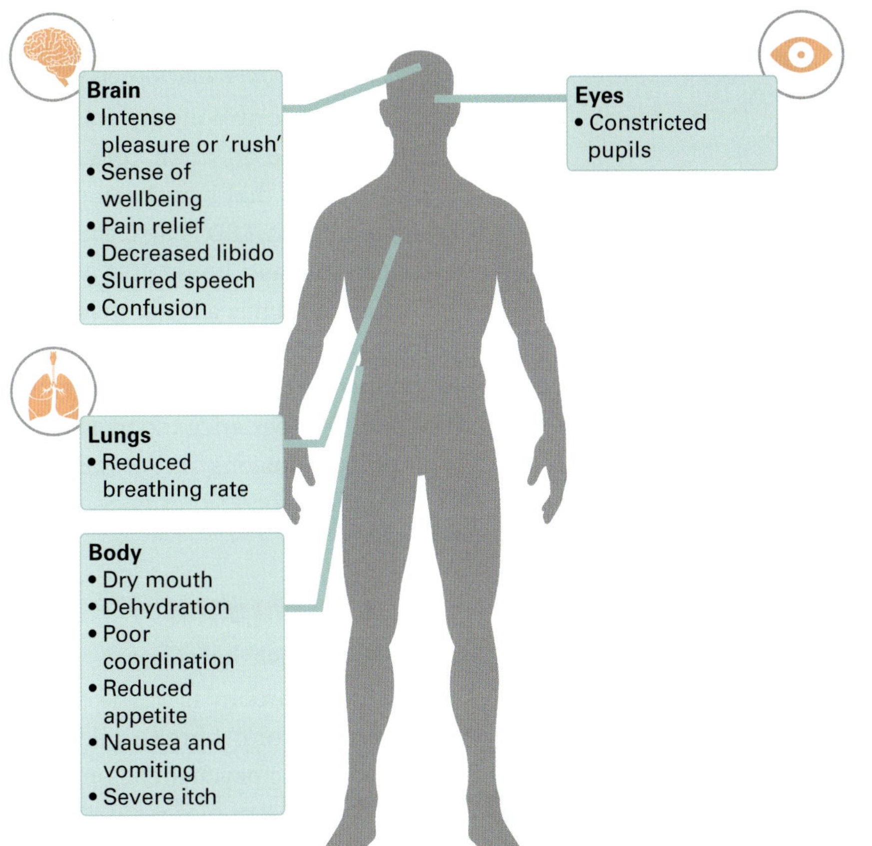

Figure 1.21 Short-term effects of heroin.

placing burning coals above it. The smoke produced from the tobacco bubbles up through a chamber of water and is then sucked into the mouth via a long flexible tube. In recent years, the smoking of shisha has grown in popularity.

Alamy Stock Photo/Agencja Fotograficzna Caro

Figure 1.22 Shisha can be as harmful as cigarettes.

Shisha No Thanks campaign (NSW)

The 'Shisha No Thanks' campaign was launched in 2019. It is a New South Wales research translation project aimed at raising awareness of the dangers associated with shisha (water pipe) smoking in young people from Arabic backgrounds.

Weblink Further information about the 'Shisha No Thanks' campaign, including resources, can be found online.

FAST FACTS

1. A 45-minute shisha session can be as harmful as smoking 100 cigarettes!
2. Shisha smoke is toxic – like cigarette smoking, it can lead to lung cancer and heart disease.
3. Australians spend more than $14 billion a year on cigarettes and tobacco. This equates to a total average individual cost of around $5237 a year.

Dreamstime.com/Bakalaeroz Photography

Figure 1.23 LSD is usually sold on colourful tabs like these.

FAST FACT

Other names for LSD include:

- Acid
- Tabs
- Trips
- Blotters
- Microdots

LSD

Drug classification: hallucinogen

LSD, or lysergic acid diethylamide, is most commonly sold on small, colourful pieces of blotting paper that have been dipped into a prepared liquid. LSD is also soaked into sugar cubes or sold as a liquid. It is odourless and colourless and has a bitter taste. The first effects of LSD occur within 30 to 90 minutes of taking the drug and are unpredictable. Users may experience mood swings, changing emotions, delusions and/or hallucinations. A 'bad trip', or bad experience, can also result in death.

Alamy Stock Photo/Everynight Images

Figure 1.24 Magic mushrooms can be dried, brewed, cooked, smoked with tobacco or eaten raw from the ground.

Magic mushrooms

Drug classification: hallucinogen

A 'magic mushroom' is a type of fungus that grows naturally in the wild. It contains a mind-altering psychedelic property that alters the perception of reality. Once the drug is consumed, the immediate effects can last between four and six hours but, as with any drug, there are dangers. There is no way of knowing whether the experience will be enjoyable or not.

Worksheet 1.6

FAST FACT

91 per cent of Australian's aged 14+ have never used hallucinogens. It is normal not to use drugs.

Source: 'Young people, "ice" and ecstasy: Sorting out fact from fiction' by Paul Dillon, Drug and Alcohol Research Training Australia, https://darta.net.au/wordpress-content/uploads/2016/02/TEACHERS-STIMULANTS-2016.pdf

FAST FACTS

Other names for magic mushrooms include:

- Shrooms
- Mushies
- Magics
- Golden tops
- Liberty caps

CASE STUDY

EFFECTS OF DRUG TAKING

Identify

Prescription medication is sold with information packs outlining common side effects. Illegal drugs do not and therefore it is difficult to predict what may happen as there are often unknown ingredients in the drug and therefore unpredictable side effects.

Understand

Preston Bridge, a 16-year-old student from Western Australia, was celebrating with his friends in a Perth hotel after his school ball. That fatal night, Preston fell to his death from the hotel balcony. He had taken a $2 LSD tab that had been purchased online. The deadly drug is thought to have been the hallucinogen known as 25I-nBOMe, a synthetic version of LSD, and 60 times stronger. Had Preston known what he was taking, he may have made an entirely different decision that night.

Sideffect Australia was founded by Preston Bridge's father, Rodney Bridge, to educate young Australians about substance use. Explore the Sideffect website before answering the questions below.

SIDEffECT Australia

Source: 'Dad's hard-hitting drug message to change young minds' by Rourke Walsh, *The West Australian*, 26 July 2017

Discuss

1 State why was the website Sideffect Australia created.
2 Identify Sideffect Australia's mission.
3 Discuss Sideffect Australia's aim.
4 Communicate Sideffect Australia's goal.
5 Determine ways you can assist Sideffect.

Weblink
Watch The Ripple Effect videos, which aim to inform young Australians about the dangers associated with drug use and peer pressure. Do you think they are successful?

Ketamine (date rape drug)

Drug classification: hallucinogen

Ketamine is a fast-acting **dissociative anaesthetic** used primarily for veterinary purposes. Once consumed, the drug can take effect within five minutes and the effects can last for four to eight hours. Ketamine has a hallucinogenic property, which means it alters the way of thinking, distorts sight and sound, and changes the perception of time and emotion. Many describe the effects of ketamine as being like an out-of-body experience, where the mind and body 'separate' from each other. When sold illegally for recreational use, ketamine usually comes in the form of a crystalline powder or a pill. It is also known as a date-rape drug, as it can easily be dissolved in a liquid.

dissociative anaesthetic a drug that produces feelings of dissociation or detachment from the mind and body and distorts the perception of sound and sight

Getty Images/Nicolas Asfouri

Figure 1.25 Research is currently being conducted into whether ketamine can be used to treat depression.

FAST FACT

Other names for ketamine include:

- Special K
- Super K
- Kitkat
- K
- Ket

Effects of hallucinogens

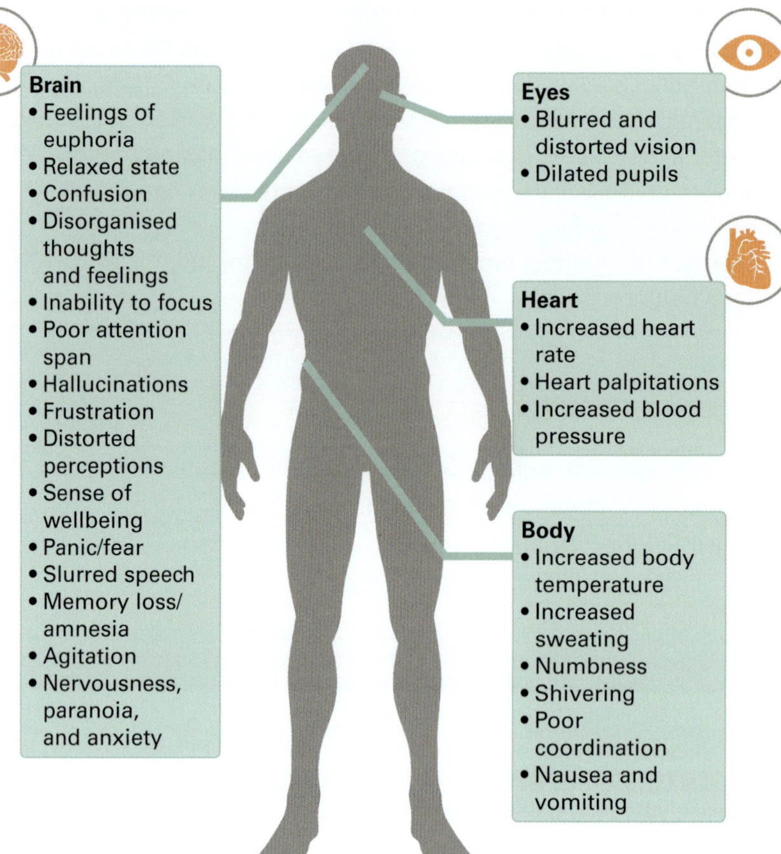

Figure 1.26
Short-term effects of hallucinogens.

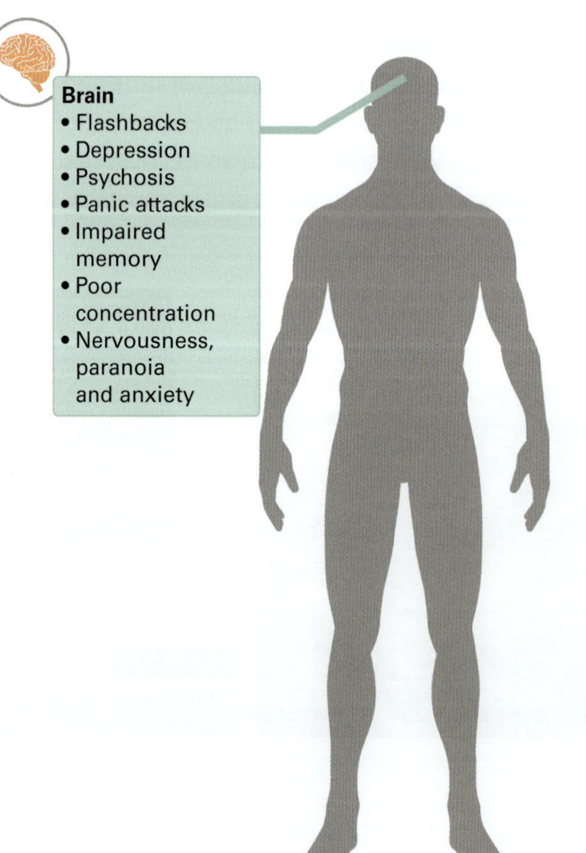

Figure 1.27
Long-term effects of hallucinogens.

FACE TO FACE Have your say

Read the following scenarios. In pairs, discuss how you would respond and provide three strategies for each scenario.

1 It's a Saturday night and you and your best friend head to a party. Your best friend has offered to be the designated driver (DD) and will be responsible for getting you home safely after the event. Halfway through the evening you discover that your DD has started drinking. What do you decide to do?

2 Your parents have gone away for a few days and your older sister decided to have a last-minute party. You have no idea how many of your sister's 'friends' are there and there are drunk people everywhere. The house is an absolute mess. What do you decide to do?

Worksheet 1.7

Drug use and the law

It is an offence to use, possess, supply and/or manufacture illegal drugs in Australia. If caught, a person could receive a criminal conviction, a substantial fine and/or possible imprisonment.

Weblink
Play the Pure Rush drug education game – your mission is to get to a music festival before the tickets sell out. You need to navigate through different Australian landscapes, avoiding illegal drugs along the way.

Class game – How do drugs affect the body?

Several signs will be placed around the classroom. Each sign will have the name of a drug that you have studied in this chapter. Your teacher will read out a fact about one of these drugs. After hearing the fact, identify the drug it relates to and move to the correct sign.

1 Describe the five main reasons people use drugs.
2 Which drugs are often used to spike drinks?
3 Describe the similarities and differences between two hallucinogens in terms of their:
   a use
   b appearance
   c effects.

1 You are a 16-year-old girl and are at your friend's eighteenth birthday party. You brought your own alcohol, but have drunk it all by 11 p.m. Your friend's older brother, who you have never met, offers you a can of pre-mix alcohol that has already been opened. What do you do? Discuss as a class.
2 Listen to the podcast on energy drinks and then answer the following questions:
   a Discuss why energy drinks are considered to be so dangerous.
   b Explain why athletes should avoid consuming energy drinks.

Weblink
ABC Radio podcast

1 Using the internet, investigate other naturally occurring hallucinogens not discussed in this chapter, and describe their effects on the human body.

2 Find out about how magic mushrooms can be used as a therapeutic treatment by visiting the ADF website.

   a Magic mushrooms are usually associated with which era?

   b What is psilocybin and what is it commonly classified as?

   c Scientists are currently researching the effects of psilocybin on the human brain. Bullet point the major findings.

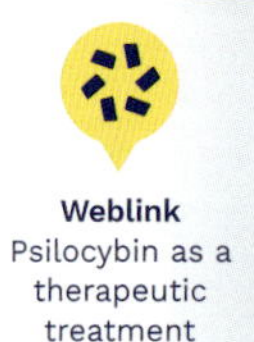

Weblink
Psilocybin as a therapeutic treatment

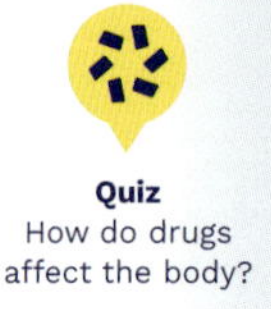

Quiz
How do drugs affect the body?

WHAT FACTORS INFLUENCE THE USE OF DRUGS?

Worksheet 1.4

It is a common misunderstanding that drug users are outcasts or misfits who care only about themselves. Drug users are generally thought of as deviant, untrustworthy and unemployed losers within their communities. However, many drug users are professional, highly skilled workers leading apparently normal lives.

Professionals are more likely to abuse the following drugs:

- ecstasy (MDMA)
- cannabis
- cocaine
- magic mushrooms.

Some reasons why busy professionals would use these types of drugs are to cope with everyday stresses and to 'treat' themselves, as a reward after a hard week at work.

Shutterstock.com/Monkey Business Images

Figure 1.28 Stereotypes don't correctly portray drug users.

PROTECTIVE/RISK FACTORS

Personal, social, environmental/cultural and economic factors play an important part in the choices young people make about drug use. The influence of these factors can either be positive or negative. Any influence that affects your choice can be either a risk factor or a protective factor.

Protective factors

Protective factors help young people avoid the need to abuse substances. These are coping strategies that help you deal more effectively with any dangers or stresses you may be challenged with. Protective factors do not eliminate the risks associated with drug use, but they can reduce the likelihood of risks occurring.

Risk factors

Risk factors increase the likelihood of a young person abusing substances. Risk factors are not predictable, and they can all have a damaging effect on an individual's health and wellbeing.

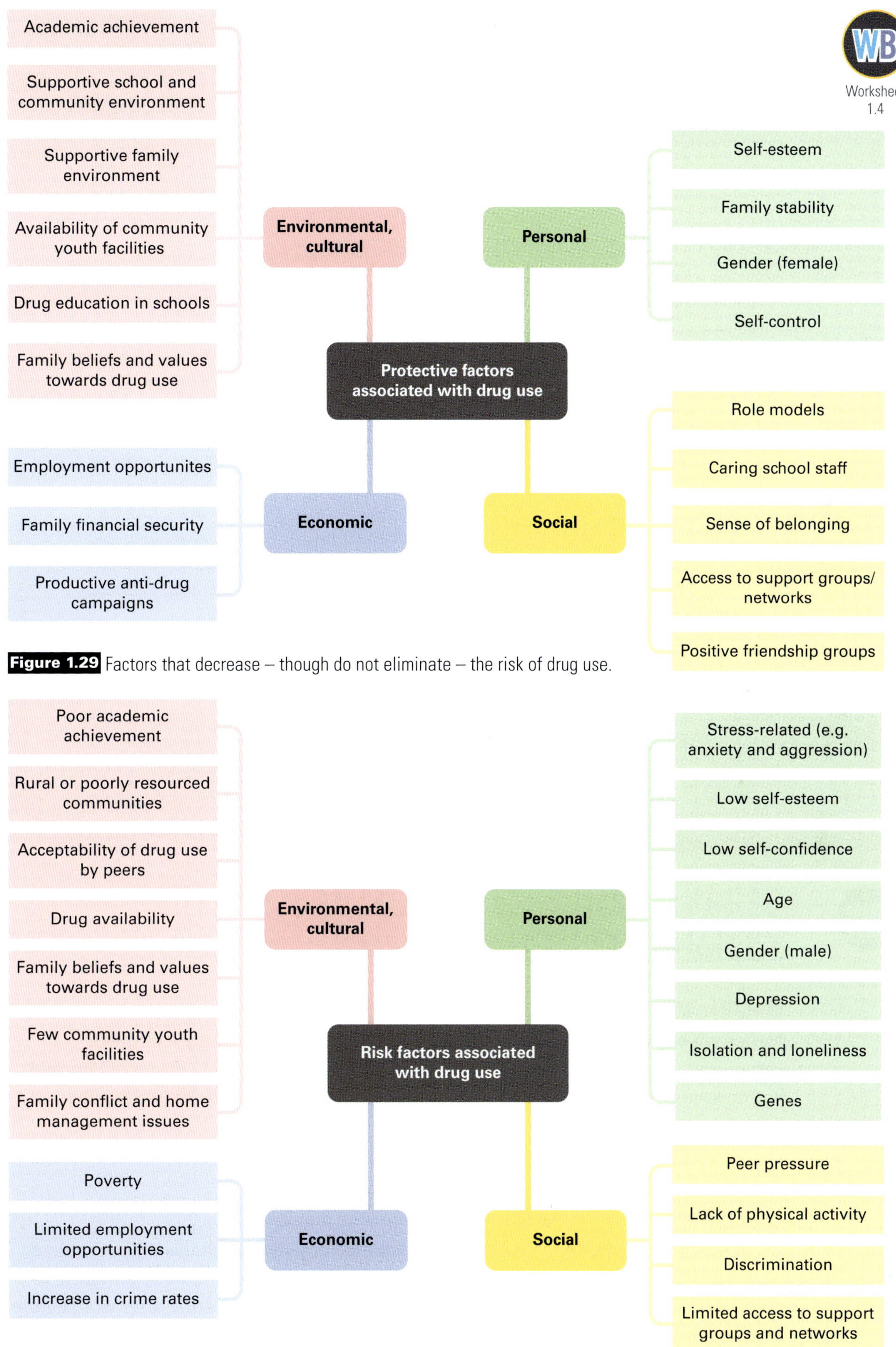

Figure 1.29 Factors that decrease – though do not eliminate – the risk of drug use.

Figure 1.30 Factors which increase the likelihood of drug use.

WB Worksheet 1.4

CASE STUDY ⇨ PROTECTIVE VERSUS RISK FACTORS

Identify

Read through the following scenarios and analyse the protective and risk factors affecting Sam and Melissa.

Understand

Sam is a 14-year-old boy who lives in a small country town, 10 kilometres from the ocean. He lives with his mum, who works part-time at a local store. Sam's mum and dad separated when he was nine years old. He complains that there is not much for kids his age to do in town and he is easily bored. Sam is an only child and attends the local high school. He used to enjoy school and was getting good grades, but in recent months he has seemed distracted and his grades have suffered. Sam has no idea what he wants to do after he finishes school. Each day, after school, he hangs out with a small group of mates at the local run-down skate park.

Melissa is a 14-year-old girl who lives in the outer suburbs of a large city. She lives with her mum, dad and younger brother. Both Melissa's mum and dad commute to work in the city every day. Melissa enjoys going to school and her grades are excellent. She hopes to go to university. Melissa plays netball twice a week for a local club. She has several friends and enjoys shopping, watching movies and spending time with her best friend.

Discuss

1. In small groups, compare the lives of Sam and Melissa.
2. Discuss the personal, social, environmental and economic factors that may influence either Sam or Melissa to abuse or avoid substances.
3. Prepare a visual summary to go with the case studies, including images of both Sam and Melissa.

Worksheet 1.8

BEHAVIOURAL FACTORS INFLUENCING THE COMMUNITY WELLBEING

Norms and stereotypes

Norms are behaviours that are considered acceptable within a group, culture or community. Norms generally refer to things that are expected and considered normal within society such as rules that guide behaviour. Examples of norms in the public setting include:

- not cutting in front of someone when standing in a queue
- sitting quietly during assemblies or parades
- offering your seat on the tram to someone who may need it.

In Australia, alcohol consumption is considered socially acceptable. It is considered a big part of Australian culture. Whether you're at a sporting event, BBQ, gathering or even a funeral, alcohol is often available. It's common to go for a drink after work before heading home. Drinking alcohol is considered a norm in most cultures.

Stereotypes are generalisations about specific types of people or individuals. These assumed generalisations, or beliefs, are often seen as inaccurate, negative and overgeneralised. Examples of stereotypes may include:

- boys wearing blue and girls wearing pink
- all obese people eat too much
- Asian people are excellent Badminton players.

It is widely known among other cultures that Australians love to drink. Such a global assumption is perceived as a stereotype. While alcohol is seen to be deep-rooted in Australian culture, the stereotype of a nation of hard drinkers does not reflect reality. In fact, more Australians are giving up alcohol.

FACE TO FACE Stereotypes

In small groups (same gender), discuss the positive and negative stereotypes associated with:

1 Teenage girls and alcohol consumption
2 Teenage boys and alcohol consumption.

Worksheet 1.9

Alamy Stock Photo/UK Sports Pics Ltd

Figure 1.31 Drinking is arguably a huge part of Australian culture.

HOW DOES DRUG USE AFFECT INDIVIDUALS, FAMILIES AND THEIR COMMUNITIES?

CRIME

Crime and drug abuse have been linked for a long time. Many people who commit a criminal act are doing so to fuel their drug addiction. Common drug-related offences carried out to support a drug habit include:

- manufacture, possession or supply of illicit drugs
- petty theft to armed robbery
- domestic violence
- drug trafficking
- prostitution
- participation in organised crime.

Criminal behaviour resulting from alcohol and/or other drug abuse includes:

- driving under the influence of alcohol or drugs
- violence, including damage to property
- sexual assault.

FAST FACT

Over 50 per cent of Australians would be in favour of having a device fitted to their car that would prevent the car from starting if their BAC was under the legal limit.

Source: 'Australian drink driving survey & statistics 2021', Budget Direct, 1 August 2021, https://www.budgetdirect.com.au/car-insurance/research/drink-driving-statistics.html

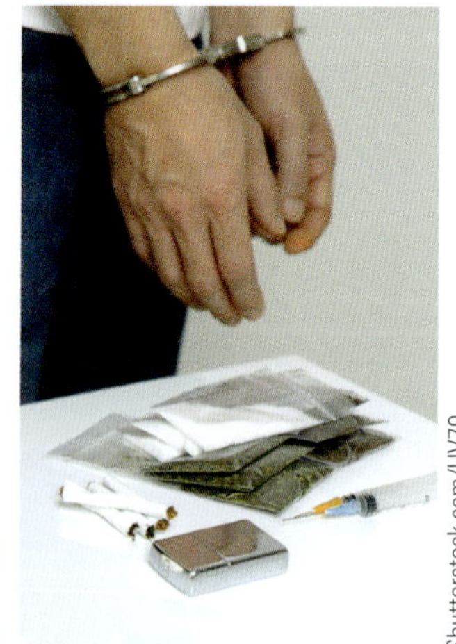
Shutterstock.com/UV70

Figure 1.32 As well as drug-related offences to *support* a drug habit, such as possession or supply of drugs, criminal activity can also *result from* drug abuse, such as drink/drug driving.

CASE STUDY ➩ CRIME AND DRUGS

Identify

All crime starts somewhere. Victoria talks about the recent changes in her 16-year-old sister Amy's behaviour, and her petty theft.

Understand

'Amy is never at school and rarely comes home. We think she is living with her druggie boyfriend. She has changed so much since she met Sean. He is 22 years old and unemployed, and I know he smokes dope! I knew something was wrong when she started taking money from my piggy bank and when Dad caught her taking money from Mum's purse. When I started asking questions she'd just yell at me and say that it was none of my business. Mum gets upset easily these days, and Dad is just not himself. She's never around anymore and her behaviour is destroying our family.'

Discuss

1. Justify why Amy's behaviour has changed since meeting Sean.
2. Put yourself in Amy's shoes. If you could wind back time to the moment you met Sean, what three pieces of advice would you give yourself?

FACE TO FACE The effects of drug use on individuals over time

Look at the four pairs of photos of people who have been damaged by drug abuse. These photos show just how much people's faces can change. In pairs, discuss the following questions:

1. Analyse the photos and describe the changes that you see in each individual.
2. Based on your analysis, which drug do you think might be responsible for these drastic changes.
3. Apply your knowledge of the possible drug responsible and justify why the changes have been so dramatic.

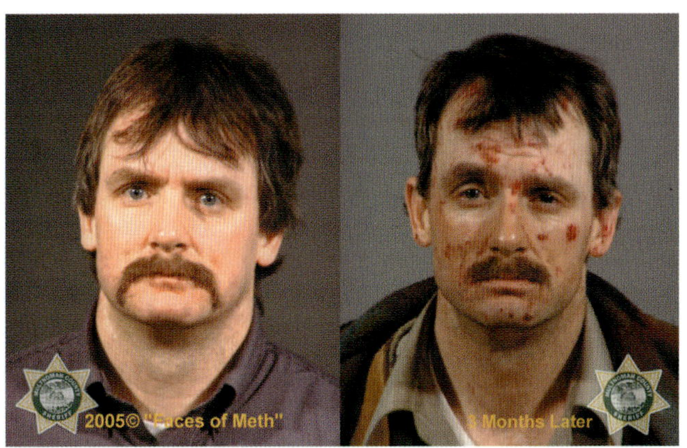

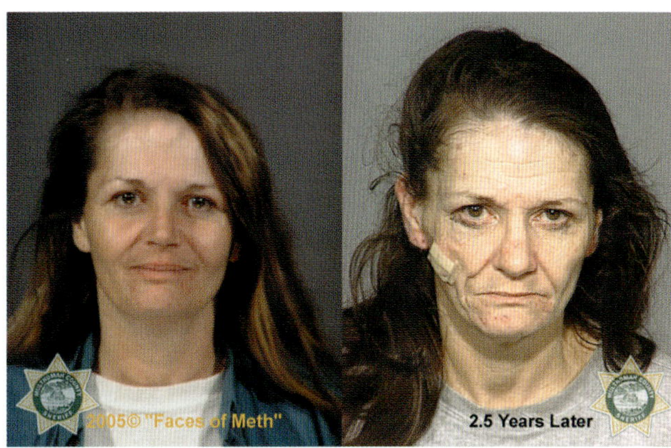

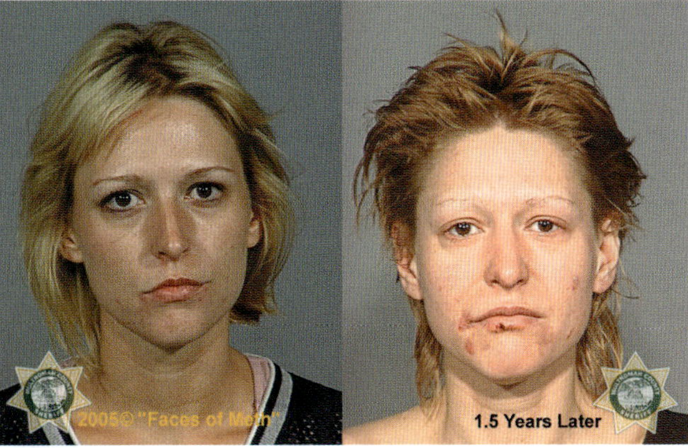

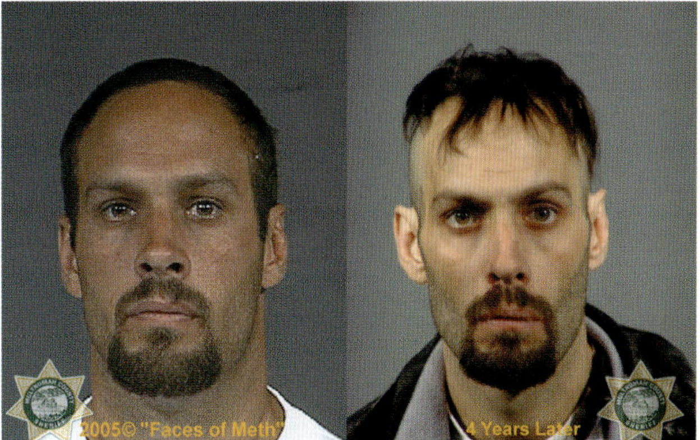

Figure 1.33 Effects of drug use over time.

DRINK/DRUG DRIVING

FAST FACTS

1. Australia has one of the highest rates of vehicle accidents caused by drink driving in the world.
2. Those who drive after being awake for 17 hours have the mental function equal to those with a blood alcohol concentration of 0.05 per cent (the legal limit for driving). This results in impaired coordination, judgement and reaction times.

Driving a car or any motorised vehicle safely is a complex task that requires total concentration and a clear mind. Even a small amount of alcohol affects a person's ability to drive. On Australian roads, reckless drink driving causes significant road trauma and death. Australia has very strict laws about drinking alcohol and driving. The legal blood alcohol concentration or content (**BAC**) limit for drivers across Australia is below 0.05. For **novice drivers**, probationary licence drivers (P-plates) and learner drivers (L-plates), the legal BAC limit is zero, meaning that these people are not allowed to drive with any alcohol in their system at all. Police can stop a driver at any time and request a driver or even a passenger to take a breathalyser test.

BAC blood alcohol concentration or content; refers to the percentage of alcohol in the blood

novice driver someone who has held a driver's licence for less than two years; this does not include the period when the driver was learning (on L-plates)

Alamy Stock Photo/Elspeth Graham

Figure 1.34 Drink driving is not worth the risk to yourself and others.

Driving under the influence

Driving under the influence (DUI) refers to driving when your ability to do so has been impaired by alcohol or other drugs. Depending on the severity of the offence committed and whether the person is a repeat offender, penalties for driving under the influence can include one or more of the following:

- fines
- demerit points on your driver's licence
- cancellation of a driver's licence
- imprisonment.

In addition to the legal consequences, your actions may also result in additional negative consequences, such as:

- the injury or death of a friend or family member; they may require hospitalisation or time off work and be left with medical expenses to pay
- emotional stress – feelings of blame and guilt affecting not only you but also your friends and family
- financial stress – the need to find the money to pay for the damage caused, legal charges and fines.

9780170463102

CASE STUDY

DRUNK DRIVERS IN OUTBACK QUEENSLAND TO FOOT $2K BILL FOR ALCOHOL INTERLOCK DEVICES

Identify

Drunk drivers in Outback Queensland are being stopped in their tracks with new interlock devices as frustrated police try to lower rates of high-risk drink driving.

Understand

The devices feature facial recognition technology and are connected to the Internet, with convicted motorists forced to foot the estimated $2,000 bill for installation.

'These devices are well worth it as they make drivers accountable for their actions', Sergeant Paul Quinlan, officer in charge of the Mount Isa District Road Policing Unit, said.

Beating the system

Up until now, convicted drivers in rural parts of Queensland could apply for a court exemption because there was no qualified technician in the area to install alcohol interlock devices. But an auto technician is due to arrive at the end of September to begin fitting the upgraded systems.

The devices are fitted to vehicles and require a breath test, and now facial verification, to start the ignition.

In 2021, the Queensland state government lowered the limit for a high-risk Blood Alcohol Concentration (BAC) from 0.15 to 0.1 state-wide. Queensland's Assistant Minister for Regional Roads, Bruce Saunders, welcomed the upgraded technology.

'The new system is absolutely brilliant, they're going to be more accurate', Mr Saunders said. 'I don't have any sympathy for drunk drivers'.

Worse than the state average

Sergeant Quinlan said, on average, 55 people were killed in Queensland each year in alcohol-related road crashes, while another 500 were seriously injured. He said the north-west Queensland district had a worse record for drink-driving than other regions.

'On average across Queensland, one in 75 drivers turns a positive breath test', Sergeant Quinlan said. 'In the Mount Isa police district, we have one in 65 drivers turning a positive breath test'.

Discuss

An alcohol ignition interlock device is a small, hand-held breathalyser for car ignitions. They are being installed to prevent drivers from being able to start their vehicle after consuming alcohol.

Should interlock devices be fitted to all Australian vehicles? Examine the case study and information from the Internet to determine your response.

FAST FACT

The DUI offence does not just apply to people who are driving a motor vehicle. You can also be charged if you are using a skateboard, riding a bike or rollerblading under the influence of alcohol or drugs. You can even be charged if you are riding a horse!

Table 1.2 How alcohol concentration affects driving ability

BAC	Driving ability
0.02–0.05	Reduced ability to see or locate moving lights as well as to judge distances Increased desire to take risks Ability to respond to several stimuli is reduced
0.05	Twice as likely to be involved in a road traffic accident
0.05–0.08	Slower reactions Sensitivity to red lights impaired Shorter concentration span Ability to judge distances reduced further
0.08	Five times more likely to be involved in a road traffic accident
0.08–0.12	Driving ability overestimated Reckless driving Peripheral vision and perception of obstacles impaired
0.12	Ten times more likely to be involved in a road traffic accident
0.15	Twenty times more likely to be involved in a road traffic accident

Source: https://www.towardszero.vic.gov.au/

Weblink
Visit the Health Engine website and use the online calculator to estimate your blood alcohol concentration through a simulated drinking session.

Weblink
'Just Over' and 'Grow Up' campaigns

Tackling drink driving

There are many national, state and territory-based campaigns aimed at raising awareness about everyday issues related to drug use that may affect you or those around you. The Road Safety Commission has launched two drink driving campaigns, 'Just Over' and 'Grow Up'. These campaigns are aimed at reducing the road toll on Western Australian roads caused by drink driving, as well as at raising awareness of the dangers and consequences associated with drink driving. What advice would you give to a younger person about drink driving? Would you be happy for them to follow in your footsteps if they saw you as their role model?

FACE TO FACE Community action initiatives

1. In small groups, choose an alcohol or other drug-related issue that you would like to promote in your school.
2. Discuss possible campaign ideas or initiatives that could have a positive influence on the health and wellbeing of others in your school. Plan how you would raise the awareness of your chosen issue.
3. Using a reliable search engine, identify drug and alcohol support services in your local community that could be used to help raise the awareness of your chosen issue.

BLOOD-BORNE VIRUSES

Worksheet 1.11

Blood-borne viruses (BBVs) are viruses that are found in blood or bodily fluids. BBVs can be transmitted from person to person via the sharing of needles associated with drug use or by risky behaviour, such as unprotected sex. Risky behaviour is often associated with alcohol and other drug use. BBVs can be transmitted bodily fluids such as semen and vaginal sex.

The most prevalent BBVs include the Human Immunodeficiency Virus (HIV), Hepatitis B (HBV) and Hepatitis C (HCV). HIV is a virus which attacks the human immune system and can lead to acquired immunodeficiency syndrome (AIDS), if HIV is left untreated. AIDS is the last stage of the HIV infection and this occurs when the body is severely damaged by the virus. There is no cure for AIDS.

The Hepatitis B virus (HBV) is a serious disease that affects the liver. Long-term health issues include inflammation of the liver, liver damage, liver failure, liver cancer and even death. Whilst HBV is considered a chronic medical condition, it can be successfully managed with medical treatment.

Hepatitis C is also a chronic medical condition that also affects the liver in a similar way to Hepatitis B. So, you may be wondering what the difference is between these two viruses. Hepatitis B is transmitted through both blood and bodily fluids whereas Hepatitis C is usually only transmitted through blood. Hepatitis B is more likely to cause death and yes, people can be diagnosed as carrying both Hepatitis B and C at the same time.

Unfortunately, there are some people who may not even realise that they have a BBV as the symptoms associated with BBVs may not present for many years. So, it is important to have regular check-ups with a health professional.

However, BBVs are preventable if you take the following precautions:

1. Wear a condom when having sex.
2. Avoid sharing needles.
3. Have vaccinations (HBV).
4. Ensure that all piercings or tattoos are received from licensed and reputable facilities.
5. Avoid sharing body jewellery such as earrings.

1. Define the following terms: BAC, DUI, drug trafficking.
2. True or false:
   - A person who looks fit and healthy can have a BBV and show no symptoms.
   - The only way to determine whether you have a BBV is to have a blood test.
   - You can contract a BBV when getting a tattoo or a body piercing.
3. Brainstorm all the crimes you can think of related to drugs and alcohol.

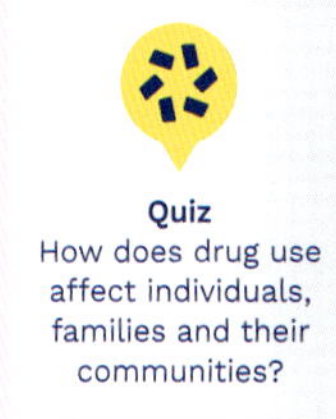

Quiz
How does drug use affect individuals, families and their communities?

1. You are a 17-year-old boy and you are about to head to Bali with a group of friends to celebrate the end of your Year 12 exams. No adults will be going. When you arrive at the airport all your friends have already checked in. Your best friend has just discovered a small packet of marijuana in his pocket. He has already checked his baggage, so he asks you to hide it in your suitcase. What do you do? Justify your decision.
2. You are a 15-year-old girl and are at a friend's party. Before you go to the party you arrange to get a lift home with your best friend. The party is winding down and it's getting late. You have looked everywhere for your best friend but can't find her. A male friend who has just got his license says he can give you a lift home. You know he has been drinking, but he doesn't appear drunk.
   a. What do you do? Discuss with a partner.
   b. What options do you have? As a class, analyse each option and discuss the consequences and possible benefits of each one.

1 Watch the *60 Minutes* interview with Cassandra Sainsbury.
   a Discuss the decisions Cassandra made that led to her serving six years in a Colombian jail.
   b Discuss the consequences of her actions, both on her and on those close to her.

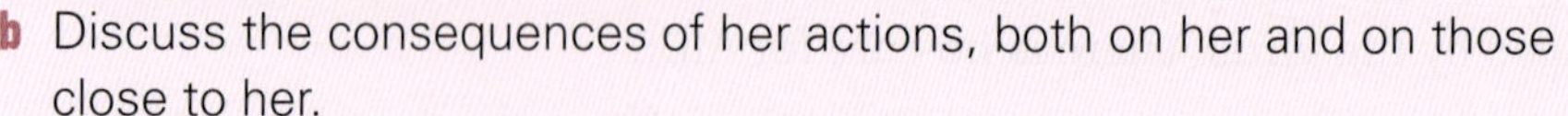

Weblink
60 Minutes: Cassandra Sainsbury

SKILLS TO DEAL WITH CHALLENGING SITUATIONS

Values and concerns

According to the 2021 National Australia Youth Survey Report, the three most valued items amongst males and females aged between 15 and 19 were friendships (other than family), family relationships and mental health. The same survey also identified the top five issues of personal concerns for young people. These were coping with stress, mental health and school and study problem, body image and COVID-19. Alcohol and drugs were ranked 16th and 17th in the list of personal concerns for both males and females.

Source: Tiller, E., Greenland, N., Christie, R., Kos, A., Brennan, N., & Di Nicola, K. (2021). Youth Survey Report 2021. Sydney, NSW: Mission Australia

Table 1.3 What young people (15–19 years) value as important in life

Male		Female	
Friendships (other than families)	81.5	Friendships (other than families)	82.3
Family relationships	77.5	Family relationships	79.7
Physical health	67.8	Mental health	71.6
Mental health	60.9	School or study satisfaction	70.7
School or study satisfaction	59.0	Physical health	63.8

Source: Tiller, E., Greenland, N., Christie, R., Kos, A., Brennan, N. & Di Nicola, K. (2021). Youth Survey Report 2021. Sydney, NSW: Mission Australia

Table 1.4 Issues of extreme personal concern to young people (15–19 years)

Male		Female	
Coping with stress	27.0	Coping with stress	59.5
Mental health	25.2	Mental health	52.8
School or study problems	23.3	Body image	47.2
Physical health	22.3	School or study problems	47.0
COVID-19	18.0	COVID-19	36.5

Source: Tiller, E., Greenland, N., Christie, R., Kos, A., Brennan, N. & Di Nicola, K. (2021). Youth Survey Report 2021. Sydney, NSW: Mission Australia

FACE TO FACE Personal values and concerns

1 As the data from 2021 National Australia Youth Survey Report was broken down into gender, divide the class into one male and one female group. As a gender group, create a list of your top five most valued items and the top five issues of concerns. Compare male and female results.
2 In pairs, consider the results of the 2021 National Australia Youth Survey and the class results and discuss reasons why there might be gender differences.
3 Consider a personal concern. Decide whether this concern may influence a young person to use alcohol or other drugs. Provide reasons for your decision.

Personal strengths

Every one of us has unique character qualities that make us who we are. Our qualities are our strengths, and these strengths are very important as they influence how we act, think and feel. They define who we are and are very important in our growth and development.

Dealing with challenging situations can be very frustrating. Facing challenges helps you to build important personal strengths that you will need throughout your lifetime, such as determination, courage, empathy, resilience and confidence. Consider every challenge you face as an opportunity to grow and use your strengths to help you manage any situation you find yourself in.

FACE TO FACE Using personal strengths to manage situations

Refer to the VIA Character Strengths webpage. Discuss the following scenario using the questions below to guide you.

Yetunde and Ed have been at a pool party and are waiting for a bus to take them home. Clancie pulls up in her car and offers them a lift. She lives very close. Yetunde wants to get in the car but Ed notices that there are already three others in the car. He knows that his parents would not be happy if he travelled home in an overcrowded car, not wearing a seatbelt. Yetunde opens the door to get in.

- How might Ed be feeling in this situation?
- Discuss any negative thoughts Ed might be experiencing.
- What choices does Ed have?
- How could Ed respond in a positive way?
- What personal strengths could help Ed manage this situation?

Weblink
VIA Character Strengths

WHAT SUPPORT STRATEGIES CAN I USE TO MAKE INFORMED DECISIONS ABOUT DRUGS AND ALCOHOL?

There are strategies that you can use to make informed decisions about alcohol and other drugs. These include assertive behaviour, support from your friends and harm minimisation.

PARTY SAFETY

Everyone enjoys a fun party or gathering. Many young people, in particular, like to party and enjoy themselves, but one bad decision could turn a fun night into a nightmare. When planning a party or just relaxing with friends, the last thing you expect is

WELLBEING CHECK IN

DEEP BREATHING

Identify

Sometimes when we're feeling bad, a big deep breath can help. Taking some time to breathe deeply is even better.

This activity has been developed in collaboration with Dr David Bakker of MoodMission

Understand

Our breathing tends to become fast and shallow when we're feeling upset or stressed. Some people use drugs and alcohol to feel calmer and reduce stress, but we can actually trick our bodies and brains into feeling calmer by breathing slower and deeper. If you are stressed, such as at a party that is getting out of control, you could use this technique to calm down so you can think more clearly about what to do.

Practise

1 Find a quiet place where you can sit down. You can do deep breathing while standing, walking or doing other things, but it might be easiest to practise while sitting with nothing else to do.
2 Place your hand on your belly, just beneath your ribs. This is where you'll be breathing from – pushing your belly out against your hand when breathing in and letting your hand fall back towards you when breathing out.
3 Set a timer for one minute.
4 Take a slow, deep breath in for a count of 4.
5 Hold the breath for a count of 2.
6 Breathe out for a count of 4.
7 Hold for a count of 2 before breathing in again.
8 Continue this 4-2-4-2 counting while breathing in and out slowly.
9 After a minute, finish up and note how you're feeling.

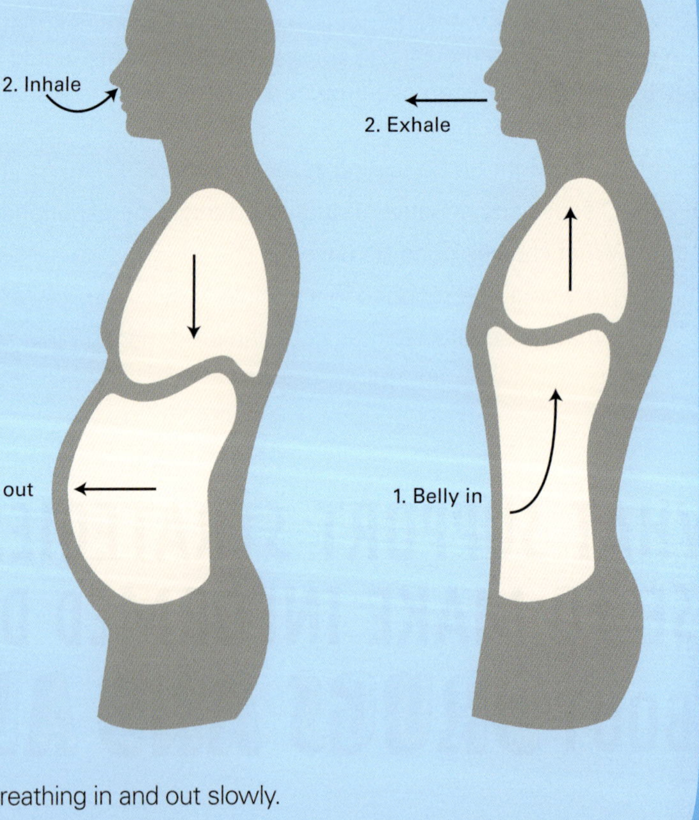

Reflect

Do you feel any more relaxed after doing this? You may feel a little dizzy; this is because your blood is getting more oxygen in it than normal. This is totally fine and safe. To reduce the dizziness, try making the out breaths slightly longer.

When might you be able to do this deep breathing?

9780170463102

for something to go wrong. However, things can go wrong, especially as a result of the following behaviours:

- binge drinking
- drink driving
- drink spiking
- underage drinking
- drug overdose
- unprotected sex
- sexual assault
- violence
- trouble with the police.

iStock.com/rez-art

Figure 1.35 How do you explain the hole in the wall or the vomit stain on the couch to your parents? What would you say to your parents if you were arrested?

CASE STUDY ⇨ PILL TESTING AUSTRALIA

Identify

Pill testing is a way to minimise the harm from drug use by allowing users to find out the exact content and purity of a substance before they take it. This means they can make an informed decision about consuming the substance, potentially reducing the likelihood of a fatal overdose from unknown contaminants. Pill testing is seen as a 'harm minimisation' approach to drug use. Australia's first legal pill testing program was set up at Canberra's 'Groovin The Moo' music festival in 2018. In addition to pill testing, users also have the opportunity to talk to a drug counsellor. Pill testing does not suggest that it is safe to take a drug, and cannot predict how that drug is going to react to an individual's body chemistry.

Weblink
Watch the SBS 'How pill testing works' video, which outlines the pill testing procedure.

Understand

The lives of seven young people were potentially saved at the weekend when a second pill testing trial at the Groovin the Moo festival in Canberra identified lethal substances.

Patrons discarded their drugs after testing alerted them that their pills contained dangerous n-ethylpentylone. MDMA was the most common substance identified followed by cocaine, ketamine and methamphetamines.

Organiser Gino Vumbaca from Pill Testing Australia hailed the trial a success and said 171 samples had been tested on behalf of 234 participants. He said there were long queues to use the pill testing service and people waited patiently for up to 20 minutes. Testing takes about 10 minutes.

About 35 volunteer doctors, chemists and counsellors worked on shifts during the trial to educate youngsters about drug use and negative effects. One doctor flew over from Perth to participate.

'I spoke to a lot of young people as they walked out and the most common thread was they know a hell of a lot more now than they did when they walked in,' Vumbaca told the *Guardian*.

'Some were very appreciative saying, "Thank God you guys are here because we know (about the recent festival deaths) in New South Wales and Queensland and other states".'

Source: *The Guardian*, www.theguardian.com

Discuss

1 In pairs, using the internet, case study and video clip, formulate arguments for and against pill testing.

2 Do you believe pill testing should be rolled out at all music festivals across the country?

Strategies when planning a party

Worksheet 1.10

Involve your parents in the planning.

Provide a 'chill out zone' for friends who need a place to relax. Make sure it is supervised by an adult.

Make the party 'invitation-only' and request RSVPs to ensure you know exactly who will be attending. Do not post details about the party on social media sites such as Facebook.

Have an emergency plan in case things go wrong. Keep emergency telephone numbers by the phone.

Indicate a definite start and finish time.

Keep the number of a taxi service in your contacts list.

Ensure you have responsible adults around who can check people arriving at the party, to avoid unwanted guests (gatecrashers).

Tell your friends to leave their cars at home if they want to drink.

Consider asking other parents if they can help out with security.

Use plastic containers only.

Provide plenty of food.

Provide only water and non-alcoholic drinks.

Figure 1.36 A suggested checklist for responsible party planning.

Strategies for when you attend a party

Use the following checklist to help you enjoy parties safely.

- Plan a safe ride home (designated non-drinking driver, parent, taxi, public transport).
- Don't assume you can invite additional people to an invitation-only party. Ask the host before you invite anyone to come with you.
- Never leave your drink unattended.

In many states around Australia, it is now against the law to serve alcohol in a private home to a person under the age of 18, unless their parents have provided consent. The law is getting tougher on underage drinking, and adults caught serving alcohol to people under 18 may face hefty fines.

Weblink
Visit the NSW Government website 'Your Room' to test your knowledge about drugs and alcohol.

9780170463102

Dealing with gatecrashers

Too often you hear about a teenage party that has become uncontrollable. Many of these parties are advertised innocently on social media sites, including Facebook. Consider what you will do if uninvited guests show up at your party. Would you let the gatecrashers in? Do you need some help to work out what you would say?

1. Act quickly by refusing the gatecrashers entry and ask them to leave immediately. Be polite and calm and use non-confrontational, yet confident, body language. Remember, you want to avoid starting a fight.
2. If the gatecrashers don't leave and become aggressive, tell them that the police will be called. Make sure you have friends around you to politely back you up.
3. Continue to explain to the gatecrashers that the party is invitation-only and that you would like them to leave. Remember, you have the right to refuse people entry to your property, and trespassing is an offence.
4. Expect a little resistance from the gatecrashers; like you, they may be just looking to have a good time ... or they may be trouble! Continue to explain that the party is private.
5. Keep an eye on your unwanted guests as they leave, but don't follow them off your property.
6. Have emergency numbers available and call the police if the situation becomes out of control.

Figure 1.37 These are a few handy hints that will help you manage this situation if it happens.

Remember, be assertive and take ownership of the situation. Trust your instincts. It's your party and you have an obligation to keep everyone safe.

If you are concerned at all, speak to your parents or the designated 'responsible adult' who is monitoring your party. And, should the situation get out of control, don't be afraid to call the police.

CASE STUDY → RIOT POLICE SHUT DOWN OUT-OF-CONTROL HOUSE PARTY

Identify

The police riot squad had to be called in after a house party quickly spiralled out-of-control in Perth.

Understand

Officers were called to a Ballajura home on Meadowview Drive after reports of juveniles and adults drinking in the street. An hour later the party was declared out-of-control. At least 400 people RSVP'd to the event advertised on Facebook.

Fairfax Syndication/Nick Moir

Figure 1.38 Riot police.

The mother of the 17-year-old who organised it insists the party was supervised and there were no more than 50 people at the rental property.

'It was not 200, I can't allow 200 people to come in house, the house is not for me,' Rachel Dweh told 9NEWS.

It is alleged while the party-goers were leaving the home, several fights began breaking out between them. A few bottles and cans were thrown towards police, and some of the people kicked street signs and bins. Some neighbours feared for their safety.

'Absolutely unsafe, yesterday was very unsafe. There was nothing we could have done if it had got out of control on the front lawns and it was scary,' Sumen Bhatia said.

Two 17-year-old boys were issued an Infringement Notice for Disorderly Conduct.

This out of control party isn't the first police have attended to in Ballajura – in the past few weeks police were forced to break up another party where residents saw party-goers smash property windows and damaged parked cars.

'People who host it need to be fined big dollars to stop it,' witness Brigitte Saleh said.

Source: 'Riot police shut down out-of-control house party' by Michael Stamp, 9news.com.au, 12 November 2017

Discuss

1 What impact does an out-of-control party have on the surrounding community?
2 How can you ensure a party does not spiral out of control as in this case study?

HOW TO HELP YOUR FRIENDS

What would you do if your friend passed out or collapsed because they had drunk too much alcohol or had a reaction to drugs they may have taken?

If anyone collapses, ALWAYS dial 000 and call for an ambulance.

There is a common misconception that if you call an ambulance, the police will be notified as well. However, ambulance officers only call the police if they are in danger or someone dies. For example, if they attend an uncontrollable party and there is violence and rioting, an ambulance officer will need to call the police.

The ambulance officers are there to help, so you will need to ensure that you provide as much information as possible, including:

- what has happened
- how much alcohol and/or drugs have been consumed
- how long ago the alcohol or drug was last consumed
- whether your friend has any existing medical conditions.

Never leave a friend who needs help just because you want to avoid trouble. Never walk away from either a friend or your personal responsibility for others.

Weblink
The Recovery Position

The recovery position

Watch the St John Ambulance video clip: The Recovery Position.

This clip shows you how to put an unconscious, yet breathing person into the recovery position if you are by yourself.

In pairs, practise placing each other into the recovery position until you are sure you can do it.

DRSABCD action plan

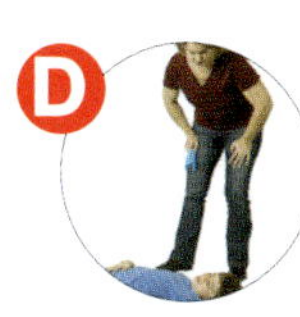

Danger Check for danger and ensure the area is safe for yourself, bystanders and the patient.

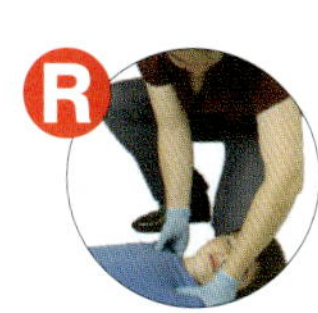

Response Check for a response: ask name and squeeze shoulders. **No response?** Send for help. **Response?** Make comfortable; monitor breathing and response; manage severe bleeding and then other injuries.

Send for help **Call triple zero (000)** for an ambulance or ask a bystander to make the call. Stay on the line. [If you are alone with the patient and you have to leave to call for help, first turn the patient into the recovery position before leaving.]

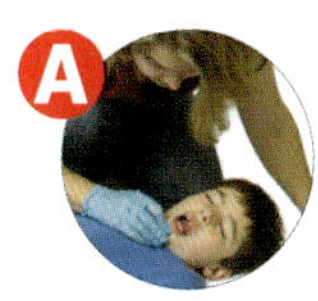

Airway Open the patient's mouth and check for foreign material. **Foreign material?** Roll the patient onto their side and clear the airway. **No foreign material?** Leave the patient in the position found, and open the airway by tilting the head back with a chin lift.

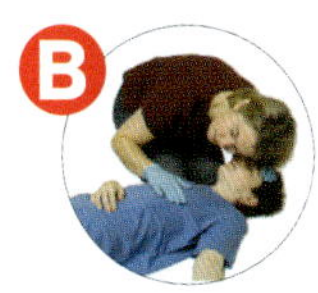

Breathing Check for breathing Look, listen and feel for 10 seconds. **Not normal breathing?** Ensure an ambulance has been called and start CPR. **Normal breathing?** Place in the recovery position and monitor breathing.

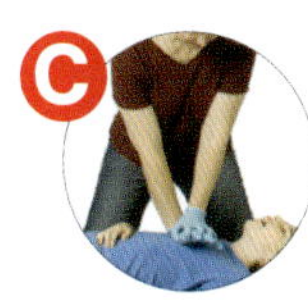

CPR Start CPR: 30 chest compressions followed by 2 breaths. Continue CPR until help arrives, the patient starts breathing, or you are physically unable to continue.

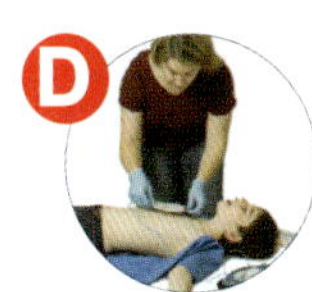

Defibrillate Apply a defibrillator as soon as possible and follow the voice prompts.

In a medical emergency call Triple Zero (000)

DRSABCD Danger ▶ Response ▶ Send for help ▶ Airway ▶ Breathing ▶ CPR ▶ Defibrillation

You could save a life with first aid training • www.stjohn.org.au • 1300 360 455

This information is not a substitute for first aid training. Formal instruction in resuscitation is essential.

Figure 1.39 DRSABCD action plan.

Worksheet 1.12

Performing CPR

Watch the video clip: How to do CPR on an adult.

This clip provides an accurate demonstration of how to perform CPR on an unconscious person who is not breathing.

In pairs, practise performing CPR on manikins if you have access to them. **Do not practise CPR on each other!**

Weblink
How to do CPR on an adult

Quiz
How are performance-enhancing drugs used in sport?

1 Describe three ways you can plan a party responsibly and safely.
2 Describe three ways you can attend a party responsibly and safely.

1 Read the following party scenarios:
   - You go to a party with a group of friends. At the party you are offered an LSD tab for the first time.
   - You've been going out with your boyfriend/girlfriend for a while and things are going really well. However, you are not ready to have sex yet. One night at a party you both go upstairs. Things are starting to move too fast.
   - You are at a friend's party and decide you want to go home because you are not feeling well. But no one else wants to leave yet.
   - You're about to leave a party with some friends. You've had a great time but it is 2 a.m. and you know you have to walk home. You are offered a lift by a friend's 19-year-old brother. You know he has been drinking but he seems okay.

   After reading the scenarios, in pairs, consider the following questions:
   a What are the risks?
   b How might you be feeling in this situation?
   c What options do you have?
   d What will you decide to do and why?
   e If you had consumed alcohol, do you think your decision would have been different?

1 Investigate other first-aid skills that might be required at a party.
   a Brainstorm three injuries you could sustain at a party while drinking.
   b Research online to find out how to treat each one.
   c Create a leaflet or poster that can be handed out/put up at school to teach students first aid for parties.

HOW ARE PERFORMANCE-ENHANCING DRUGS USED IN SPORT?

FAST FACT
The World Anti-Doping Authority (WADA) has banned more than 192 performance-enhancing substances and methods. New performance-enhancing drugs are constantly being developed, as are the methods of detection.

The use of performance-enhancing drugs in sport is a controversial and topical issue. Drugs have been used in sport for hundreds of years. The early Greek Olympians used a variety of herbs and mushrooms to enhance their sporting performance. Today, sport is a multi-billion-dollar industry, and the difference between winning and losing can have major financial implications for top athletes who relentlessly push themselves to their limits. The development of professional athletes takes years of training, hard work and dedication. Some athletes consider using

9780170463102

illegal performance-enhancing drugs to help achieve that 'winning edge' and fulfil their ultimate dream. What do you think it means to be 'the best'?

Why take performance-enhancing drugs?

There are many different types of illegal performance-enhancing drugs available to athletes. The type of drug used by an athlete will depend on their needs. Drugs may be taken for different reasons, such as to:

- help an athlete recover from an injury more quickly
- develop greater physical performance, such as speed, strength or endurance
- help athletes train harder and longer
- help athletes be the very best.

Table 1.5 describes various illegal performance-enhancing drugs.

AAP Image/AP Photo/Michel Euler

Figure 1.40 Australian showjumper Jamie Kermond was provisionally suspended by Equestrian Australia for testing positive for the stimulant cocaine.

Table 1.5 Illegal performance-enhancing drugs

Drug	Target group	Advantages	Disadvantages
Anabolic steroids	Sports requiring strength and power (e.g. weightlifters, endurance athletes)	- is a synthetic substance - increases muscle mass - increases strength - reduces recovery time - allows athletes to train harder	- depression - aggression - liver damage - acne - excessive hair growth
Polypeptide hormones (human growth hormone – HGH)	Sports requiring strength and power (e.g. weight lifters, endurance athletes)	- increases muscle mass - increases strength - burns fats - reduces recovery time - improves endurance - is difficult to detect	- swelling of the brain - joint pain - increased blood pressure - heart enlargement - acne - excessive hair growth
Diuretics	Sports where ideal weight is important (e.g. horseracing [jockey], boxing)	- helps make the desired weight - increases urine production - is a **masking agent**	- weakness, dizziness - joint pain - muscle cramps - diarrhoea

masking agent
a substance or drug designed to hide or mask the effects of another substance or drug that may have been taken

Drug	Target group	Advantages	Disadvantages
Beta blockers	Any sport that involves a target (e.g. archery)	slows heart rate blocks effect of adrenaline reduces anxiety and provides a sense of calm	tiredness difficulty breathing weakness, dizziness
Narcotic analgesics (codeine, morphine, opiates)	Injured athletes who need to continue competing	masks pain increases sense of wellbeing allows continuation of performance when injured	weakness false sense of wellbeing dependence increased risk of injury worsening
EPO (erythropoietin) (hormone naturally found in the human body)	Endurance athletes	increases number of red blood cells increases oxygen-carrying capacity improves VO_2max	blood clotting heart failure strokes death
Blood doping (some blood is removed from the body, the plasma and red blood cells are separated, then the red blood cells are later injected back into the body)	Endurance athletes	increases number of red blood cells increases oxygen-carrying capacity improves VO_2max	infections blood clotting heart failure strokes death

Worksheet 1.13

AUSTRALIAN SPORTS ANTI-DOPING AUTHORITY (ASADA)

The use of drugs by athletes has raised a number of ethical issues, mainly about whether the practice provides an unfair advantage. How would an athlete feel if they knew that other athletes could outperform them because they were taking substances to enhance their performance?

The Australian Sports Anti-Doping Authority (ASADA) is a government statutory authority based in Canberra. ASADA's vision is to be a 'driving force' for pure performance in sport. ASADA has the prime responsibility for the implementation of the World Anti-Doping Code (the 'Code') in Australia.

Alamy Stock Photo/dpa picture alliance

Figure 1.41 Sun Yang is a three-time Olympic swimming champion. He has been banned from competing for refusing to provide blood and urine samples to doping officials.

What is the World Anti-Doping Code?

The World Anti-Doping Code is a document that provides a worldwide framework for anti-doping policies, rules and regulations across all sports, sporting organisations and public authorities.

If a sport or country does not comply with the Code, it may be subject to sanctions from the International Olympic Committee (IOC) and other sporting organisations. Sanctions include losing the right to host the Olympic Games. The Code is, therefore, a very powerful document.

Worksheet 1.14

Part of ASADA's role is to undertake random testing on athletes. Athletes can be selected for testing anywhere or at any time, in or out of competition. Athletes are either randomly selected or targeted by ASADA if they have received a 'tip off' that the athlete may be using drugs.

There are a number of high-profile athletes and countries who have been 'named and shamed' for using performance-enhancing drugs, including:

- American cyclist Lance Armstrong, who was banned for life in 2013 for using blood doping, EPO and testosterone. He was stripped of his seven Tour de France titles.
- Russian tennis player Maria Sharapova, who was banned for two years in 2016 after failing a drug test at the Australian Tennis Open. Maria tested positive to the banned substance meldonium, which is used to help improve circulation and oxygen supply in the brain.
- British track sprinter CJ Utah tested positive for a banned substance at the 2020 Tokyo Olympics resulting in Great Britain being stripped of their silver medal in the men's 4x100m relay.
- Chinese three-time Olympic champion swimmer Sun Yang, was banned for four years for refusing to submit blood and urine samples to doping officials. Sun Yang will be able to compete in the Paris 2024 Olympics.
- Russia was banned from global sports competitions for four years, first in 2015 and again in 2019, due to state-sponsored doping. Athletes not implicated in the scandal 'competed at the 2020 Beijing Olympics (which took place in 2021) under a neutral flag'.

Weblink
Find out more information at the Australian Sports Anti-Doping Authority (ASADA) and World Anti-Doping Agency (WADA) websites.

1 Explain ASADA's role in Australian sport.
2 Describe four reasons why sportspersons take performance-enhancing drugs.

1 Ryan has been playing basketball since he was nine years old. He currently plays state basketball and competes in major tournaments around Australia. The day before a major competition Ryan was practising with his teammates. After jumping up for a rebound, he landed awkwardly on his left ankle. Ryan was desperate to compete in the tournament the following day, so his coach gave him codeine to mask the pain.

In small groups, discuss whether Ryan should compete having taken codeine. Provide evidence to support your answer.

Weblink
Russian state-sponsored doping scandal

1 Investigate the state-sponsored Russian doping scandal and prepare a short report detailing the events. You can use some of the weblinks on Nelson MindTap to help you.

Quiz
What factors influence the use of drugs?

CHAPTER 1 REVIEW

1 What are the three most commonly abused pharmaceutical drugs?

2 Identify three tips to use medicine safely.

3 Which alternative therapy uses essential oils and how can they be applied?

4 What is the definition of a psychoactive drug?

5 Speed, ice and base are all forms of methamphetamine. How do they differ from one another?

6 Explain why it is dangerous to mix energy drinks with alcohol.

7 There are three drugs that are associated with sexual assault and 'date rape'. Identify these drugs and compare their short-term effects.

8 Binge drinking is a prevalent issue in today's society. Discuss the dangers of excessive drinking over a short period of time.

9 LSD is a dangerous hallucinogen. Explain the effects of LSD on the human body.

10 The smoking of shisha has increased in popularity around the world. Outline the dangers associated with smoking shisha.

11 Discuss four factors that can increase the likelihood of a young person abusing substances.

12 Your parents have agreed to let you have a party to celebrate your sixteenth birthday. How would you ensure that your party is safe and enjoyable for all?

13 You have a friend who has collapsed and is unconscious, yet breathing. What would you do to ensure their safety?

14 Do you think Australian authorities are doing enough to stop doping in sport? Discuss your answer.

9780170463102

NUTRITION AND SUSTAINABILITY

2

IN THIS CHAPTER

You will explore ways to be healthy, safe and active, including ways to communicate and interact with others for health and wellbeing. You will also explore how to contribute to healthy and active communities.

Shutterstock.com/Nicetoseeya

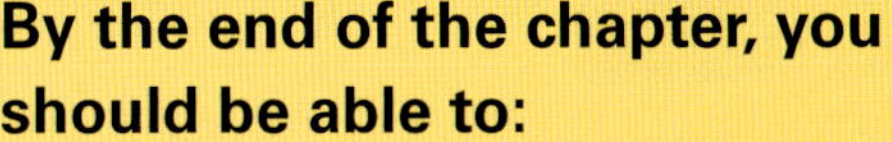
By the end of the chapter, you should be able to:

- understand energy and nutrition requirements for healthy living and performance
- consider and manage factors that influence eating habits
- make informed and justifiable decisions about eating
- analyse eating habits and propose strategies for improvement
- comprehend the trends and consequences of eating habits in Australia and the cultural and contextual factors that shape these trends.

WHAT IS A HEALTHY DIET?

Quiz
Pre-chapter

Before you start, take the pre-chapter quiz to find out how much you already know.

Worksheet 2.1

Nutrition is the process of obtaining the food necessary for health and growth. In Australia there are excellent resources and guidelines about nutrition and healthy eating. It is important to use these resources and guidelines to inform decisions around eating and drinking habits.

GUIDELINES

The key publications available in Australia are the Australian Guide to Healthy Eating and the Australian Dietary Guidelines, produced by the National Health and Medical Research Council (NHMRC). Used together, these two documents provide reliable and accurate advice to help people make informed decisions about healthy eating. Both documents are readily available on the internet.

These publications are trustworthy because they are not produced by companies that have an interest in selling their own products; they focus only on the health of Australians.

Publications of this nature tend to take many years of research and testing to produce. They are updated every few years to reflect new scientific understanding and new trends in communities. Australia's healthy eating guidelines aren't the same as those from other countries. The following illustrations show some of these differences; this will help to reinforce your knowledge of the current Australian guidelines.

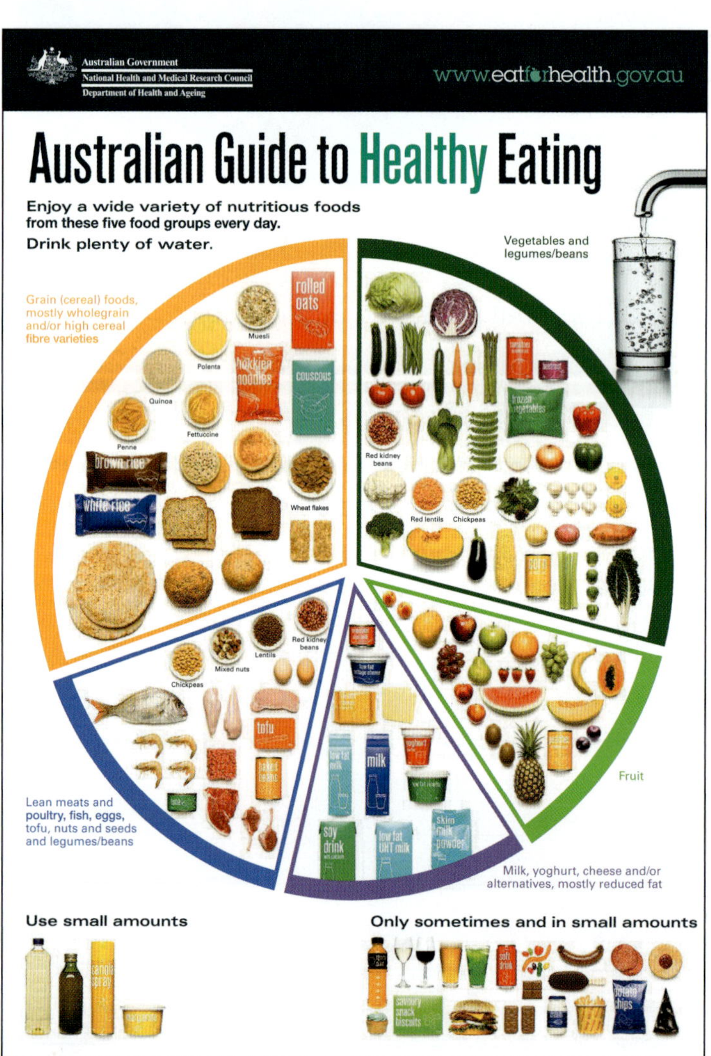

Figure 2.1 The Australian Guide to Healthy Eating

Canada's food guide
Eat well. Live well.
Eat a variety of healthy foods each day
Have plenty of vegetables and fruits
Eat protein foods
Make water your drink of choice
Choose whole grain foods
Discover your food guide at
Canada.ca/FoodGuide
Health Canada Santé Canada
Canada
Canada's food guide
Eat well. Live well.
Healthy eating is more than the foods you eat
Be mindful of your eating habits
Cook more often
Enjoy your food
Eat meals with others
Use food labels
Limit foods high in sodium, sugars or saturated fat
Be aware of food marketing
Discover your food guide at
Canada.ca/FoodGuide
Health Canada Santé Canada
Canada

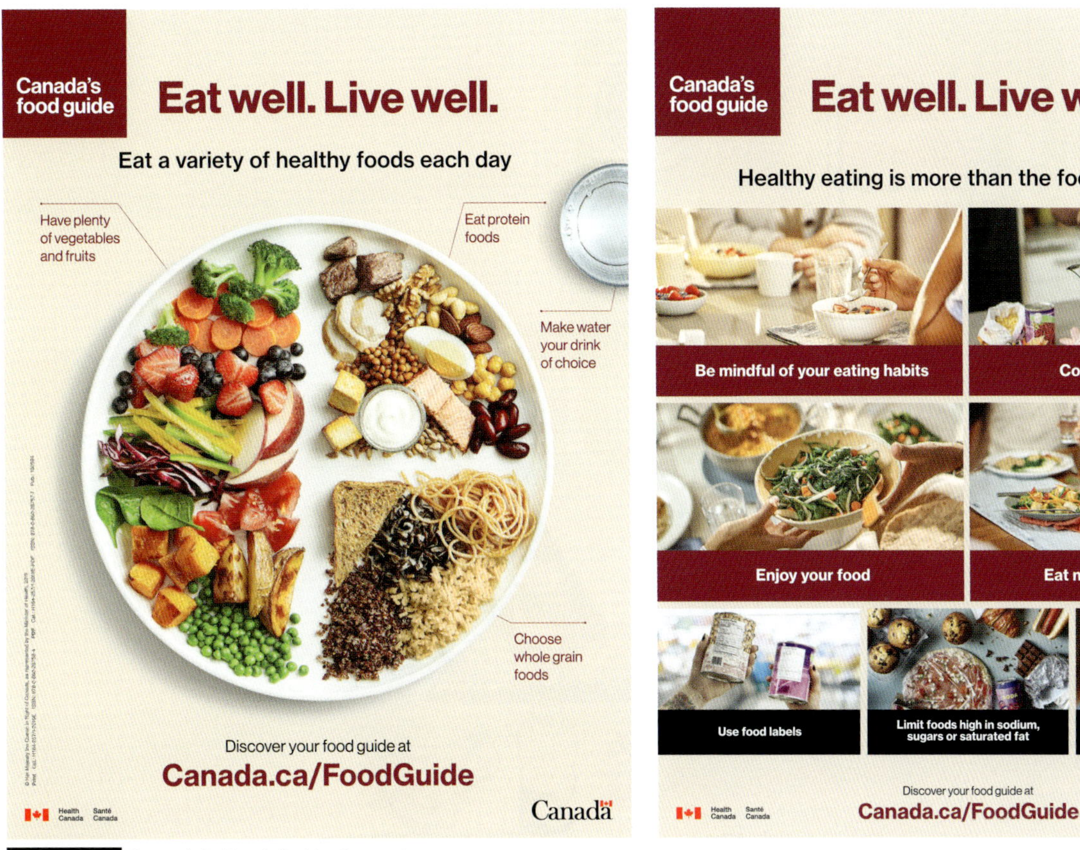

Figure 2.2 Canada's Food Guide Snapshot

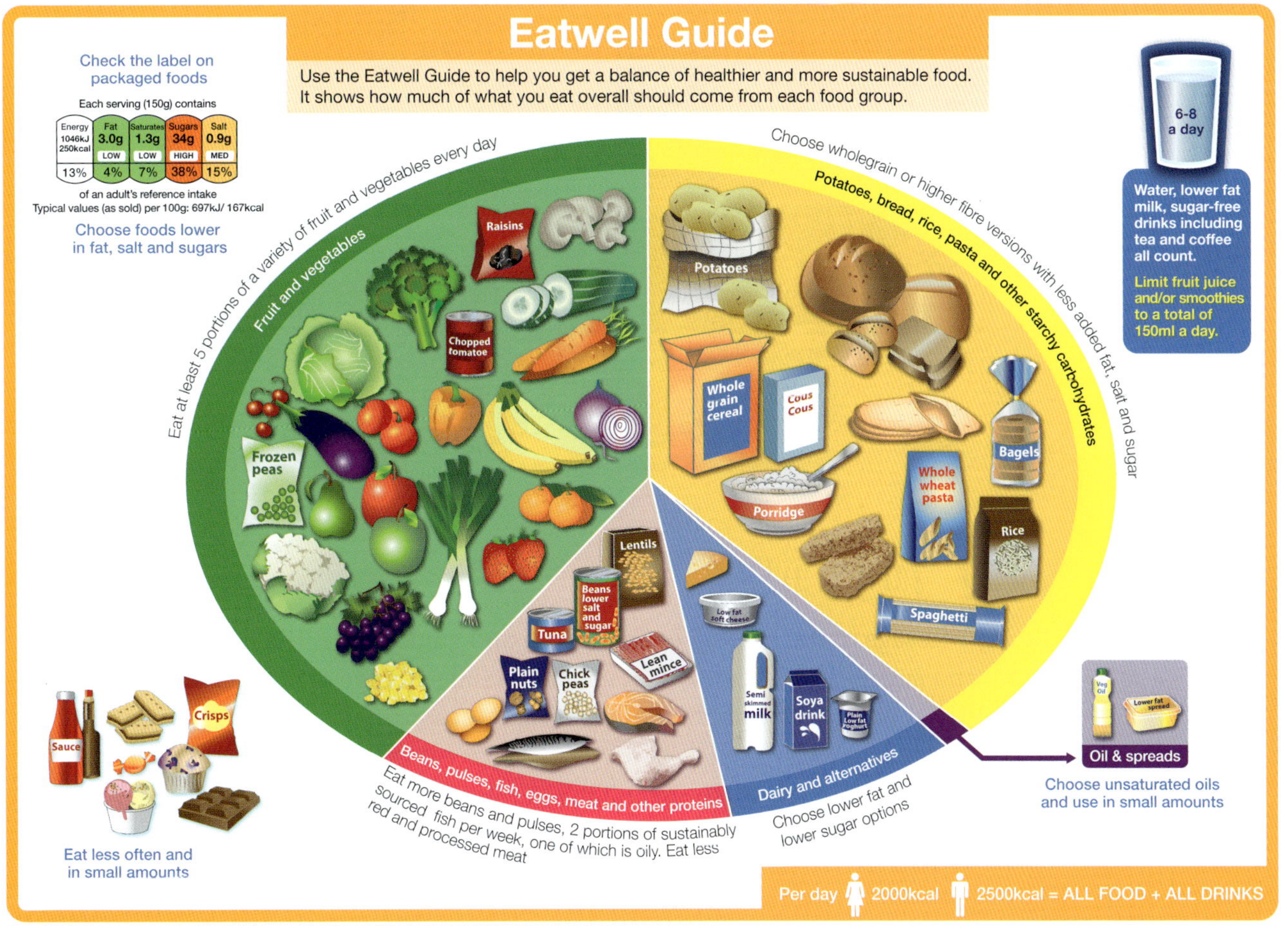
Eatwell Guide
Use the Eatwell Guide to help you get a balance of healthier and more sustainable food. It shows how much of what you eat overall should come from each food group.
Check the label on packaged foods
Each serving (150g) contains
Energy 1046kJ 250kcal 13%
Fat 3.0g LOW 4%
Saturates 1.3g LOW 7%
Sugars 34g HIGH 38%
Salt 0.9g MED 15%
of an adult's reference intake
Typical values (as sold) per 100g: 697kJ/ 167kcal
Choose foods lower in fat, salt and sugars
Eat at least 5 portions of a variety of fruit and vegetables every day
Fruit and vegetables
Raisins
Chopped tomatoes
Frozen peas
Choose wholegrain or higher fibre versions with less added fat, salt and sugar
Potatoes, bread, rice, pasta and other starchy carbohydrates
Potatoes
Whole grain cereal
Cous Cous
Bagels
Porridge
Whole wheat pasta
Rice
Spaghetti
Lentils
Beans lower salt and sugar
Tuna
Plain nuts
Chick peas
Lean mince
Beans, pulses, fish, eggs, meat and other proteins
Eat more beans and pulses, 2 portions of sustainably sourced fish per week, one of which is oily. Eat less red and processed meat
Low fat soft cheese
Semi skimmed milk
Soya drink
Plain low fat yoghurt
Dairy and alternatives
Choose lower fat and lower sugar options
6-8 a day
Water, lower fat milk, sugar-free drinks including tea and coffee all count.
Limit fruit juice and/or smoothies to a total of 150ml a day.
Veg Oil
Lower fat spread
Oil & spreads
Choose unsaturated oils and use in small amounts
Sauce
Crisps
Eat less often and in small amounts
Per day 2000kcal 2500kcal = ALL FOOD + ALL DRINKS

Source: Public Health England in association with the Welsh Government, Food Standards Scotland and the Food Standards Agency in Northern Ireland

Figure 2.3 United Kingdom guidelines

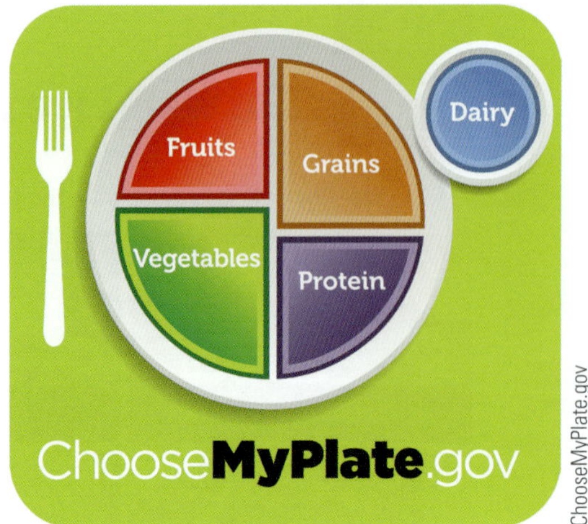

ChooseMyPlate.gov

Figure 2.4 American guidelines

© Swiss Society for Nutrition SSN, Federal Food Safety and Veterinary Office FSVO/2011

Figure 2.5 Swiss guidelines

VARY YOUR DIET

The key message in the healthy eating advice and guidelines is the need for *variety*, *balance* and *moderation*. A healthy diet should include a range of different foods, an even and proportional distribution of foods, and reasonable limits to the amounts of food consumed.

As young people progress from childhood though adolescence to adulthood, they gain more freedom of choice and self-responsibility. There isn't always going to be a parent or guardian or teacher looking over your shoulder and guiding you in your eating and drinking choices.

Making healthy decisions about nutrition isn't about 'never eating chocolate' or 'eating only vegetables and whole grains'. It is about being suitably informed, self-aware and respectful of our bodies. No single food item is completely forbidden; however, there is a 'correct' amount that should be consumed, and that amount is different for each type of food.

FACE TO FACE Discussion questions

Share your responses to these questions with a classmate:

1 Make a determination on how well the Australian Guide to Healthy Eating reflects the concepts of variety, balance and moderation. Justify using a specific example from the guide to these three terms.

2 Reflect and evaluate about how you have applied each of the concepts of variety, balance and moderation in your decisions around eating and drinking over the past few days.

HOW MUCH, HOW OFTEN

Nutrition and healthy eating isn't just about what is eaten, but also *how much* and *how often*. Serving size tells us how much of each type of food *should* be eaten, and portion size helps us track how much of each type of food is *actually* eaten. Serving size is also included in the nutrition information panels on all packaged food and drink in Australia. This topic will be explained later in the chapter.

Everyone's nutritional needs are different. These needs are based on age, sex, metabolism, body size, health conditions, lifestyle and activity levels, so it isn't possible to provide one set of recommendations that suits everybody. That's where portion sizes come in. This approach enables larger and more active people to take their extra energy requirements into consideration (this can be handy for growing teenagers too).

Shutterstock.com/Tish1

© iStock.com/zoomstudio

Figure 2.6 Serving size refers to what you should eat; portion size refers to what you actually eat.

UP AND MOVING Guess how much

This activity could be performed with the whole class or in groups of five or six. It will require some preparation at home before your next lesson. You will also need at least one set of kitchen scales to verify your results.

Each person in your class or group is to bring in an item of food (note, this might not be able to be eaten afterwards, as everyone will be handling it). Make sure you discuss what you can bring with your parents or guardians before you raid the pantry. Try not to bring anything that is particularly perishable. An example of an appropriate item to bring is a partly empty box of breakfast cereal.

One person in each group should record each group member's suggested food item and make sure no one doubles up.

Each person is to take a turn at guessing how much of each food item is considered a 'serve', based on the nutrition information panel on the packaging. For example, with breakfast cereal, you would guess how many handfuls you think should make a serve. Once everyone has guessed, use the scales to weigh the amount guessed and see how close each person was. Then answer the following discussion questions.

1. Evaluate how successful you were at matching your idea of a serve to the suggested serve size on the packaging. Was it difficult or easy? Justify why you think this is so.
2. Comparing your guesses and the guesses of others, determine if there were any trends of overestimating or underestimating appropriate serve sizes.
3. Obesity levels in Australia are rising – more than 60 per cent of all adults and 25 per cent of adolescents and children are **overweight** or **obese**. Can you infer, with justification, that overestimating portion sizes compared to serve sizes could be a contributing factor to this increase?
4. There is a perception among some people that healthy food is more expensive. Indeed, a common strategy among fast-food retailers is to offer upsized portions that represent better value. Make an argument for or against the idea that if we adhered to more appropriate portion sizes, healthy meals and snacks wouldn't be particularly expensive.
5. You may have heard about people around the world who don't have enough to eat, or that the continuing global population increase will mean the planet will not be able to produce enough food to support everyone. Hypothesise the benefits of populations in developed countries such as Australia being more careful about portion sizes. Could global food shortages be reduced?

overweight in adults, a BMI from 25 to 30

obese in adults, a BMI of 30 or higher

Weblink
Watch the ABC's The *Checkout* clip titled 'Serving Size Me' on YouTube. Discuss with your class the strategies being used by food product manufacturers to make their products seem healthier in an effort to sell more of them.

Some people find it difficult to limit their portion sizes. This can sometimes be managed by trying to eat smaller meals and snacks, and giving the body a chance to feel satisfied. Here are some tips:

- Eat slowly, so you know when you feel full and can stop eating. Chew each mouthful completely, savour the taste and feel of your food, put your cutlery down between mouthfuls, sip on some water.
- Don't skip meals, particularly breakfast – if you feel ravenous, you are more likely to overeat.

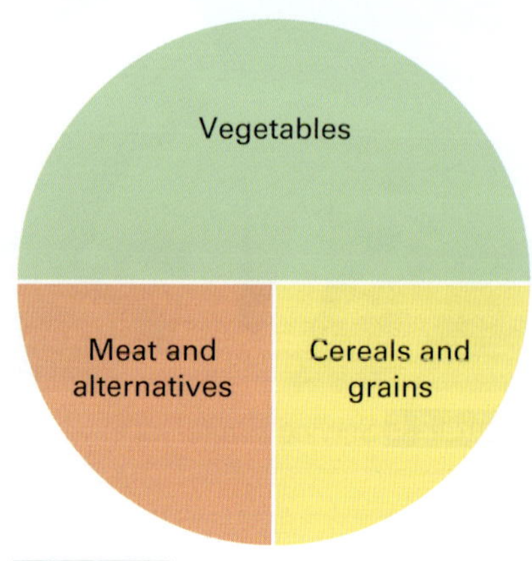

Figure 2.7 Food on your plate should be roughly proportioned in this way.

- Eat a range of fresh vegetables; this helps to fill you up without making you feel stuffed or bloated.
- Use smaller plates.
- Order entree sizes at restaurants if you know the main sizes are quite large.
- Share a dessert or snack with a friend.
- Avoid eating meals and snacks in front of the TV or when distracted.

1 List the food groups, and recall the main foods that each group is based upon.
2 Order the food groups according to what you should eat most of, and explain why this variance exists.
3 Recognise some examples of 'discretionary foods', and explain why these foods aren't included in any of the food groups.

1 Compare and contrast the models for healthy eating provided for Australia, Switzerland, Canada, the UK and the US. Note some similarities and differences between them. You can also compare the guidelines for Spain, South Korea, Japan and Greece. Use a table like the one below to help you complete your assessment.

Country	General design (e.g. pyramid)	Number of food groups	Largest food group	Smallest food group	How food item or group is represented as being significantly different in importance from the Australian model
Australia	Wheel/ plate/pie	5	Grain	Dairy, fruit	–

a Adjudicate each country's model and how similar they are in terms of the advice about how much of each food group should be consumed. Describe some similarities and differences.

b People across the world function very similarly and require essentially the same nutrition. Is this reflected in the different healthy eating models from the different countries? Explain why you think this may be the case.

c Critique the different models. Decide which model you think is best and justify why.

d The UK and the US previously used a pyramid but have moved to a circular design that represents a plate. Theorise why they made this change.

Weblink
Other countries' food guidelines
Nutrition Australia

2 Nutrition Australia have regularly produced a Healthy Eating pyramid since 1982. Review the previous versions of the Healthy Eating/Living Pyramid (1982, 1986, 1999 and 2004). Select one of these older models and compare it with the current version. Use a **Venn diagram** to note the similarities and differences. Features that are common to both models go in the intersection area of the two circles; features not shared go into either side.

Venn diagram a diagram used to represent the similarities and differences between two or more sets of data

1 Now that you have considered a range of past and current healthy models, try designing your own. Carefully consider:
- how you will develop your model
- how you will portray appropriate nutritional advice (be able to justify this)
- making your model visually appealing.

Figure 2.8 The 2015 Healthy Eating Pyramid

The Healthy Eating Pyramid © Copyright The Australian Nutrition Foundation Inc

WHAT ARE MY NUTRITIONAL NEEDS?

Eating habits may have more influence on general health, resistance to disease, physical function, athletic performance, state of mind, mood and concentration than any other daily activity. The role of a healthy diet cannot be overstated. Eating healthily doesn't have to be difficult, inconvenient or hard to tolerate, especially when the benefits are considered.

WHAT IS NEEDED

Bodies need a range of **nutrients** to maintain a healthy state. These nutrients come from different food sources.

nutrients the compounds we consume in food and drinks that are essential for life and cannot be produced by the body alone

- Most of the nutrients required to keep you healthy cannot be stored in the body. Small amounts of each nutrient need to be consumed every day.

- It is possible to have both too much or too little of each nutrient, and either can cause health problems.
- Typically in Australia, nutrition-related health problems are due to overconsumption of certain **macronutrients** and underconsumption of certain **micronutrients**.
- It isn't possible to eat a single food item to provide each individual nutrient – you wouldn't be able to eat this much food! Try to include food items in your diet that provide a wide range of nutrients.

macronutrients nutrients that are needed in relatively large amounts; examples are water, fats, protein and carbohydrates (including fibre)

micronutrients nutrients that are needed in relatively small amounts; examples are vitamins and minerals

WHAT HAPPENS WHEN NOT ENOUGH NUTRIENTS ARE OBTAINED?

An unhealthy diet can cause a range of health conditions and issues. Apart from medical conditions, our day-to-day general health and performance can also be affected. A deficiency or excess of any particular nutrient (or an imbalance in nutrient intake) can contribute to health disorders or bodily dysfunction. Usually this is a gradual process that occurs over time, but not always.

The majority of diet-related health issues in Australia are due to excessive intake of carbohydrates and fats. Usually this is in the form of energy-dense but nutrient-poor foods. The common result is obesity, which brings with it a huge range of further health complications, or conditions, referred to as **lifestyle diseases**. It is easy to think of these types of medical conditions as problems that only older people need to worry about, but increasingly this is not the case. The dietary habits young people develop tend to continue through adulthood, so it is beneficial to develop the right habits now.

lifestyle diseases diseases that can potentially be prevented by changes in lifestyle (e.g. diet, activity levels, smoking)

Being overweight or obese can lead to some types of cancer, type 2 diabetes, Alzheimer's disease, cardiovascular disease (including arteriosclerosis, strokes and heart attacks), dental cavities, osteoarthritis and sleep apnoea. There are also related emotional and social issues such as poor self-esteem, social exclusion and **stigma**.

stigma unfair disapproval or discontent with a person on grounds that distinguish them from cultural norms

Getty Images/Design Pics

Figure 2.9 You may struggle to concentrate in class if you didn't have a nutritious lunch.

INVESTIGATION

HOW ARE LIFESTYLE DISEASES AFFECTED BY OUR LIFESTYLE CHOICES?

Purpose

To examine the relationship between lifestyle choices, particularly diet and body weight, and three highly prevalent diseases in Australia.

Method

1 Select one of the Nelson MindTap sources on either type 2 diabetes, cancer or Alzheimer's disease, then respond to the questions below.
2 Feel free to search the web for additional sources (making sure they are reliable).

Weblink
Type 2 diabetes Cancer Alzheimer's disease

Discussion

1 State the prevalence of the disease you have chosen to examine in Australia. Is this number increasing and/or expected to increase?
2 Describe how poor diet and/or unhealthy body weight contributes to the rising prevalence of this disease.
3 Consider how the author of the article knows this to be true. Outline the evidence provided in the article.
4 Other than diet and body weight, identify any risk factors for this disease. List them.
5 Summarise the advice suggested in the article to reduce the risk of developing this disease.

FAST FACT

Over the past 20 years, the average weight of Australian adults increased by around 0.5 to 1 kilogram per year.

FACE TO FACE Causes of death across the globe

The following graphs from the World Health Organization (WHO) depict the 10 greatest causes of death from around the world, accounting for more than half of all deaths worldwide.

As a class, refer to Figure 2.10 and consider which of the 10 listed causes are directly linked to being overweight or obese, and the type of diet that would create these disorders. (Hint: these causes add up to about half (16.8 million) of the 30.7 million total deaths.)

Then compare the causes of death in wealthy, developed countries such as Australia with poor, developing countries as shown in Figures 2.11 and 2.12.

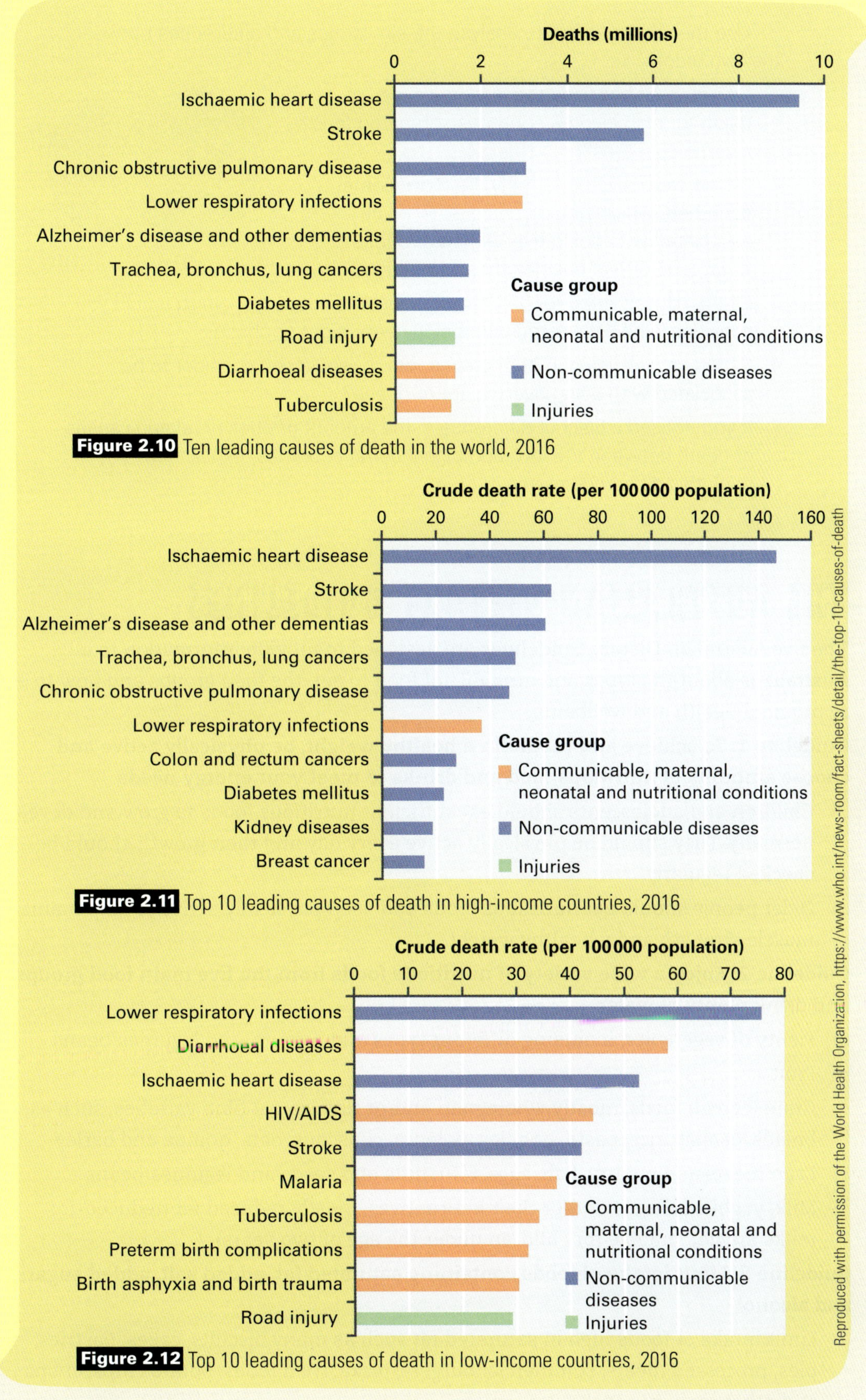

Figure 2.10 Ten leading causes of death in the world, 2016

Figure 2.11 Top 10 leading causes of death in high-income countries, 2016

Figure 2.12 Top 10 leading causes of death in low-income countries, 2016

Use the Venn diagram to analyse the similarities and differences between the two graphs.

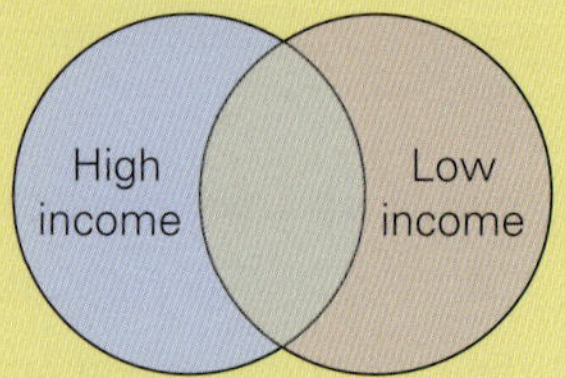

1 Differentiate between the leading causes of death for high income countries compared to low income countries.
2 Of the causes listed for only the low-income countries, which would you consider to be associated with being overweight or obese? What type of diet enables these conditions?
3 Of the causes listed for only the high-income countries, which would you consider to be associated with being overweight or obese?
4 Of the causes listed in both graphs, which would you consider to be associated with being overweight or obese?
5 If you noticed a distinct difference in the amount of obesity-related causes of death between the low- and high-income countries, explain why you think this difference exists.

THE AUSTRALIAN DIETARY GUIDELINES

Worksheet 2.2

The five Australian Dietary Guidelines are designed to provide information to all Australians about the types and amounts of food to eat, the food groups and how to eat to promote health and wellbeing.

Guideline 1: To achieve and maintain a healthy weight, be physically active and choose amounts of nutritious food and drinks to meet your energy needs

- Children and adolescents should eat sufficient nutritious foods to grow and develop normally. They should be physically active every day and their growth should be checked regularly.
- Older people should eat nutritious foods and keep physically active to help maintain muscle strength and a healthy weight.

Guideline 2: Enjoy a wide variety of nutritious foods from the five main food groups and drink plenty of water

- Plenty of vegetables, including different types and colours, and legumes/beans
- Fruit
- Grain (cereal) foods, mostly wholegrain and/or high cereal fibre varieties, such as breads, cereals, rice, pasta, noodles, polenta, couscous, oats, quinoa and barley
- Lean meats and poultry, fish, eggs, tofu, nuts and seeds and legumes/beans
- Milk, yoghurt, cheese and/or their alternatives, mostly reduced fat (reduced-fat milks are not suitable for children under the age of two years)

Guideline 3: Limit intake of foods containing saturated fat, added salt, added sugars and alcohol

a Limit intake of foods high in saturated fat such as many biscuits, cakes, pastries, pies, processed meats, commercial burgers, pizza, fried foods, potato chips, crisps and other savoury snacks.
   - » Replace high-fat foods that contain predominantly saturated fats such as butter, cream, cooking margarine, coconut and palm oil with foods that contain

predominantly polyunsaturated and monosaturated fats such as oils, spreads, nut butters/pastes and avocado.

  - Low-fat diets are not suitable for children under the age of two years.

b Limit intake of foods and drinks containing added salt.

  - Read labels to choose lower sodium options among similar foods.
  - Do not add salt to food in cooking or at the table.

c Limit intake of foods and drinks containing added sugars such as confectionery, sugar-sweetened soft drinks and cordials, fruit drinks, vitamin waters, energy and sports drinks.

d If you choose to drink alcohol, limit intake. For women who are pregnant, planning a pregnancy or breastfeeding, not drinking alcohol is the safest option.

Guideline 4: Encourage, support and promote breastfeeding

Guideline 5: Care for your food; prepare and store it safely

While many people can follow these guidelines quite easily, there are also many who can't or don't. In fact, a range of studies have shown that Australians are not very good at following nutritional guidelines. This also tends to be the case in other modern, well-developed countries, where there is a continual rise of obesity and lifestyle diseases attributed to excessive energy consumption and inadequate micronutrient consumption. Why do you think people don't follow nutritional guidelines?

FAST FACT

Only one third of middle-aged Australian women meet at least half of the dietary guidelines. Of 10 561 women surveyed, only two met all 13 points and sub-points in the guidelines.

Census time

There could be any number of reasons why people don't follow the advice in the Australian Guide to Healthy Eating and the Australian Dietary Guidelines. Consider the following questions, for example:

- Do they know these publications exist?
- Have they heard of them but don't know what they entail?
- Do they think they know what is contained in the publications, but are actually wrong?

It is also possible that they know exactly what is advised in these publications, but are unable to follow them for some reason. Perhaps they perceive the guidelines as too hard to follow, too costly, too inconvenient or demanding of time or effort. Perhaps there are too many distractions, influences or temptations.

Working in small groups, design a survey using the above questions as a basis to develop your own specific questionnaire. Negotiate the scope of your survey with your teacher – will you survey members of the wider community or just your school and perhaps family members? Decide on the time frame, too, and how many respondents you will need in order to gain some meaningful data. You may make hard-copy surveys to distribute, use a free online survey service such as Survey Monkey or develop one using Google Forms.

Remember that when analysing the results, multiple choice, yes/no or rating 0–5 questions are much easier to work with than questions that require full sentences to answer.

Alamy Stock Photo/Richard G. Bingham II

Figure 2.13 Surveys enable us to gather relevant and timely data.

Gather your responses into tables of data, and use these to generate some graphs, either by hand or using spreadsheet software.

1 Were you able to establish a significant or observable trend in your results? Can you distinguish a particular stand-out reason behind people not adhering to the nutritional guidelines?
2 Hypothesise a reason for this.
3 Were your findings consistent with those of other groups? Propose what this means for the validity of your results.
4 Evaluate your survey design for any flaws or issues. Recommend any refinements for if you were to conduct this survey again.
5 If you were to take the results of your survey to the Department of Health and Ageing or the National Health and Medical Research Council, what course of action would you propose they take based upon your findings?

INTAKE VERSUS EXPENDITURE

BMR the energy used when at rest to maintain life (e.g. breathing, brain function, heart function and cell repair)

There are three components of energy expenditure:

- the basal metabolic rate (**BMR**)
- the process of eating and digesting food
- the levels of physical activity performed.

The physical activity component is the only one of the three that can be easily changed. The BMR can be increased to a degree, but this is achieved by physical activity also.

Apart from increasing physical activity, which is always a good idea, there are other ways to balance energy intake and expenditure. Make sure you include a lot of vegetables in your meals. They have a high nutrient density and low energy density, so they help make you feel full without overloading your kilojoule intake. Look at the following examples of more filling choices. Note the size in grams of each snack for the exact same amount of energy (every food item shown contains 600 kilojoules).

9780170463102

INVESTIGATION

THE RELEVANCE OF BMI – DO WE NEED A BETTER WAY OF MEASURING BODY FAT LEVELS?

Purpose

While the Body Mass Index remains a standard measurement when assessing body fat levels and the associated health risks, it has its limitations and is far from perfect. We need to investigate if we should still be using the BMI or if there are alternative methods that are more effective, and if so, what those alternative methods are.

Materials

This will be a research investigation. Materials will be text and online publications (from reliable sources). Consider how you can access a range of publications, so we can consider multiple perspectives and hopefully discover some common findings. Some sources are available on Nelson MindTap to get you started.

Weblink
Follow the weblinks to learn more about BMI.

Method

1 Find a range of sources about the topic. If you are using a web search, consider carefully how to word your questions. Establish the credentials of the authors of your sources and the potential motives behind their perspective.
2 Summarise the perspective of each source and the reasons used by the author to support their perspective.
3 Compare the perspectives of each source. Are there commonalities or consensus?
4 Develop your responses to the discussion questions below.

Discussion

1 List the advantages of using BMI over other methods.
2 State any limitations of the BMI. Who may it not be applicable to?
3 Consider if we should be completely disregarding the BMI. Why or why not?
4 State two other methods of determining whether someone is of a healthy weight range. Are these more reliable than the BMI?
5 Select the 'gold-standard' of measuring body fat and determining healthy weight. Why don't we use this as an everyday measure?

Your guide to filling up on lower-kilojoule foods

Each of the food portions pictured provides **600 kilojoules**. It's easy to see that the foods on the left will fill you up faster (for the same number of kilojoules) than the foods on the right.

GETTING ENERGISED

Most people, at some stage in their life, wish they had more energy. Perhaps you are studying for an exam, making your way to sports training after a long day at school, heading out with friends, getting through some household chores or getting ready for a big game on the weekend.

What you really want is to feel more energetic. This is different from needing more energy – one kilogram of body fat contains 37 000 kilojoules of energy, which is enough to last the average adult four days. Everyone has plenty of energy contained within their blood, muscles, liver and fat. However, your body can be more efficient in the way it accesses and uses its available energy, and this will make you feel much more energetic.

From top, left to right: Shutterstock.com/Yeti studio; Shutterstock.com/tetiana_u; Shutterstock.com/science photo; Shutterstock.com/Andrei Kholmov; Shutterstock.com/baibaz; Shutterstock.com/baibaz; Shutterstock.com/Jan Martin Will; iStock.com/pamela_d_mcadams; Shutterstock.com/SOMMAI; Shutterstock.com/Nataliia K; Shutterstock.com/bergamont; Shutterstock.com/Valery121283; Shutterstock.com/Maks Narodenko; Shutterstock.com/paulista; Shutterstock.com/amenic181; Shutterstock.com/Hong Vo

Figure 2.14 Different food portions for the same kilojoule value

Shutterstock.com/hlphoto

Figure 2.15 Energy AND nutrient dense

Tips to help the body feel more energetic include:

- Eat a complete and proper breakfast – never skip or snack when it comes to breakfast.
- Make sure to have three complete meals that include a wide range of foods.
- Wait until you are hungry before you eat, but don't wait until you are ravenous.
- Develop a regular routine of eating times and portion sizes.
- If your snacks don't belong to one of the five food groups, this means they are classed as 'discretionary foods'. Limit these to weekly treats, not daily regulars. Swap to healthier snacks instead.
- Carry a water bottle with you whenever possible. Having all the energy in the world means nothing if you are not adequately hydrated.
- Complex carbohydrates such as those found in whole grains will keep you feeling ready for action much more effectively than simple carbohydrates (sugars). Do not rely on sugar hits to try and pick you up when you are feeling flat.

Energy comes from three main nutrients – carbohydrates (17 kilojoules per gram), protein (17 kilojoules per gram) and fats (37 kilojoules per gram). Fibre is a type of carbohydrate that can only be partially digested, so it yields only 8 kilojoules per gram. Alcoholic drinks yield a lot of energy – 29 kilojoules per gram of alcohol.

The body prefers to use carbohydrates as its energy source, particularly during physical activity. Carbohydrates serve only as a fuel source, whereas fats and protein play other important roles in the body as well as providing fuel.

Glycaemic index

Not all carbohydrates are equal; the speeds at which they are digested and utilised vary. Foods that contain carbohydrates that cause blood glucose (sugar) levels to rise rapidly are considered high-GI foods. Foods that contain carbohydrates that cause blood glucose levels to rise gradually are considered low GI. Low-GI foods tend to provide sustained energy, as they don't cause rapid spikes in blood glucose levels that drop away quickly.

Shutterstock.com/Kostsov

Figure 2.16 Always carry a water bottle to help ensure you drink enough water.

Worksheets 2.3 and 2.4

Video
Energy: What do you consume when you want to feel more energetic? Are you making the right choices when consuming foods to feel more energetic? Watch the video and start the discussion!

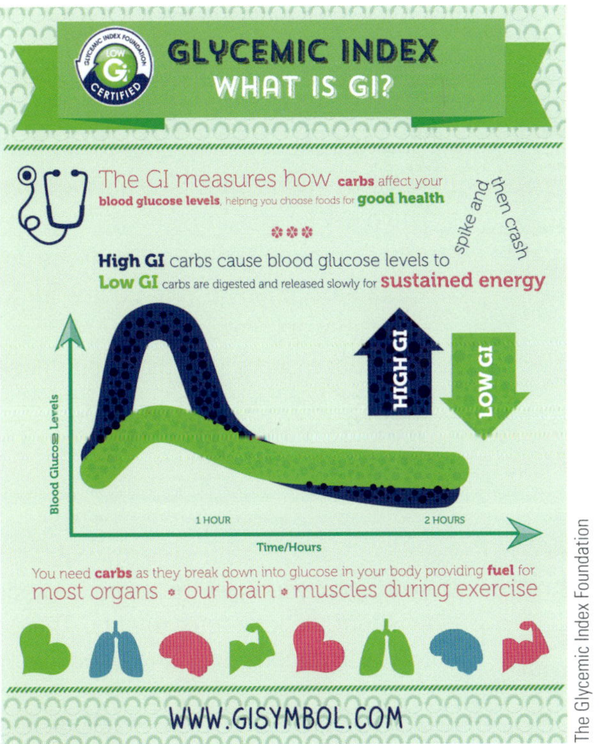

The Glycemic Index Foundation

Figure 2.17 High-GI foods cause blood glucose levels to rise and fall rapidly, whereas foods with a low GI value have a slower impact on these levels.

In general, sugary or sweet foods, including many fruits, tend to be high GI, as do white bread, rice, potatoes and most highly processed foods. Most vegetables are low GI, as are wholegrain foods, pasta and some types of rice.

The GI symbol is found on many food labels and packaging, but not all. If you are eating mainly from the five food groups, limiting your intake of discretionary foods and following the principles of variety, balance and moderation, you probably don't need to track the GI of the foods you eat.

HOW TO COMFORT EAT RIGHT

This activity has been developed in collaboration with Dr David Bakker of MoodMission

Identify

Ever find yourself comfort eating? Eating certain foods can help us feel better for a moment, but different foods might help us feel better for longer.

Understand

Comfort eating is when you crave something tasty and calorie-rich to eat to make you feel better. However, as covered in this chapter, these high-GI foods often create a powerful but brief 'sugar high'. Sometimes we might feel sad and guilty after eating them. This can start a vicious cycle; we sometimes reach for more comfort food to help us cope with the bad feelings that arise in the aftermath.

Practise

1. Create a list of the comfort foods that you most often eat to feel better.
2. Now, using the principles outlined in this chapter, list some healthier alternative foods you could eat for comfort.
3. Make sure the healthy alternatives are available at home. Add them to the shopping list or stock them in your pantry. Try to make them more easily available than the unhealthy options.
4. Consider making sure that the unhealthy foods aren't easily available to you. You could leave them off the shopping list, put them on a high shelf, or even put a parent or family member in charge of them.

Reflect

Do you think you'll get the same immediate good feelings from the healthier comfort foods? How about the long-term effects? There's nothing wrong with eating to improve your mood, but choosing foods that have good long-term effects is always preferable.

The nutrition information panels located on food packaging always list the amount of carbohydrates, protein, fat, fibre and total energy.

The manufacturers of the foods work out how much energy is contained in each item by performing some simple calculations. The label provided shows the following nutrition information. Using the 'Per 100 g' column enables the calculations shown in the following table.

Energy from protein (17 kJ per gram)	9.3 g × 17	=	158.1
Energy from fat (37 kJ per gram)	1.2 g × 37	=	44.4
Energy from carbs (17 kJ per gram)	62.9 g × 17	=	1069.3
Energy from fibre (8 kJ per gram)	21.2 g × 8	=	169.6
Total sum of these energy sources		=	1441.4
This matches the amount of energy listed on the nutritional information panel.			

Nutrition Information

Servings per package – 16
Serving size – 30 g (2/3 cup)

	Per serve	Per 100 g
Energy	**432 kJ**	**1441 kJ**
Protein	2.8 g	9.3 g
Fat		
Total	0.4 g	1.2 g
Saturated	0.1 g	0.3 g
Carbohydrate		
Total	18.9 g	62.9 g
Sugars	3.5 g	11.8 g
Fibre	6.4 g	21.2 g
Sodium	65 mg	215 mg

Ingredients: Cereals (76%) (wheat, oatbran, barley), psyllium husk (11%), sugar, rice, malt extract, honey, salt, vitamins.

Based on material provided by the National Health and Medical Research Council

Figure 2.18 Nutrition information panel

INVESTIGATION ⇨ KNOW YOUR NUTRIENTS

Purpose

In order to develop a clear understanding of the role and importance of each nutrient, we will need to conduct some research. You need to know which food sources provide each nutrient, the function of each nutrient in the body and what happens if you have too much (excess) or not enough (deficiency) of each nutrient.

Materials

Conduct your research online using web searches, and use a table such as the one provided to record your findings.

Method

1 You will need to ensure you are using reliable sources of nutritional information; be mindful of the credibility of the sources you use. Here are some quick tips for credible online sources:
   - Refer to large institutions such as universities or government organisations – the domain name can be a good indicator.
   - Watch out for commercial biases – companies looking to make profit from selling their own products.
   - What makes the author of this website an expert?
   - Check the sources used by the author of the website – there should be references.
   - If the design of the site appears amateurish, or the writing style casual or error-riddled, this denotes a lack of credibility.

2 Present your research findings clearly, using a table. You shouldn't need whole paragraphs of text – just a few words for each section of your table may even suffice. See the example for protein below.

Macronutrients	Food sources	Function	Deficiency	Excess
Protein	Meat, dairy, eggs, seeds, nuts, legumes	Rebuild and repair cells (hair, muscle, blood, skin, etc.)	Muscle wasting, impaired growth, anaemia	Liver and kidney stress, weight gain
Water				
Fats				
Simple carbohydrates (sugars)				
Complex carbohydrates (starches)				
Dietary fibre				

Fats	Food sources	Function	Deficiency	Excess
Monounsaturated				
Polyunsaturated				
Trans				
Omega-3 fatty acids				
Omega-6 fatty acids				
Saturated				

Vitamins	Food sources	Function	Deficiency	Excess
Vitamin A				
Vitamin B1 (thiamin)				
Vitamin B2 (riboflavin)				
Vitamin B3 (niacin)				
Vitamin B12				
Vitamin C				
Vitamin D				
Vitamin E				
Vitamin K				

Minerals	Food sources	Function	Deficiency	Excess
Calcium				
Magnesium				
Potassium				
Sodium				
Iron				
Iodine				

Discussion

1 Can you identify any nutrients that did not have a health consequence associated with deficient or excessive consumption of that nutrient? What message discussed earlier in this chapter does this support?
2 Which food sources did you list the most? What can you infer about how often you should be eating these food sources?
3 List any nutrients that should ideally be avoided entirely and state why.
4 What food sources are typically associated with nutrients that should be avoided entirely?
5 What nutrient do we mainly receive by absorbing it from our environment instead of eating it?

Weblink
To get an idea of exactly how much of each nutrient you should be getting daily, enter a couple of your details into the Eat for Health Nutrient Calculator.

1 List three examples of each of these types of nutrients: macronutrients, fats, vitamins, minerals.
2 Identify three lifestyles diseases that are considered leading causes of death in Australia.
3 What does BMI stand for, and how is it calculated?
4 Identify each nutrient that contributes to the total energy we consume.
5 Name three high-GI and three low-GI foods.

1 Use the following nutrition panels to calculate the energy from each of the sources and then the total energy. The table will help you with your calculations. Which product is the best source of energy?

Nutrition information		
Servings per can: 2		
Serving size: 210 g		
	Average quantity per serving	Average quantity per 100 g
Energy	895 kJ	425 kJ
Protein	10.8 g	5.1 g
Fat: Total	1.2 g	0.6 g
Saturated	0.2 g	0.1 g
Carbohydrate	33.7 g	16.1 g
Sugars	15.5 g	7.4 g
Dietary fibre	11.9 g	5.7 g
Sodium	1300 mg	620 mg
Potassium	650 mg	310 mg
Iron	2.7 mg	1.3 mg

Nutrition information
Servings per package: 9.5 (17 slices + 2 crusts)
Average serving size: 74 g (2 slices)

	Avg. quantity per serving	% daily intake and per serving	Avg. quantity per 100 g
Energy	773 kj	9%	1044 kj
Protein	6.1 g	12%	8.2 g
Fat, Total	2.2 g	3%	3.0 g
- Saturated	0.6 g	2%	0.8 g
Carbohydrate	33.2 g	11%	44.8 g
- Sugars	3.0 g	3%	4.0 g
Dietary fibre	3.0 g	10%	4.0 g
Sodium	324 mg	14%	438 mg

Energy from protein	g × 17	=	
Energy from fat	g × 37	=	
Energy from carbs	g × 17	=	
Energy from fibre	g × 8	=	

1 During the upcoming school holidays, you will be coaching your soccer team in a major tournament. The team will play two games per day for four days. This is a lot more consecutive game time than the players are used to, so nutrition and hydration will be especially important. Develop a set of guidelines that you can use to advise your players on what to eat and drink for their main meals and for snacks before, during, between and after games to maintain their energy levels and hydration.

Check your responses with your teacher or by using the Sport Dietitians Australia website.

Weblink
Sports Dietitians Australia

2 You are participating in the school swimming carnival tomorrow and hope to win your favourite events, the 50-metre freestyle and 100-metre individual medley. You are also in a couple of relay teams. This is your schedule for the day:

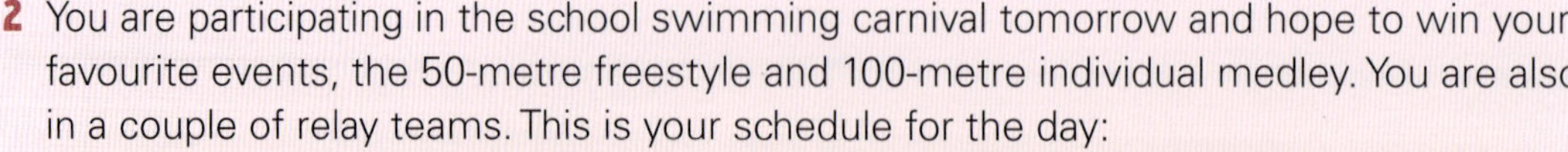

Time	Activity	Advice
7.00 a.m.	Breakfast at home	
9.00 a.m.	Warm up and race 50-metre freestyle heats	
10.00 a.m.	Break	

Time	Activity	Advice
11.00 a.m.	Warm up and race 50-metre backstroke	
11.30 a.m.	Recover, warm up and race 50-metre freestyle final	
12.00 p.m.	Lunch	
1.30 p.m.	Warm up and race 100-metre medley	
3.00 p.m.	Break	
4.00 p.m.	Warm up and race 4 × 50-metre freestyle relay	
4.30 p.m.	Recovery, warm up and race 4 × 50-metre medley relay	
6.00 p.m.	Dinner at home	

a Provide some advice for each time period throughout the day. It could be as simple as sipping on water or describing a nutritious meal.

b Check your responses with your teacher or by using the Sport Dietitians Australia website.

Quiz
What are my nutritional needs?

WHAT STRATEGIES WILL SUPPORT ME HAVING A HEALTHY, BALANCED DIET?

It is easy to forget or disregard the official advice in the Australian Guide to Healthy Eating and the Australian Dietary Guidelines, especially when out with friends or busy with school or sport.

It can be handy to have some quick tips and tricks that are easy to remember and use when making nutrition decisions.

Worksheet 2.5

MODIFYING RECIPES

You don't have to be a gourmet chef to take a hands-on approach to managing your diet. Sometimes it is just about making small changes to your favourite meals or snacks to make them healthier. This usually means making them less energy dense and more nutrient rich.

Custom cooking

Your challenge here is to modify the recipe for one of your favourite meals or snacks so that it becomes a healthier food choice. You will need to find a suitable recipe, modify it to make it a healthier meal or snack, and then cook and eat your modified version. You can make a meal or snack healthier by changing the ingredients, the cooking or preparation method, or both.

Fairfax Syndication/Robert Pearce

Figure 2.19 What is your favourite meal? How could you make it healthier by changing the ingredients, the cooking or the preparation method?

There are several options that you could consider and negotiate with your teacher:

- bringing your modified recipes to school to be judged by your teacher or someone else at the school
- preparing and cooking your meal at home, taking photos of the process and the finished product. You might also need an independent taste tester such as a family member.
- preparing and cooking your meal at home, taking photos of the process and finished product, then wrapping or packaging it appropriately and bringing it to school to share
- negotiating to use some time in a cooking, food technology or home economics class (if possible) to prepare and cook your meal or snack at school
- nominating a recipe or two from the class to be cooked by the staff in your school's canteen or cafeteria.

Bring your recipe, modifications, photos and/or finished product to school to present to your class.

Follow the weblinks on Nelson MindTap for some ideas.

Make sure you discuss your options with your parents.

Weblink
Eat for Health
FoodSwitch

Weblink
Dietitians Association of Australia

1. Upon reflection, how well did your meal or snack turn out? Was it successful? Why or why not?
2. Describe the modifications you made to the ingredients and/or the cooking or preparation method to make your meal or snack healthier.
3. Evaluate how your meal or snack become healthier. What additional health benefits do you get now that you wouldn't have before?
4. Propose are some reasons why you might not continue to use your new method for this recipe. Taste? Cost? Difficulty?

WELLBEING CHECK IN ⇨ SAVOURING FOOD

This activity has been developed in collaboration with Dr David Bakker of MoodMission

Identify

We can change how much we enjoy something by focusing on it differently. Food is a great example of this.

Understand

Have you ever been to a fancy restaurant where they serve smaller amounts of food, carefully presented? One reason for this is to encourage diners to savour the food. Savouring is a skill that can allow people to focus on the good parts of something. Food is a great way to practise savouring, but it can be used with anything really.

Practise

1 Get a small amount of food that you like. It could be a piece of chocolate, a strawberry or a teaspoon of peanut butter, for example.
2 Make sure you don't have any distractions, like TV, or other people talking to you.
3 Before eating the food, take a close look at it. Notice all the colours and textures. You can also take a moment to smell the food.
4 Put a very small amount of the food into your mouth and roll it across your tongue. Resist the urge to chew and swallow the food – just enjoy the flavours as they are.
5 When you're ready, chew and swallow the food in your mouth. It might take a full minute or so to fully savour the mouthfull before swallowing it. Repeat with the rest of the food, taking care to savour the experience.

Reflect

What did you notice when you ate the food slowly? Did you enjoy it more? Is there anything else, besides eating, that you could take this slow, savouring approach to?

SELECTING HEALTHY OPTIONS FOR SNACKS AND MEALS

As adolescents move towards adulthood, they gain more freedom of choice and influence over their diets. Children under the age of seven or eight rarely have any input when it comes to their meals. Teenagers usually choose some of their weekly meals, while adults make their own decisions. There often comes a time when adults are also responsible for the diet of people under their care.

This increased choice and independence comes with a level of responsibility. It takes planning and organisation to manage a week's worth of meals and snacks. Not only are you trying to follow the healthy eating dietary guidelines, but there are also budgets to stick to and tastebuds to satisfy.

Worksheet 2.6

Getty Images/Klaus Tiedge

Figure 2.20 Who is responsible for the food choices in your house?

The word 'diet' has been used a lot throughout this chapter, but it means something different to the kind of diet that you might hear about a celebrity following, or that a family member or friend is trying out. 'Diet' in the context of this chapter refers to usual overall eating and drinking practices, decisions and habits. 'Dieting', on the other hand, usually refers to the practice of eating food according to a regulated or restricted system to cause a change in body weight.

Dieting is often used as a desperate measure or last resort to shed body weight. Such strict controls over food choices and eating habits makes dieting unsustainable, and many people quickly return to their old habits and choices, which means their body weight also returns to its pre-dieting state. This is called 'yo-yo dieting'. This sort of dieting tends to be quite expensive, and often limits the intake of some essential nutrients.

CASE STUDY  DIET FADS

Identify

Fad diets might achieve short-term results, but they are difficult to sustain in the long term. They often eliminate entire food groups, which means they're unlikely to provide adequate amounts of key nutrients that are essential for our health and wellbeing. Fad diets and rapid weight loss can also increase the risk of serious health problems such as gall bladder disease and gallstones.

Understand

When assessing whether a diet is a fad, ask yourself, does the diet:

- contradict advice from qualified health professionals?
- promote or ban specific foods or whole food groups?
- promote a one-size-fits-all strategy?
- promise quick, dramatic or miraculous results with minimal effort?
- focus only on short-term results?
- promote 'miracle' pills, supplements or products touted to 'burn fat'?
- make claims based on personal testimonials or one random study?

If the answer to two or more of these questions is 'yes', it's probably a fad.

So, how do today's popular diets measure up?

Blood type diet

The blood type diet has been around for some time. It's based on the idea that your blood type is a key factor in predicting your body weight, nutritional requirements, risk of chronic disease, and overall wellbeing.

According to this diet, those with blood type A should follow what resembles a vegetarian diet. Type Os are supposed to limit carbohydrates and increase their protein intake. Type Bs should avoid chicken, corn, wheat, lentils, tomatoes, peanuts, and sesame seeds; while type ABs should avoid caffeine, alcohol, and cured meats.

But a comprehensive review of 16 studies found there is no scientific literature to back up this list of dos and don'ts.

Gluten-free diet

Gluten is a protein naturally found in wheat, rye and barley, plus some food additives. People with diagnosed coeliac disease must eliminate gluten from their diet to avoid serious damage to their gut, but many people choose to avoid gluten as a weight-loss strategy.

Eliminating gluten does not automatically reduce your kilojoule intake or induce weight loss. But some gluten-containing foods such as pizza, bread, pasta and cakes are energy-dense, so removing them completely will reduce your total energy intake, which may lead to weight loss. Gluten-free alternatives can be just as high in kilojoules as the gluten-containing version, and sometimes can be higher in kilojoules.

Removing gluten-containing foods without considering what foods will replace them can also reduce your intake of important nutrients such as fibre, folic acid and other B vitamins. Recent studies discourage unnecessary gluten-free diets due to the reduced intake of beneficial whole grains, which are key to a healthy diet and are associated with lower heart disease and cancer risk.

Mediterranean diet

The Mediterranean diet has a strong focus on intake of core foods in addition to olive oil, coffee and wine, and low intake of meat, sugar and highly processed foods.

While the main focus of the Mediterranean diet is not weight loss, when combined with a kilojoule restriction, it can be effective for weight loss.

Among studies that did not prescribe an energy restriction, following the Mediterranean diet was not associated with gaining weight. The Mediterranean diet has also been shown to improve components of metabolic syndrome, even without weight loss.

Source: 'Blood type, Pioppi, gluten-free and Mediterranean – which popular diets are fads?', Clare Collins, Lee Ashton and Rebecca Williams, 23 October 2018, *The Conversation*, https://theconversation.com/blood-type-pioppi-gluten-free-and-mediterranean-which-popular-diets-are-fads-104867

Discuss

Compare the analysis in this article with *The Conversation* article looking at three other types of diet. Follow the weblink provided on Nelson MindTap.

Weblink The Conversation diet analysis Healthy Eating Quiz

1. A range of popular diets are specifically mentioned. Identify four diets from either article that are considered to be ineffective, unhealthy or 'fads', and describe why the authors have found this to be the case.
2. Both articles provide a checklist for determining if a diet is a 'fad' or unlikely to lead to long-term weight loss. Compare these two checklists and select the four recommendations you think would make the best advice for anyone looking to lose weight.
3. Why do you think so many different 'fad' diets exist (there are hundreds)? Why are they so appealing to people?
4. What advice would you offer a family member or friend who decided to commit to a 'fad' diet? How would you go about persuading them with healthier nutritional advice?
5. Have a go at the 'Healthy Eating Quiz' as developed by the University of Newcastle. How healthy is your diet?

Better weight loss results can be obtained by making gradual, long-term, sustainable changes in diet and physical activity levels that give your body time to adapt and adjust. This approach means your body still accesses the full range of nutrients; some forms of dieting eliminate certain nutrients completely.

Worksheet 2.7

1 List five different ways a recipe can be modified to make a meal healthier.

2 List all the ingredients and the main cooking methods used in your favourite meal. Propose two modifications to these so the meal is healthier.

3 On average, how many times a week do you get to choose the meals you eat? When you are an adult, you will be responsible for all of your meals. Identify the challenges to maintain a healthy diet when you will have to make all of the decisions about what you eat.

1 Working in small groups, your teacher will assign each group a different diet. Popular diets include Atkins, Keto, South Beach, 5:2, Paleo, The Master Cleanse, Macrobiotic, Cabbage Soup, Dukan, Beverly Hills and Hay. Research your assigned diet and develop a presentation to your class. Make sure you cover the following points:
- name of the diet
- the diet's key elements
- reasons the diet is supposed to work
- criticism or flaws of the diet
- reasons the diet could be considered unhealthy or unsustainable
- ways the diet could be considered healthier than the Australian Dietary Guidelines.

There are some links on Nelson MindTap to start off your research.

Weblink
Better Health Channel FUSE resource: Contemporary food fads, trends and diets

2 There is no doubt that many diets would cause weight loss, particularly some of the more extreme ones. If someone followed the Australian Dietary Guidelines and was physically active, would they ever be in a situation where they needed to lose weight? Justify your response.

3 Of the different dieting strategies studied by your class, would you consider any of them to be healthier than the diet recommended in the Australian Dietary Guidelines? Why or why not?

1 'Determinants' are factors that influence (determine) our health. They are referred to as 'risk factors' if they have a negative effect or 'protective factors' if they have a positive effect. An example of a risk factor for cardiovascular disease is a diet high in sodium or saturated fats. An example of a protective factor for type 2 diabetes is a healthy body weight. Contemporary research suggests that some of the best protective factors for young people to develop healthy eating habits is for them to be involved in as many stages of food planning and preparation as possible. Examples might include helping to write the weekly grocery shopping list, or helping to peel and chop vegetables for evening meals.

   a Develop a list of all the different ways young people can be involved in food routines and activities in the home. You should be able to come up with at least 10. Ask your classmates or conduct a web search if you need some ideas.

   b Perform a self-evaluation: how many of these food routines or activities do you participate in at home?

   c Devise some ways to involve yourself more in these food routines or activities.

   d Food preparation doesn't have to be a chore; it can be joyous, rewarding and fun. Propose some ways you and your family could make meal times healthier and more enjoyable.

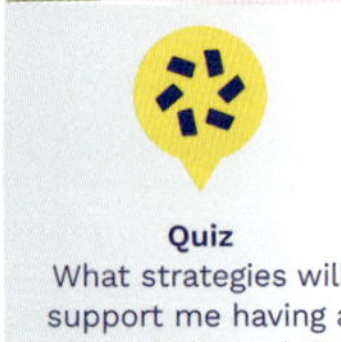

Quiz
What strategies will support me having a healthy, balanced diet?

9780170463102

HOW IS MY DIET INFLUENCED?

There are many factors that influence food choices and eating habits. Some are positive; others are negative. Positive influences tend to help people stick to a healthy diet, whereas negative influences become barriers to this. Some of these influences are intrinsic – they come from within, such as motivation or self-discipline. Some influences are extrinsic – these factors are external, such as time, money or friends.

These influences are with you in some form every day of your life and they affect every decision you make about eating or drinking. Understanding these influences will help you to manage them.

OUR DIETARY HABITS

There are correlations between both geographical areas and **socio-economic** groups, and the prevalence of overweight and obesity.

socio-economic status the social standing of an individual or group, measuring a combination of education, income and occupation

In 2017–18, about one in three children and adolescents from regional areas were overweight or obese, compared to about one in four in major cities. There is a similar difference in the figures for adults, too.

Rates of obesity (not overweightness) in children and adolescents were two and a half times higher among those in the lowest socio-economic areas compared to those in the highest. For adults, almost three in four Australians in the lowest socio-economic areas are overweight or obese, compared to less than two in three in the highest socio-economic areas.

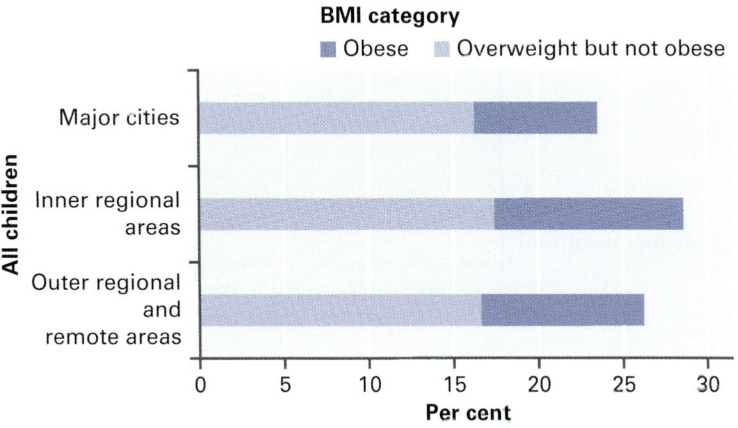

Figure 2.21 Proportion of overweight and obese children and adolescents aged 2–17 by remoteness area

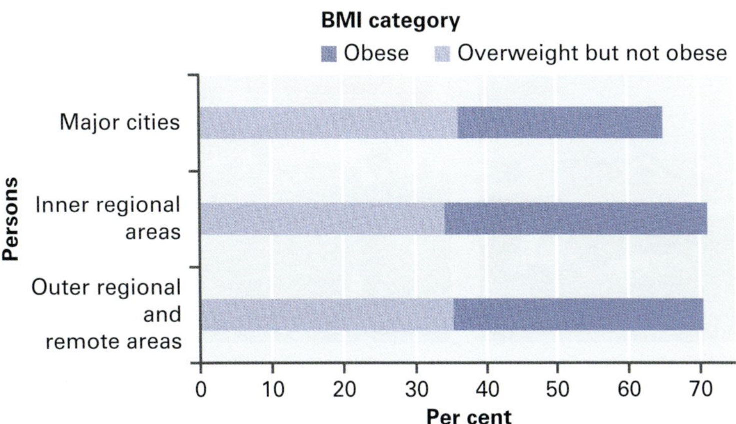

Figure 2.22 Proportion of overweight and obese adults by remoteness area

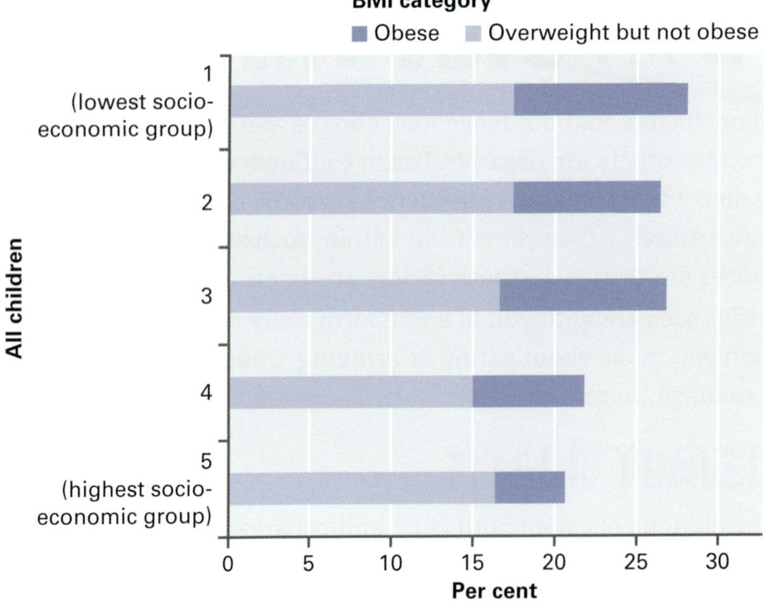

Figure 2.23 Proportion of overweight and obese children and adolescents aged 2–17 by socio-economic status

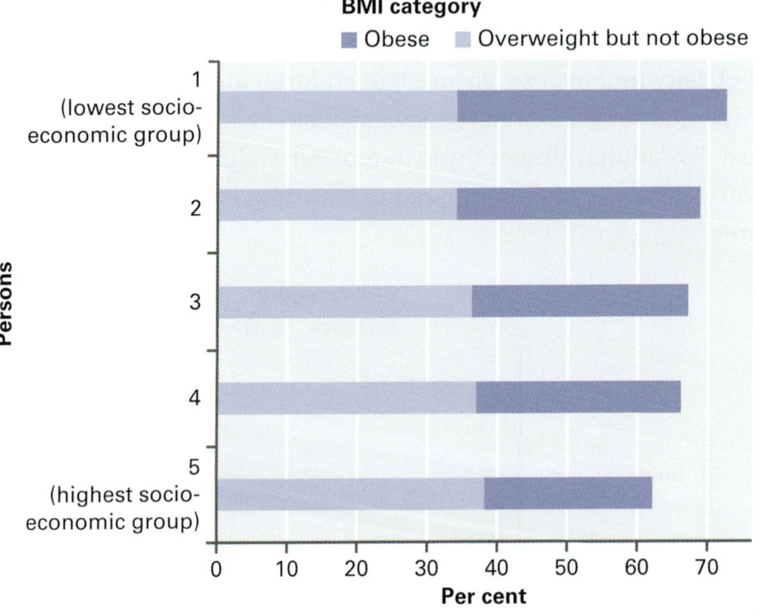

Figure 2.24 Proportion of overweight and obese adults by socio-economic status

Figure 2.25 Country towns tend to have fewer fast-food outlets than cities.

Socio-economic groups can influence food and nutrient intake as well as physical activity levels. People living in high socio-economic areas are more likely to consume more fruit, grains, meat and dairy products, have lower intakes of sugar and higher intakes of fibre, and are more likely to meet recommended physical activity levels.

FAST FACT

45 per cent of adults and 63 per cent of children eat sufficient serves of fruit (approximately two portions), but only 9 per cent of adults and 9 per cent of children ate sufficient serves of vegetables (approximately 4–6 portions, depending on age).

Worksheets 2.8, 2.9 and 2.10

Weblink

Access the ABC's The Checkout episodes titled 'Healthy Labelling', 'Superfoods', 'Seeing Stars' and 'Energy Crisis' on YouTube for examples of clever marketing ploys used by food and drink manufacturers to entice consumers to buy their products.

THE MEDIA

Everyone is constantly exposed to advertising; you can see it and hear it everywhere. There's an obvious reason for this – it works! Advertising is used very effectively by food and drink manufacturers in order to increase sales of their products. Television and social media advertisements can cost thousands of dollars per second, depending on the timeslot when they are broadcast. For most manufacturers and marketers, it is money well spent, because it helps to generate more revenue.

When we compare nutrition to other health issues discussed in this publication, it becomes evident how often we are exposed to a wide range of messages encouraging us to make unhealthy dietary decisions. These messages are cleverly developed by marketing and advertising professionals and cost thousands and millions of dollars.

FAST FACT

Of adolescents who try a new food or drink product, 55 per cent do so because they saw it advertised.

SOCIAL MEDIA & 'INFLUENCERS'

CASE STUDY

THE SCANDAL THAT SHOULD FORCE US TO RECONSIDER WELLNESS ADVICE FROM INFLUENCERS

Identify

Former social media influencer and 'wellness guru' Belle Gibson first caught public attention after claiming she cured herself of terminal cancer by rejecting conventional medicine in favour of a healthy diet and lifestyle. Her story was documented on a blog and social media, which became the basis for a successful book and app, featuring lifestyle advice and healthy recipes.

Understand

In 2015, however, Gibson was exposed as a fraud. It was revealed that she never had cancer and failed to donate the proceeds from her app to charity, as promised. Now, she has been summoned to appear in Federal Court following her failure to pay a $410,000 penalty for misleading health claims.

Beyond the psychological factors motivating Gibson's deceit, the scandal raises important questions about the cultural and technological conditions that enable lifestyle gurus to flourish.

The rise of lifestyle gurus

Claims about how to heal illness through diet and alternative therapies are far from novel. What is new is the unprecedented speed and scale afforded by online transmission. Social media also enables bloggers to monetise their following through advertorials, affiliate programmes and blog shops. The influencer economy has become a billion dollar industry, resulting in a surge in the number of 'uncertified' bloggers competing to achieve lifestyle guru status.

Although Gibson's story is seemingly unique, the narrative upon which it was scripted is common to lifestyle gurus. Lifestyle gurus define themselves in opposition to experts. Selectively, they combine elements from science, esoteric systems of knowledge, self-help and positive thinking. The advice given, which often comes at a commercial premium, appeals to common sense. But practical recommendations to eat more fruit and vegetables, exercise regularly and reduce alcohol consumption are generally followed by pseudoscientific detox products, cleanses and online services that offer quick fix solutions to complex problems. While some influencers claim to be nutritionists, few have the credentials required to give medical advice. Instead, their fame and credibility is derived from a series of techniques. These include carefully constructed personas and narratives of self-transformation, documenting their journey from illness to self-recovery. The personal improvements they document online rest mostly on anecdotal evidence and photographs which reveal their transformation into attractive, ostensibly happier and healthier people.

There is no commitment to independent testing procedures and results by objective, scientific methods. Rather, online metrics (such as followers, likes and shares) validate their status. Lifestyle gurus connect and inspire their followers through disclosing their struggles and vulnerability. Each life crisis, confession and revelation shared online results in more likes and followers.

Social media has altered how we are influenced. Engineered around the quest for visibility and attention, influence is measured by follower counts and engagement. An expert may have credentials and years of experience, but they are unlikely to be as compelling as an attractive lifestyle guru who is 'instafamous', with a highly curated social media feed to verify their advice. The issue here is not merely about the risk of misinformation, but the techniques used to influence us to decide what information to trust and who to believe.

'The scandal that should force us to reconsider wellness advice from influencers', The Conversation, *21 May 2019, https://theconversation.com/the-scandal-that-should-force-us-to-reconsider-wellness-advice-from-influencers-117041*

Discuss

1. Summarise three strategies used by social media influencers, as described in the article, to generate revenue.
2. Explain the ways in which social media has 'altered how we are influenced'.
3. Hypothesise some reasons why some people are more receptive to wellness advice from unqualified, unscientific, instafamous lifestyle gurus than they are from trained and experienced professionals who are qualified to offer medical advice.

INVESTIGATION ➡ ADVERTISING TECHNIQUES

Purpose

To discover the range of advertising techniques used to promote discretionary food or drink products, and healthy food or drink products (as advised by the Australian Guide to Healthy Eating).

9780170463102

Materials

- Access to advertisements online, through printed materials or television.
- Use word processing or presentation software to collate your findings.

Method

Look for six food or drink advertisements from a range of printed or digital sources (e.g. newspapers or magazines, TV, screen grabs from online) and gather them together for comparison. Examine the advertisements and list the techniques utilised by the marketers to make the advertisement more effective.

Discussion

1. Compare the advertisements you found for healthy foods or drinks to those you found for discretionary foods or drinks. Which techniques were most prevalent?
2. Appraise the effectiveness of these techniques.
3. Why do you think these sneaky advertising techniques are used so frequently to promote discretionary products?
4. Should there be rules in place to limit the way advertisers go about generating more sales, particularly if it is at the expense of consumers' health?

1 This is a three-stage process to organise the factors that influence our eating habits.
  a Brainstorm different factors that can influence someone's diet and nutritional choices. List them on your classroom whiteboard. You should be able to come up with at least 20 different factors.
  b Next, use a technique called 'mind mapping'. This will help organise all your brainstorm ideas. You can do this by hand on paper, preferably A3, or use computer software or an online tool such as Mindmeister. You could use the terms Personal, Social, Economic and Cultural as the main 'branches' of the mind map.
  c Then take the influencing factors on a hot-air balloon ride. Design an eye-catching and colourful poster, preferably on A3 paper, that depicts a hot-air balloon with you in it.
    - In the far-off distance is the objective you are aiming for in your balloon. Label it 'healthy diet'.
    - Floating around inside the hot-air balloon and helping to keep you moving towards your objective are all the positive influences on your diet that you have identified.
    - Hanging around the outside of your balloon's basket or gondola are bags of ballast, which can be cut away to make the balloon rise more easily. Each bag needs to be labelled with a negative influencing factor that you have identified. You might be able to show yourself cutting away the negative factors that are holding you back from your objective.

2 Knowing your positive and negative influencing factors is one thing, but knowing how to 'cut away' the negatives that are stopping you from achieving a consistent healthy diet is another story. Propose some strategies that could be used to minimise the negative influences you identified.

3 In what ways can you support a family member or friend to achieve healthy eating habits?

1 As quickly as you can, think of three advertisements for food or drink products. You might even remember their jingle or slogan. Compare your responses with a partner – were any of your answers the same as theirs? Share any products that both you and your partner had in common. These could be listed on the whiteboard.

a Are there any brands that appear more often in people's memories?

b How many of the brands would be classified as junk food, fast food or discretionary food?

c What does this tell you about the power of advertising?

d What does this tell you about the relationship between expensive, effective advertising and junk food, fast food or discretionary food?

e What is the part of the advertisement that sticks in your mind? Why?

1 Go online and find four advertisements for Coke or Pepsi products. Take note of the images, words (or lack thereof) and messages they are trying to send.

a Do any of the advertisements endorse the product in any way? In other words, does the advertisement describe how well the product will work?

b What messages do the advertisements try to communicate to consumers? How do you know this?

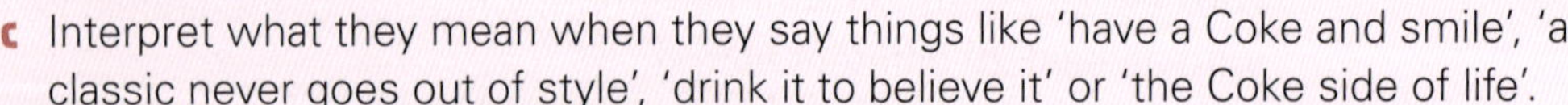

c Interpret what they mean when they say things like 'have a Coke and smile', 'a classic never goes out of style', 'drink it to believe it' or 'the Coke side of life'.

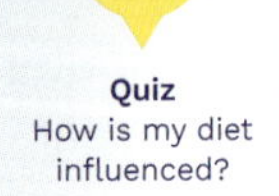

Quiz
How is my diet influenced?

d Do these advertisements make you feel like drinking a cola? Why or why not?

e These advertisements would have cost millions of dollars each to design, produce and place in the media. What is the justification for this?

HOW CAN I MAKE SUSTAINABLE FOOD CHOICES?

Producing enough food to meet the demands of the world's growing population is a serious issue. While more food is currently produced in Australia than Australians need, the production of food is resource intensive and will not necessarily always be possible due to the limited availability of water, soil nutrients climate change and other raw materials.

MINIMISE!

Shutterstock.com/smereka

Figure 2.26 How can we produce food more sustainably?

All Australians can make a difference to the demands and impacts placed on the environment by food production. Australians should look to minimise three aspects of their dietary habits – travel, production and waste.

Travel

The further away food is produced and needs to travel to reach the point of consumption, the more resource intensive that food becomes. Sometimes, distance travelled also has impacts on the freshness, taste and nutrient density of the food. Buying locally from small stores and farmers' markets, and growing your own food at home or in a school or community garden are good ways to minimise 'food miles'. It is also best to read food labels and try to buy foods manufactured or grown in Australia.

Production

iStock.com/immature

Figure 2.27 Food that is imported has to travel from other countries to get here, so isn't as sustainable as veggies you might grow at a school ag plot.

Try to choose foods that are less processed. The more effort that goes into manufacturing food, the greater the environmental cost. Looking for foods with minimal packaging is an easy way to achieve this. Organic farmers aim to minimise the use of chemicals that can increase the impact of food production on the environment.

Purchasing and eating seasonal foods is also a good way to know you are consuming fruit and vegetables that have used minimal resources in their production.

Farmers producing food crops are subject to weather, pests, disease and many other factors. Human-induced climate change is predicted to increase the severity of extreme weather events, such as droughts, floods and cyclones.

This doesn't just impact farmers; it impacts the availability of food for all Australians, and can also make some foods unaffordable. When Cyclone Yasi hit North Queensland in 2011, it devastated banana plantations and ruined the crop. Far-north Queensland supplies 90 per cent of Australia's bananas, costing the industry $400 million.

Waste

Try to minimise food wastage and overconsumption of food. Consuming more food than is needed is not only bad for our health, but also puts extra stress on the

environment. Food wastage in Australia is huge, with about three million tonnes of edible food (worth about $8 billion) going to landfill every year.

A third of the food thrown out is fresh (such as fruit and vegetables). By composting fresh food and scraps at home, you can look after your own garden better and reduce landfill. Composting food scraps limits the amount of greenhouse gases released into the atmosphere as the food scraps break down.

In a recent US study, of the thousands of tonnes of food waste in a year, production accounted for 20 per cent, processing 1 per cent and distribution 19 per cent. The remaining 60 per cent of waste occurred at the consumer level.

FAST FACT

Ten per cent of Australia's greenhouse gas emissions are produced by agriculture, such as livestock including cattle and sheep.

Figure 2.28 Seasonal produce guide

Waste warrior

Try composting for a week so you can understand just how much waste is food-based. (If you have a compost bin at home, you probably have a good idea of this amount already!)

Ask your parents or guardians to help you choose a suitable container you can use for a week. It could be an old bin, tub, bag or drum. You will need everyone in your household to help you achieve this. For an entire week, every bit of food waste your household generates goes into your new compost bin. You might want to keep it down the back of the garden or in the shed, if possible, so the house will not smell.

Take daily photos so you can track how it grows. By the end of the week you will probably find you have quite a collection of organic waste. If you have a garden, you can dig your compost into the soil to return the nutrients.

Even better, you could start a compost bin to use permanently. You could also lobby your school to get some compost bins for food scraps and garden waste.

1 Describe three advantages of using a composting system. Consider both landfills and your own garden.

FAST FACT

On average, each Australian produces 261 kilograms of food waste per year.

Weblink
GoodFish, an Australian Marine Conservation Society website, has guidelines and a smartphone app to help choose seafood wisely.

Foodwise has a seasonal produce guide

Worksheet 2.11

2 Most people assume that organic matter such as food waste breaks down naturally in landfills, but the reality is that the decomposing process in landfills is very inefficient. This is because there is so much organic matter in the one place, and there is a lot of methane (a greenhouse gas) produced. How does composting minimise this?

3 Propose some reasons why people might choose not to compost their organic waste.

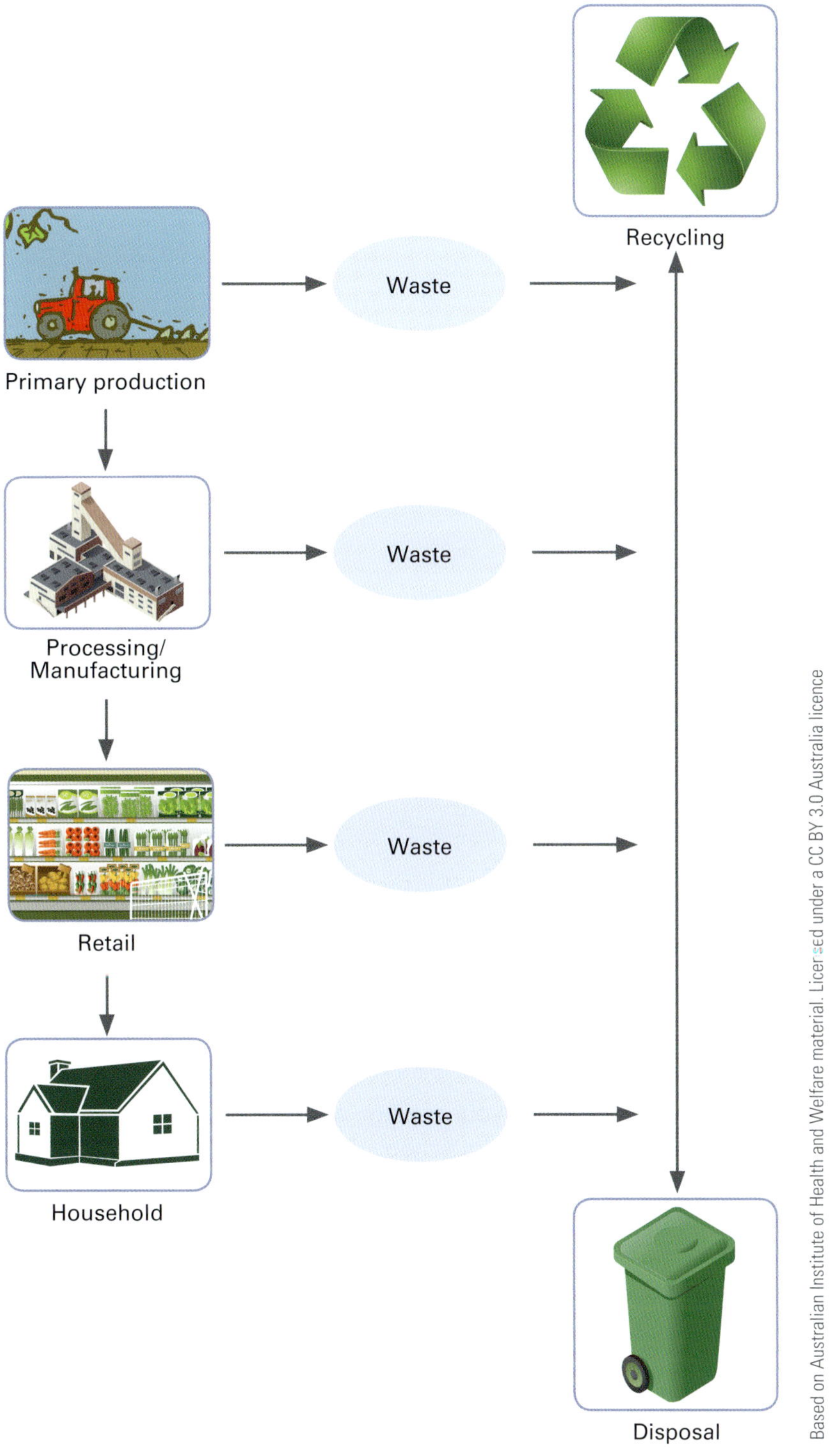

Recycling

Primary production → Waste →

Processing/ Manufacturing → Waste →

Retail → Waste →

Household → Waste →

Disposal

Figure 2.29 Sources of waste in the food supply chain

Weblink
Check out the Foodwise infographic on waste.

iStock.com/Photo_Concepts

Figure 2.30 You can be a waste warrior!

Other sustainable options

School gardens are becoming increasingly popular across Australia – does your school have one? Your class could work as a team to establish a school garden, but you will need to develop a plan. What steps would you need to consider to get one started? There are plenty of online resources to guide you through this process. You could also establish a garden at home, and use your own freshly grown vegetables, fruit and herbs in your kitchen.

If you already have a school garden, explore some options for developing it further – the following investigation will give you some ideas.

INVESTIGATION ⇨ SUSTAINABILITY PROJECTS

Weblink
Follow the weblinks for inspiration to create a sustainability project for your school.

Worksheet 2.12

Purpose

Whether your school already has some sustainability programs or projects in place, or you are starting from scratch, there are always more exciting ideas we can explore. This investigation will enable you to develop those ideas.

Materials

A planning template for tracking the goals and requirements of your project and your progress. There are lots of online links on Nelson MindTap to assist you.

Method

1 Working in a team, and in collaboration with your teacher, determine what sustainability projects or programs are already in place at your school. Will you be looking to build upon these, or to start something new? For example, you may already have a school garden, but find that manually watering it is wasteful and inefficient. Your project could focus on capturing rainwater and developing an automated drip watering system.
It can be difficult to know what ideas and options you have. Use the web to find out about all the different projects you could undertake. You can use the links above and/or conduct your own search.

Here are some ideas that have proven to be successful in schools already:

- Build a greenhouse for winter growing or propagating seeds.
- Convert your school garden to organic – no synthetic chemicals.
- Establish a hydroponics system inside or outside.
- Make a fruit stand for your school community to share, swap or sell their own fresh produce, as well as the school's.
- Transform to a 'no-dig' garden to minimise effort and make efficient use of garden bed space.
- Start composting organic waste (from school or from home) that can then be used to feed your school garden, or can be provided to your school community.
- Go high tech – moisture sensors, pH testing and automatic watering systems are just some of the ways that we can use technology to make our gardens more efficient and productive.
- Build worm farms to promote healthy soils.
- Explore the principles of 'companion planting' to maximise plant health and productivity.
- Create your own 'bush tucker' garden, using native plants that First Nations Australians have relied upon for food for thousands of years.
- Access grants or conduct fundraising to make any of the above possible.

2 Once you decide on an idea, complete a planning document that outlines: goals, stakeholders, consultation, communication, resources, timeline, review.

You can use the template in your workbook or develop your own.

Discussion

1. Being able to share the benefits of your proposed project can help to get others on board. What outcomes do you hope to achieve? List as many benefits as possible for individuals, the school or the wider community.
2. Describe the world we would live in if every school could implement a project like yours.
3. Anticipate your main barriers or challenges? What can be done to overcome these?
4. Consider the longevity of your project. Who will continue to run or maintain it in the future months and years? How will you pass on important information?

Shuttestock.com/Juice Flair

Figure 2.31 Students working as a team on a sustainability project

1 Name three aspects of our diet that we should be looking to minimise to support our natural environment.
2 Investigate how many tonnes of edible food go to landfill in Australia every year.
3 Describe the benefits of composting organic waste, as opposed to it going to landfill.
4 Explain what is meant by eating foods that are 'in season'.

1 Use the 'Uncover Food Waste' website to develop and review your knowledge of various stages of food waste – production and processing waste, retail waste and consumer waste. Complete the various activities contained within.
2 Access the Foodwise 'Food Waste Toolkit' cheat sheets and spend a week implementing one (or more) of the strategies (Smart Shopping, Storage or Shelf Life).
   a Did you find that your household wasted less food as a result of these strategies?
   b Were there other benefits you noticed after a week of using these strategies?
   c What was difficult about following these strategies?
   d Make a prediction about whether or not your household will be able to implement these strategies on a more permanent basis.

Weblink
Uncover Food Waste
Food Waste Toolkit cheat sheets

1 Use the 'From Paddock to Plate' resource from ABC Education to further extend your understanding of the processes involved in food production. Negotiate with your teacher about working in pairs or small groups. Delegate responsibility for each area, such as 'Growing crops' or 'Farming animals'. Complete the various activities contained within and share with your partner or group. There are eight challenges in the four 'Assessment' panels – how many can your team achieve?

Weblink
From Paddock to Plate

Quiz
How can I make sustainable food choices?

9780170463102

CHAPTER 2 REVIEW

1. Name two trustworthy publications with reliable nutritional advice that are available to Australians.
2. Describe the key message from three of the Australian Dietary Guidelines.
3. Select a healthy eating model from a country other than Australia. Describe one key difference and one key similarity to the Australian model.
4. Explain what the terms 'variety', 'balance' and 'moderation' mean to you and your eating choices.
5. List four macronutrients. For each nutrient, provide a healthy food source and a not-so-healthy food source.
6. Name three causes of death occurring in high-income countries such as Australia that are linked to the dietary choices people make.
7. Approximately how many kilojoules of energy do you need to consume per day to maintain a healthy weight range?
8. How many tonnes of food goes to Australian landfills each year?
9. How can the use of foods that are 'grown' instead of 'manufactured' contribute to a sustainable environment?
10. Describe three different strategies that advertisers use to try to persuade consumers to buy more of their product.
11. Why is it important to rely on the nutritional information panels instead of health claims or images on the rest of the packaging when trying to determine how healthy an item of food is?
12. Describe some strategies that could be used by the government to limit the rising levels of obesity in Australia.
13. What percentage of Australians have an inadequate vegetable intake?
14. What does the term 'yo-yo dieting' mean?
15. Select a commonly used weight-loss 'diet'. Why do diets such as these tend not to produce long-lasting weight loss?
16. What does the term 'glycaemic index' refer to?
17. List five ways the recipe for a meal or snack can be modified to make it healthier.
18. What does eating seasonally mean? For each of the four seasons, list three foods that are in season.

HEALTHY PEOPLE, HEALTHY COMMUNITIES

3

IN THIS CHAPTER

You will look at the social, cultural and environmental influences on your individual health and the health of communities. You will also investigate information and strategies that can have a positive impact on both the individual and the community as a whole.

Shutterstock.com/Nicetoseeya

By the end of the chapter, you should be able to:

- identify how societal norms and stereotypes shape the perception of appropriate physical activity behaviours
- identify the role physical activity, recreation and sport play in the lives of Australians in the past, present and future
- evaluate the influence of personal, social and cultural factors on your decision to be physically active
- evaluate and critique health information from different sources (e.g. the media, web-based, community and medical centres)
- take greater responsibility for decisions about the impact of physical activity on your health
- create, promote, implement and critique strategies to enhance the health and wellbeing of your local community
- examine social, cultural and socio-economic factors that influence the health behaviours of people in your community
- identify the health resources available in your community and evaluate how accessible they are for different groups in your community.

WHAT DOES THE AVERAGE 'AUSSIE' LOOK LIKE?

Quiz
Pre-chapter

Before you start, take the pre-chapter quiz to find out how much you already know.

Is there such a thing as an 'average' Australian? Often the **stereotype** of a 'typical Aussie' being bronzed and athletic is nothing like the reality. Sometimes stereotypical images can be offensive and highlight negative aspects of a culture; other times, the stereotype can find all the positive things that should, ideally, be seen in an 'average' Australian.

stereotype a commonly held belief about a group or type of individual

Worksheet 3.1

Vanessa Hunter/Newspix

Figure 3.1 Is this a true reflection of an average Australian?

FACE TO FACE The typical Aussie

In pairs, discuss who you think a 'typical Aussie' is.

1. Consider if the image of surf lifesavers or a 'bronzed Aussie' is true reflection of average Australians.
2. Generalise what the average Australian looks like to you.
3. State what image comes to your mind when someone says 'Australian'.
4. What three words would you use to describe the stereotypical Australian?

The case study on the opposite page shows the statistical average for Australians. The problem with these averages is that they show that Australians are getting heavier. The average Australian male has a BMI (Body Mass Index) of 27; the average BMI for an Australian female is 25. A person with a BMI of between 25 and 29.9 is considered overweight, and this greatly increases the health risks for that person.

Worksheet 3.2

CASE STUDY ⇨ HOW HEALTHY IS THE TYPICAL AUSTRALIAN?

Identify

According to the 2016 Census, the typical Australian is a 38-year-old female who was born in Australia and is of English ancestry. She is married and lives in a couple family with two children and has completed Year 12. She lives in a house with three bedrooms and two motor vehicles.

Understand

The typical Australian is a non-smoker and has never smoked, does 42 minutes of exercise every day, is overweight or obese and does not eat enough vegetables.

Data released today by the Australian Bureau of Statistics' (ABS) National Health Survey 2017–18 shows that more than half of Australians (56 per cent) thought they were in excellent or very good health, while 15 per cent were feeling in fair or poor health.

ABS Director of Health Statistics, Louise Gates, said the typical Australian male weighed 87 kg and stood 175 cm tall and was therefore overweight while the typical female weighed 72 kg and was 161 cm tall and was also overweight.

'On average, we were doing 42 minutes of exercise every day, which mostly consisted of walking for transport or walking for exercise (24.6 minutes), however we didn't participate in sufficient strength and toning activities', Ms Gates said. 'In addition, 44 per cent of us spent most of our work day sitting'.

'More than half of us were eating the recommended daily intake of fruit but not enough vegetables, with only 7.5 per cent of adults eating the recommended daily serves of vegetables."

As good news, while 79 per cent of us consumed alcohol in the last year, we did so at safe levels.

Fewer than half of Australians (48 per cent) consumed either sugar sweetened or diet drinks and 47 per cent of Australians had at least one chronic health condition.

NOTE : Please add : The average male BMI is 27 and for females this is 25.

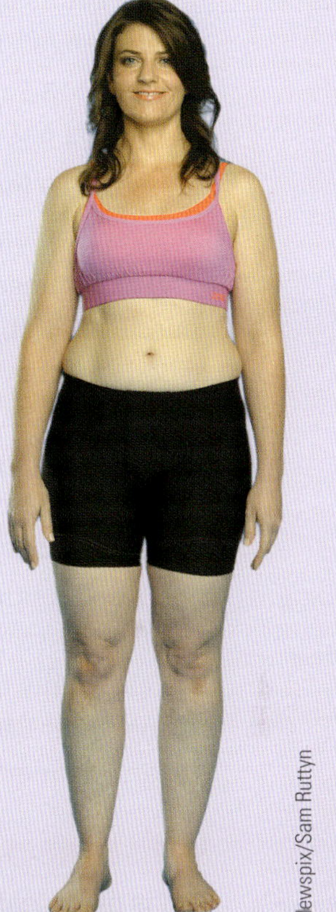
Newspix/Sam Ruttyn

Figure 3.2 An 'average' Australian

Source: *Australian Bureau of Statistics 2018 'How healthy is the typical Australian?', https://www.abs.gov.au/websitedbs/d3310114.nsf/Home/Attributing+ABS+Material, ABS website, accessed 04 June 2022. Licensed under Creative Commons Attribution 4.0 International licence, https://creativecommons.org/licenses/by/4.0/legalcode*

Discuss

Discuss how the idea of 'typical' masks the diversity and differences of Australians. Examine the word 'stereotypical' and consider how this links to making generalisations.

INDICATORS OF HEALTH

Body Mass Index (BMI)

Body Mass Index is a calculation that is used to estimate a person's total amount of body fat. It is an indicator of a healthy weight range. BMI is calculated by dividing your weight (kg) by your height squared (m^2):

$$BMI = kg/m^2$$

For example, for a 40-year-old female who is 173 centimetres tall and weighs 72 kilograms, their BMI would be 24:

$$BMI = 72 \div (1.73)^2 = 24$$

For adults aged 18 to 74 years, a BMI can be used to determine a person's weight range:

- under 18.5 means that the person is regarded as underweight
- 18.5–24.9 is a healthy weight range
- 25.0–29.9 indicates a person is overweight
- over 30 means that the person is classed as obese.

The BMI has some limitations, however, because it does not differentiate between muscle and fat. For some people it will overestimate or underestimate the amount of body fat they have, and may indicate that the person is outside the healthy weight range when, in fact, they are not. Athletes, body builders, people with physical disabilities, pregnant women, the elderly and children must be cautious when using BMI calculations. BMI calculations for children must be compared to appropriate age and sex **percentile** charts, and should only be used as a guide.

percentile the percentage of the total number of individuals that would be expected to score at or below that measure

The BMI ranges for children are different from those for adults. This is because adults have stopped growing, while children and adolescents have not! As you grow, and especially as you go through puberty, the amount of body fat you have changes, and so does your BMI. BMI calculations for children and adolescents take into account both age and gender.

Weblink
Use the calculator on the Heart Foundation website to calculate your BMI. Make sure someone you feel comfortable with measures your waist circumference, as more accurate readings are likely to be taken if you are comfortable.

Waist measurement

BMI is one measure of a person's weight and related health risk. A better predictor of health risk is waist circumference. Having a 'pot belly', or fat around the abdomen, regardless of body size, is closely linked to obesity-related health risks such as some cancers, type 2 diabetes, hypertension, high cholesterol and cardiovascular disease. Measurements of the risk of developing chronic health problems in children and young people have not yet been developed.

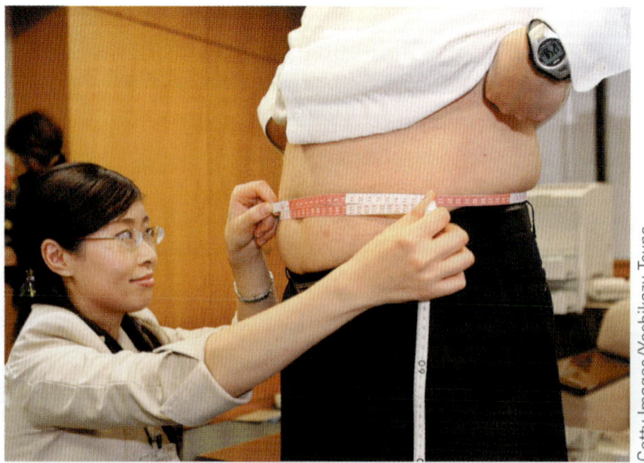

Getty Images/Yoshikazu Tsuno

Figure 3.3 Measuring the circumference of a person's waist

The waist circumferences for men and women used to indicate an increased health risk are as follows:

- Men:
  - 94 centimetres or more – increased risk
  - 102 centimetres or more – substantially increased risk.
- Women:
  - 80 centimetres or more – increased risk
  - 88 centimetres or more – substantially increased risk.

Thinking back, how does the stereotypical image you had of the average Australian fit with the reality? Sometimes ideas about what is considered the 'norm' in society can influence the way individuals think about their bodies, the physical activity that they do and the sports they play.

PHYSICAL ACTIVITY AND SEDENTARY BEHAVIOUR

Two of the biggest causes of being overweight or obese are inactivity and sedentary behaviour. Inactivity is defined as undertaking insufficient physical activity to achieve health benefits. Sedentary behaviour is associated with low levels of energy expenditure, such as sitting for long periods of time or not moving around. It is thought that reducing the time spent in sedentary behaviours may be as important to your health as increasing the time spent being active! It is worth noting that many professional athletes are highly active during their training days, but also sedentary for a large part of the day while they are recovering.

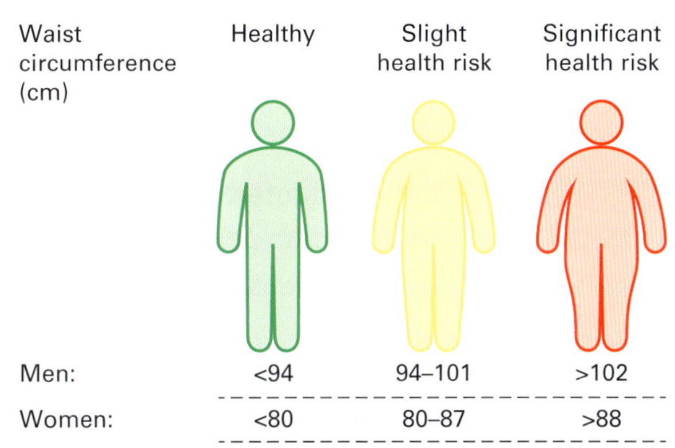

Figure 3.4 Waist circumference and health risk

FAST FACT

Physical inactivity is one of four leading risk factors in cardiovascular disease. For the first time in history, children today are projected to have a shorter life expectancy than their parents; experts say this is due to physical inactivity. Physical inactivity is estimated to cost the Australian healthcare system $800 million per year!

The Australian Government has produced a series of guidelines about the amount of physical activity needed for health. Australia's Physical Activity and Sedentary Behaviour Guidelines provide the minimum levels of physical activity required to gain a health benefit. These have recently been updated and are now known as the Australian 24-hour Movement Guidelines for Children and Young People (5–17 Years), which include sleep guidelines as well. They also suggest ways to include **incidental physical activity** in everyday life. There are many benefits of participating in regular physical activity, including a decrease in the risk of cardiorespiratory disease, diabetes, some cancers and osteoporosis. These are just the physical health benefits! There are also emotional, social and mental health benefits from physical activity. Exercise has been shown to reduce the risk of and effectively treat clinical depression and anxiety.

incidental physical activity activity that results from completing another task where the intention was not physical activity (e.g. doing the housework)

Guidelines exist for all age groups. The Australian 24-Hour Movement Guidelines for Children and Young People (5–17 years) now take into consideration the importance of sleep for healthy growth and development, and are summarised as:

Physical activity

Worksheet 3.4

- Accumulating 60 minutes or more of moderate to vigorous physical activity per day involving mainly aerobic activities.
- Several hours of a variety of light physical activities.
- Activities that are vigorous, as well as those that strengthen muscle and bone should be incorporated at least three days per week.
- To achieve greater health benefits, replace sedentary time with additional moderate to vigorous physical activity, while preserving sufficient sleep.

Sedentary behaviour

- Break up long periods of sitting as often as possible.
- Limit sedentary recreational screen time to no more than 2 hours per day.
- When using screen-based electronic media, positive social interactions and experiences are encouraged.

Sleep

- An uninterrupted 9 to 11 hours of sleep per night for those aged 5–13 years and 8 to 10 hours per night for those aged 14–17 years.
- Have consistent bed and wake-up times.

INVESTIGATION

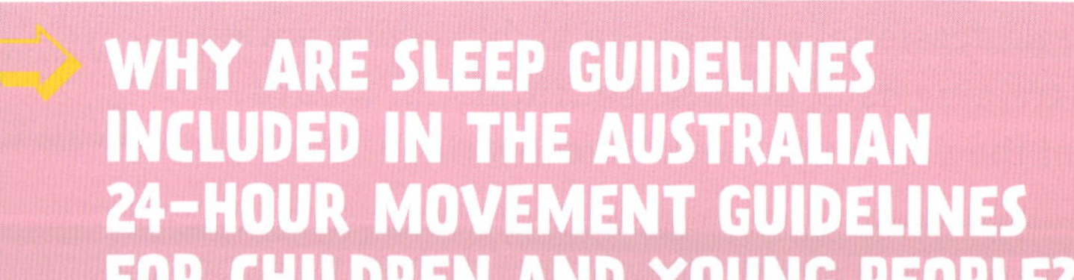

WHY ARE SLEEP GUIDELINES INCLUDED IN THE AUSTRALIAN 24-HOUR MOVEMENT GUIDELINES FOR CHILDREN AND YOUNG PEOPLE?

Purpose

Survey your class to discover their sleep behaviours. This information will be used to draw comparisons to the Australian 24-Hour Movement Guidelines.

Method

1 Design a survey to investigate sleep patterns in your class. You can use a free digital survey such as SurveyMonkey, SurveyPlanet or Google Form. These allow you to send surveys via email to all classmates, and also automatically collate the results.
When designing a survey it is important to think about what it is you are trying to find out. In this case, you could ask:
   - What time do you usually go to sleep?
   - Do you go to sleep at the same time every night?
   - If you ever have 'bad sleep', what tends to cause this?
   - How do you behave the next morning/day when you had a bad night's sleep?

2 Brainstorm other questions you could ask your classmates about their sleep habits, then create your survey.

Discussion

1 Present your results as either summary graphs or tables.
2 Summarise any trends that appear when considering the amount of sleep your classmates get.
3 State what percentage of your classmates get into 'sleep routines' and go to sleep around the same time every night.
4 Reflect on your own sleep and list three factors that cause you to experience 'bad sleep'.
5 Identify at least two consequences of having a bad night's sleep.
6 Considering the results of your survey, why do you think the Australian Government decided to add sleep guidelines to the Australian 24-Hour Movement Guidelines?

AUSTRALIA'S PHYSICAL ACTIVITY AND SEDENTARY BEHAVIOUR GUIDELINES FOR ADULTS (18–64 YEARS)

The Australian Government also provides guidelines for adults.

Physical activity

- Doing any physical activity is better than doing none. If you currently do no physical activity, start by doing some, and gradually build up to the recommended amount.
- Be active on most, preferably all, days every week.
- Accumulate 150 to 300 minutes (2 ½ to 5 hours) of moderate intensity physical activity or 75 to 150 minutes (1 ¼ to 2 ½ hours) of vigorous intensity physical activity, or an equivalent combination of both moderate and vigorous activities, each week.
- Do muscle strengthening activities on at least two days each week.

Sedentary behaviour

- Minimise the amount of time spent in prolonged sitting.
- Break up long periods of sitting as often as possible.

Tips for staying active

There are a number of 'tips and ideas' provided by the Australian Government for increasing the amount of physical activity and decreasing the amount of sedentary behaviour of adults.

Build activity into your day

- For short trips, walk or cycle and leave the car at home.
- For longer trips, walk or cycle part of the way.
- Use the stairs instead of the lift or escalator.
- Get off the bus one stop earlier and walk the rest of the way.
- Park further away from your destination and walk.

Active and safe

- If you are new to physical activity, have a health problem, or are concerned about the safety of being (more) active, speak with your doctor or health professional about the most suitable activities for you.
- Protect yourself from the sun – you should wear sun-protective clothing, including a hat, and apply sunscreen regularly.

Active at work

- Take the stairs instead of the lift.
- Walk to deliver a message to your colleague, rather than emailing.
- Leave your desk at lunch time and enjoy a short walk outside.
- Organise walking meetings.

Active indoors

Don't let the weather stop you!

- Body weight exercises like squats, push-ups, sit-ups and lunges, can all be done indoors.
- Try indoor activities like
  - dancing
  - indoor swimming
  - yoga or pilates
  - martial arts
  - squash, or
  - indoor rock climbing.

Tips for reducing sedentary behaviour

- Get up to change the channel on the TV instead of using the remote.
- When tidying up, put things away in multiple small trips rather than one big haul.
- Preset the timer on your TV to turn off after an hour to remind you to get up and move more.
- Walk around when talking on your mobile phone.
- Stand up and move during your favourite TV shows.
- Instead of sitting and reading, listen to recorded books while you walk, clean, or work in the garden.
- Stand on public transport and get off one stop earlier than your destination.

If you work in an office:

- Stand while you read at work.
- Move your rubbish bin away from your desk so you have to get up to use it.
- Use the speakerphone for conference calls, and walk around the room during the conference.
- Ask your boss for a 'walk and talk' meeting rather than a sit-down meeting.

Why not turn off the TV during the day and get out in the garden?

Set an alarm on your computer to remind you to stand up and move more often.

How about delivering the message in person, instead of by email?

Make your move – Sit less – Be active for life!

Source: © 2014 Commonwealth of Australia as represented by the Department of Health and Aged Care

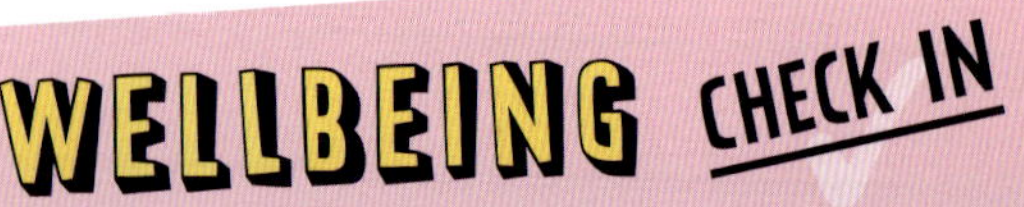

Identify

Exercise isn't just good for physical fitness. It has positive effects all over the body, including the brain.

This activity has been developed in collaboration with Dr David Bakker of MoodMission

3

Understand

Exercise can benefit mental health in many ways, two in particular. Firstly, it can promote the release of feel-good chemicals in the body called endorphins. Endorphins work in the brain to reduce perception of pain, including emotional pain, and can trigger positive, euphoric feelings, sometimes referred to as a 'runner's high'. Secondly, it can give us a sense of achievement and satisfaction, which translates into confidence and optimism. Psychologists call this 'behavioural activation', because we have activated ourselves in the direction of a goal. The feel-good rewards we get from the exercise will motivate us to do more. Exercise doesn't need to be long and strenuous to get these benefits. A short floor workout can be a great step in the right direction.

Practise

1. Start with some push-ups. Lie on your belly on the floor. Keeping your stomach tensed and your back flat, push yourself up, keeping your hands flat against the floor.
2. If you can't do a full push-up on your toes, do a half push-up on your knees.
3. Do as many as you can before you have to stop – it might be 2, 5, 10, 20 or more depending on your level of fitness.

Getty Images/MOAimage

Getty Images/MOAimage

Figure 3.5 A push-up

4. Next, let's try some sit-ups. Lie on your back with your knees bent and your feet flat on the ground. Lift your head off the ground first, followed by your shoulders, using your abdominal muscles to bring yourself up to a sitting position. In a controlled movement, slowly lie back down the starting position.
5. Do as many sit-ups as you can before you have to stop – it might be 2, 5, 10, 20 or more depending on your level of fitness.
6. Now go back to the push-ups and do another set, before doing another set of sit-ups.

iStock.com/kieferpix

iStock.com/kieferpix

Figure 3.6 A sit-up

Reflect

How do you feel after doing this short exercise? You might feel a bit hot and sweaty, but you might also feel a little more engaged. You will certainly feel different to when you started. This is helpful to notice, as this is another way that exercise can help us feel better when we're feeling down or anxious – it can make us change states, even if it's just a quick and easy exercise!

INVESTIGATION

Purpose

You are going to design a body weight circuit that is safe and enables muscle and bone strengthening.

Method

1 Research a series of exercises that only involve using your body weight as resistance (no free or machine weights). Try to come up with a couple of different exercises to target the following areas:

a arms

b chest

c back

d legs

e core.

One of the exercises should be fairly easy to complete, whereas another should be more difficult, and should place the muscles and bones under greater tension/force.

Your circuit should ideally include about 8–10 different activities.

Discussion

1 What determined how many repetitions of each exercise you completed?

2 Consider why it is a good idea not to stress the same body part in consecutive activities during the circuit.

3 Reflect on how your could 'overload' the activities, i.e. make them harder.

4 Explain why you might consider researching new activities to include in the circuit for each body part, and incorporate these after a couple of months.

1 In your own words, describe the 'average Australian'.

2 Compare the differences in the Australian 24-Hour Movement Guidelines for Children and Young People (5–17 Years), and Australia's Physical Activity and Sedentary Behaviour Guidelines, which are for adults.

1 Consider two more tips for each of the following areas suggested by the Australian Government:

a Build activity into your day

b Be active indoors

c If you work in an office

1 Do you meet the recommendations in the 24-Hour Movement Guidelines for Children and Young People? Create a weekly schedule to help you meet them. Make sure you plan out:

a daily moderate and vigorous physical activities

b muscle/bone strengthening activities
c light physical activities
d a sleep schedule
e scheduled sedentary behaviour limits (e.g. TV watching).

2 Research the physical activity recommendations for older Australians (65 years and older) and then create a weekly plan for an older Australian that will help them meet all recommendations.

Worksheet 3.3

WHAT SPORTS AND PHYSICAL ACTIVITIES DO AUSTRALIANS ENGAGE IN?

Australia and sport have always been closely connected. Sport has often been seen as part of our cultural identity. This section further explores that connection.

THE PAST, PRESENT AND FUTURE OF AUSTRALIAN SPORT

Getty Images/Gareth Fuller – PA Images

Figure 3.7 In 2019, Australia's men's and women's cricket teams both won the Ashes series against England.

FACE TO FACE Sport as cultural identity

'Australia loves sport. It has always been and will continue to be part of our cultural identity. From playing catch in the backyard through to the Olympic and Paralympic podiums the majority of Australians play, watch and enjoy sport.'

Source: SA Hajkowicz, H Cook, L Wilhelmseder, N Boughen, *The Future of Australian Sport: Megatrends shaping the sports sector over coming decades*, CSIRO, Australia, 2013, p. 4.

In pairs, discuss the following questions:

1 Is this statement the stereotype or the reality?
2 What factors have shaped Australia's sporting identity?
3 What are the likely influences on the future of sport in Australia?

PARTICIPATION IN SPORT

Table 3.1 Top five sports young people participate in nationally

Age	Male	Female
9–14	Football Swimming Basketball Australian Football Cricket	Swimming Netball Dancing (recreational) Gymnastics Basketball
15–24	Fitness/Gym Running/Athletics Walking (Recreational) Football Basketball	Fitness/Gym Walking (Recreational) Running/Athletics Swimming Netball

Source: Sport Australia AusPlay April 2019. Reproduced with the permission of the Australian Sports Commission.

INVESTIGATION

HOW DOES OUR CLASS PARTICIPATE IN SPORT AND RECREATIONAL ACTIVITIES?

Purpose

Survey your class and list the top five sports/recreational activities your class members participate in. Use this information to draw comparisons to the broader participation trends revealed in Table 3.1.

Method

1 Design a survey to investigate which sport and recreational activities your classmates participate in. You can use any of the free digital surveys such as Survey Monkey, Survey Planet or Google Form, which allow you to send surveys via email and also automatically collate the results. When designing a survey it is important to think about what it is you are trying to find out. In this case you could ask:

- Rank your top five sports/recreational pursuits.
- Provide three reasons why you participate in your top two sports.
- What sports/recreational activities would you like to participate in, but currently don't?
- List three factors that stop you from participating in activities you would like to participate in, but currently don't.

Discussion

1 Present your results as either summary graphs or tables, showing the top sports your class participates in.
2 Compare your results to the data in Table 3.1 for your age group.
3 Summarise any trends you see in your class data. Can you spot any differences between different groups?
4 State the top three reasons given for why people in your class choose to participate in sports/recreational activities.
5 When considering sports/recreational activities people would like to participate in, what are some of the common barriers?

CASE STUDY

ONGOING IMPACT OF COVID-19 ON SPORT AND PHYSICAL ACTIVITY PARTICIPATION

Identify

The ability for Australians to be active in their communities has been interrupted since COVID-19 was first confirmed in Australia in late January 2020.

Australia's most popular physical activity, recreational walking – which continued to be popular during COVID-19. Although walking is a main contributor to increases in high frequency participation, when we consider all participation (i.e. at least once per year), other activities that were accessible in a COVID-19 environment or able to be performed socially distanced also experienced increases from 2019 to 2020. The activities below experienced statistically significant increases in their estimated number of participants nationally year-on-year.

Figure 3.8 COVID-19 impacted on they ways we exercised.

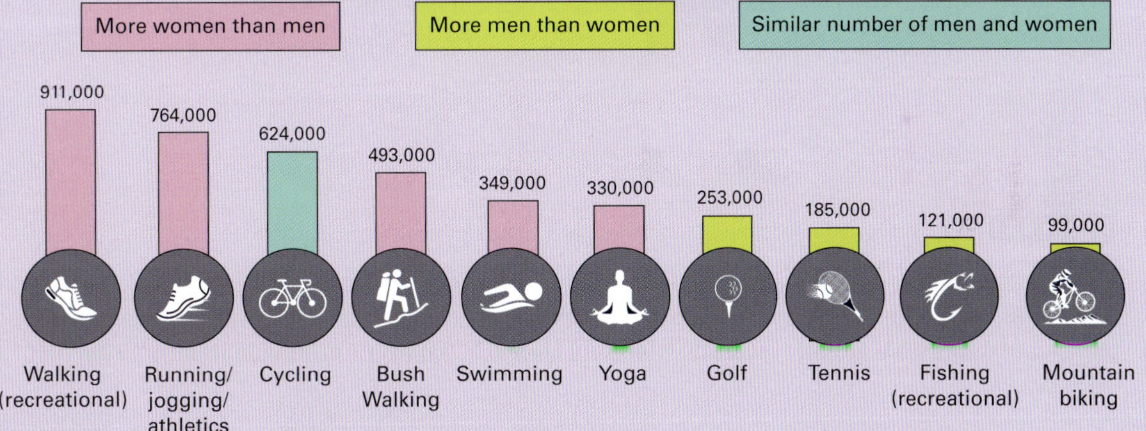

Figure 3.9 Top 10 physical activities during 2019–2020

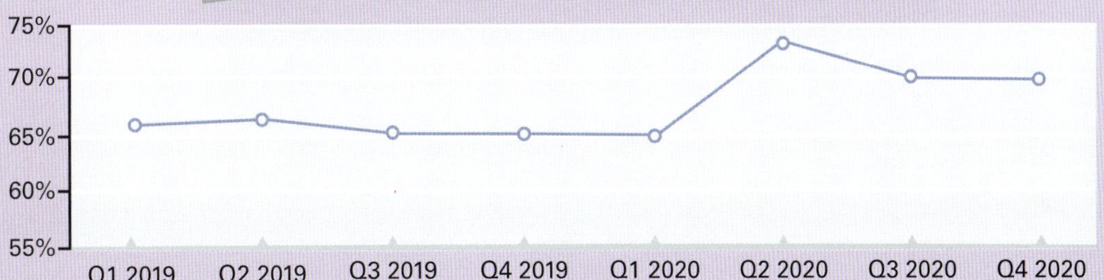

Figure 3.10 Participation in outdoor endurance/health activities 2019–2020

Some Australians had even taken up a new organised sport by March 2021. Most were already parti-cipants in other sports, but some had not played any organised sport in the 12 months before COVID-19.

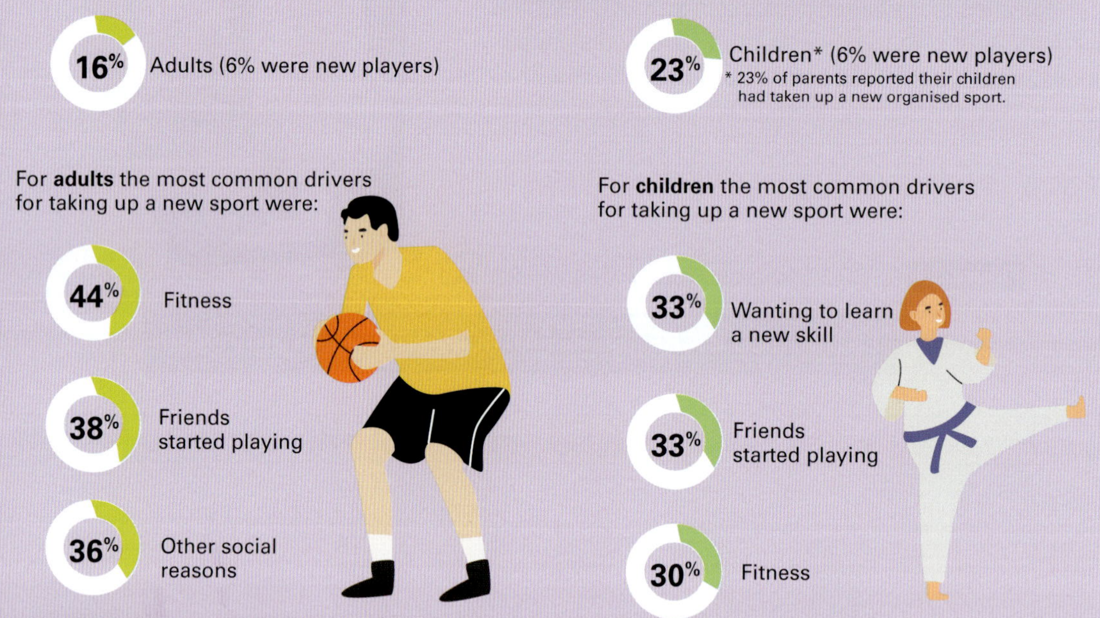

Figure 3.11 Factors involved in take up of new organised sports for adults and children

The vast majority of children's organised outside of school participation is through club sport, which was restricted during the pandemic. Therefore, the overall organised and sport-related participation rates are already revealing statistically significant declines at frequencies of 1+ and 3+ per week.

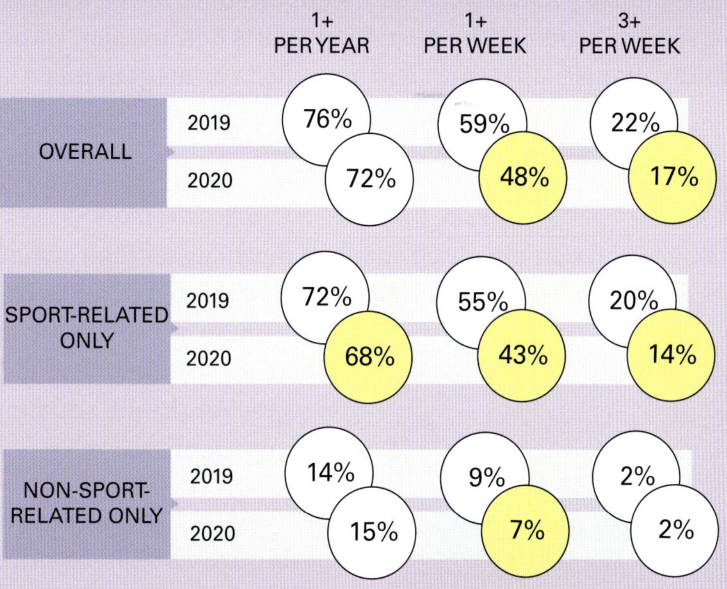

Figure 3.12 Children's participation in sport-related and non-sport-related physical activity during COVID-19

Source: AusPlay: A focus on the ongoing impact of COVID-19 on sport and physical activity participation — June 2021 update, SportAus, https://www.clearinghouseforsport.gov.au/__data/assets/pdf_file/0004/1012846/AusPlay-COVID-19-update-June-2021.pdf

Discuss

1. Reflect on your own participation in sport and physical activity during COVID-19 in 2020 and 2021. Contrast your participation levels in recreational activities to organised sport and summarise any differences.
2. Consider how your physical activity levels might have impacted on your emotional health during COVID-19. Try to generalise any links between physical activity levels and your 'mood'.
3. Explain differences between adult and children motivations for taking up a new sport when it came to 'fitness' during 2019–2020.

Future trends in participation patterns can influence policy and investment decisions by governments, industry and community groups. Sport can help improve the health and wellbeing of individuals and the community through improved physical and mental health, as well as social inclusion of marginalised groups. It is a multi-billion-dollar industry in Australia, so it's important to have an understanding of where Australian sport is heading. The Australian Sports Commission commissioned research into the future of Australian sport and identified six '**megatrends**'.

megatrend a substantial shift in social, economic or environmental conditions into the future with implications for decision making

Megatrends shaping Australian sport

A perfect fit

Participation rates in non-organised sport are much higher than organised sport, with its fixed times and higher level of commitment. It is expected that this trend will continue in the future as people look for activities that require less commitment and have a focus on recreation, health and fitness rather than on competition.

Thinkstock/istock.com/Dirima

Figure 3.13 Participation rates in non-organised sport in Australia are expected to increase.

From extreme to mainstream

Sports such as skateboarding, freestyle BMX, rock-climbing and snowboarding are examples of 'extreme' sports. Participation rates in these types of activities are increasing, and are expected to continue growing in popularity. The inclusion of BMX cycling in the 2008 Olympic Games and the push by other extreme sports to also be included is bringing these sports into the mainstream. For the first time ever, the 2020 Tokyo Summer Olympics (rescheduled for 2021) will feature surfing, skateboarding and rock climbing. There will also be new events for baseball, softball and karate. These changes are an effort to attract a younger audience by the Olympic Movement, but also indicative of increased participation in these pursuits.

Worksheet 3.5

Fairfax Syndication/Glen Watson

Figure 3.14 There are social benefits as well as health benefits to being involved in sport at the community level.

More than sport

Sport is more than just playing games! The benefits of sport and the objectives that can be met through sport are very widespread. Sport is seen as a means to tackle the rising rates of obesity and diabetes, improve mental health, reduce crime rates and work as an effective preventative health strategy; it also builds international relations and achieves social and developmental objectives.

9780170463102

3

> **FAST FACT**
> Sport participation at school can accurately predict lower adult BMI and increase the likelihood of participating in sport as an adult!

Dreamstime.com/Rob Van Esch

Figure 3.15 There are opportunities for people with disabilities to participate in sport.

Everybody's game

The changing demographic in the Australian population will change the type of sports that are played. The ageing population and the diverse range of cultures in Australia mean that sport in the future will need to cater to different sporting preferences. Low current participation rates for people with a disability mean that there is also a need to provide more opportunities and increase access to sport and recreation for these individuals.

UP AND MOVING — Sitting Volleyball

Sitting Volleyball has been included in the Paralympic games since 1980. It is played in more than 75 different countries. Currently, the USA and Iran sit at the top of the rankings in the female and male world leagues respectively. The rules are based on rules for able-bodied volleyball, though sitting volleyball is played on a smaller court and with a smaller net. There are six team members on pitch at any one point, and at all times, a part of a person's torso must touch the ground. In sitting volleyball, the torso includes the buttocks – hence a player must be sitting or lying down at all times.

Watch the video explaining the game, and then play a game yourselves. After playing the game, discuss the following questions:

1 What did you like the most about the game?
2 Did the game give you a better appreciation for some of the challenges disabled athletes face?
3 What did you find the most challenging about the game?

Weblink
Watch the A–Z video explaining Sitting Volleyball

UP AND MOVING — Modified netball

Try playing netball without anyone being able to speak during the game to simulate what it might be like to play the game if you were deaf. Or, if you wanted to experience what it might be like playing the game with impaired vision, try playing with one eye covered (this reduces the ability to focus, and also limits peripheral vision). As a class, discuss the same questions used for Sitting Volleyball.

FACE TO FACE — Accessible sports

Discuss how mainstream sports in Australia change to have greater appeal for a more diverse population.

Getty Images/Nick Laham

Figure 3.16 China has a whole country support system for elite athletes.

New wealth, new talent

Shifts in the world economy have seen both income and population growth in Asia, which will have an impact on the sporting sector in Australia. China has had rapid success in elite sporting competitions, especially the Olympics. China returned to Olympic competition in 1984 and has placed in the top four in every Games except one. The Chinese sport system has 'whole country support for the elite sport system' and, with such a large population, has a vast pool of athletes to draw from.

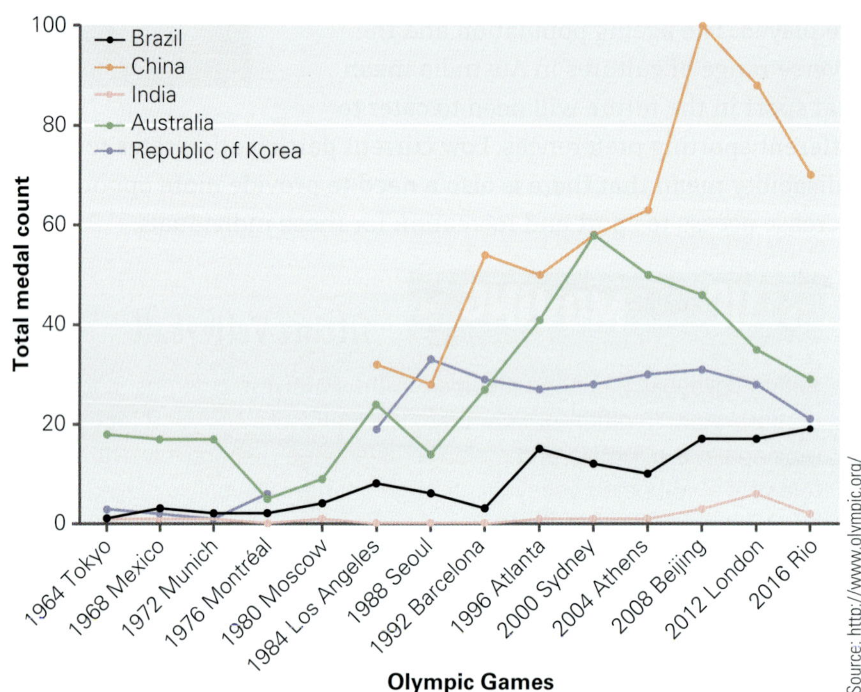

Figure 3.17 Olympic Games medal counts by country. Suggest reasons for the low medal count for India: think about population levels, economy and sports participation.

FACE TO FACE Debate

As a class, debate the statement 'Organised sport at the community level is essential for Australian success at the elite level.'

Tracksuits to business suits

Sport is big business – from television broadcasting rights, sponsorship and spectator interest to coaches, management and players. The sports industry in Australia is growing, and sporting organisations are becoming more and more like businesses, requiring professional administration. This means that the number of people employed in the sports sector will continue to grow. Many grassroots sporting clubs rely heavily on volunteers to function. In a time-poor society that fears litigation, there are barriers to volunteers in sporting clubs.

Weblink
Review the full report on the future of Australian sport

The previous six subsections are summarised from SA Hajkowicz, H Cook, L Wilhelmseder, N Boughen, *The future of Australian sport: megatrends shaping the sports sector over coming decades*, CSIRO Australia, 2013, https://publications.csiro.au/publications/#publication/PIcsiro:EP131275

9780170463102

FACTORS AFFECTING PARTICIPATION

There are many factors or reasons why individuals choose to participate in physical activity. Some may appear obvious, like the fact that some activities are enjoyable. Other influences on the choice to be physically active may be less obvious. The impact of regular physical activity on the health of the individual and the community should not be underestimated. The benefits go far beyond the physical, social, emotional and cognitive health benefits, and extend to the overall wellbeing of the local and broader communities.

Deciding to participate in regular physical activity is a good choice for overall health and wellbeing. Three of the main influences on the decision to be physically active are personal, social and cultural factors.

Video
Why don't we move?: What barriers are there to physical activity for people in your community? How may these barriers be overcome? Watch the video and start the discussion!

Personal factors

There are a number of personal factors that can increase or decrease the likelihood of an individual participating in physical activity. Table 3.2 summarises some of the main personal factors.

Table 3.2 Summary of personal factors influencing participation in physical activity

Personal factor	Explanation
Age	Generally, physical activity participation rates decrease with age; this is shown in Table 3.3.
Sex	10 years ago, males were generally more active than females, however, this is no longer true for most age groups.
Socio-economic status	People from higher socio-economic backgrounds are generally more active than people from lower socio-economic backgrounds.
Geographic location	Where you live influences the amount of physical activity you do.
Other personal factors	These include fundamental motor and sport-specific skills, knowledge, beliefs, abilities/disabilities, motivation.

socio-economic status a person's or group's position in society depending on their occupation, level of education, income, wealth and place of residence

Age and gender

Complete the case study on the following page to gain an understanding of participation rates in sport and physical activity by age and gender.

Socio-economic status, geographic location, education and employment

The socio-economic status (SES) of an individual is a determinant of physical activity. There are a number of interrelated factors related to an individual's SES that influence levels of physical activity. Income and education levels are both positively related to activity levels. Research has shown that those individuals with a higher income and who have attained a higher level of education are more likely to be active. Children living in areas of low socio-economic status are 9 per cent less likely to participate in organised sport.

iStock.com/jonya

Figure 3.18 How can communities cater to the physical activity needs of an ageing population?

CASE STUDY: HOW DOES GENDER IMPACT ON SPORT AND PHYSICAL RECREATION PARTICIPATION RATES?

Identify

The SportAus AusPlay survey provides national data on participation in sport and physical activity. The survey has run annually since 2015. The following data represents the 2020-21 survey results.

Understand

The highest participation rate in sport and physical recreation is for people aged 15 to 17 years. Participation rates in sport and physical recreation generally decrease as people get older. People aged 65 years and over have the lowest participation rate of all age groups across the lifespan.

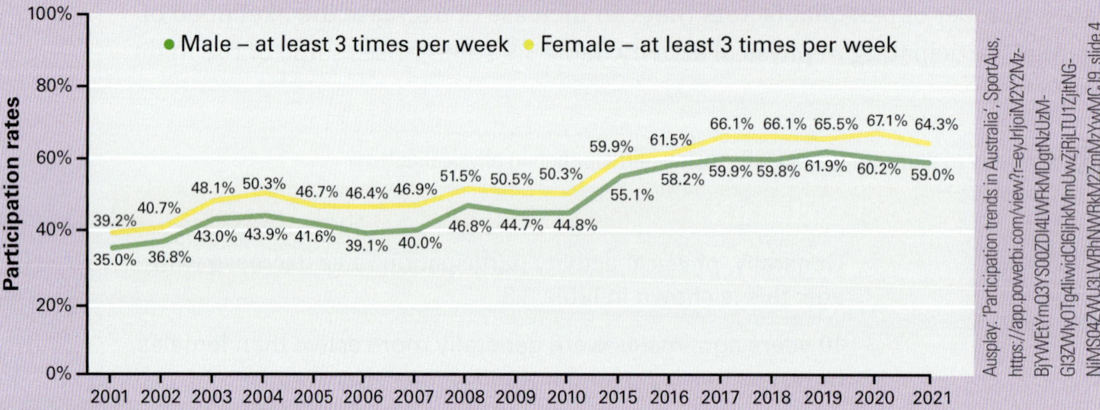

Ausplay: 'Participation trends in Australia', SportAus, https://app.powerbi.com/view?r=eyJrIjoiM2Y2Mz-BjYWEtYmQ3YS00ZDI4LWFkMDgtNzUzM-GI3ZWIyOTg4IiwidCI6IjhkMmUwZjRjLTU1ZjItNG-NiMS04ZWU3LWRhNWRkM2ZmMzYwMCJ9, slide 4

Figure 3.19 Participation in sport and physical activity over a twenty year for males and females

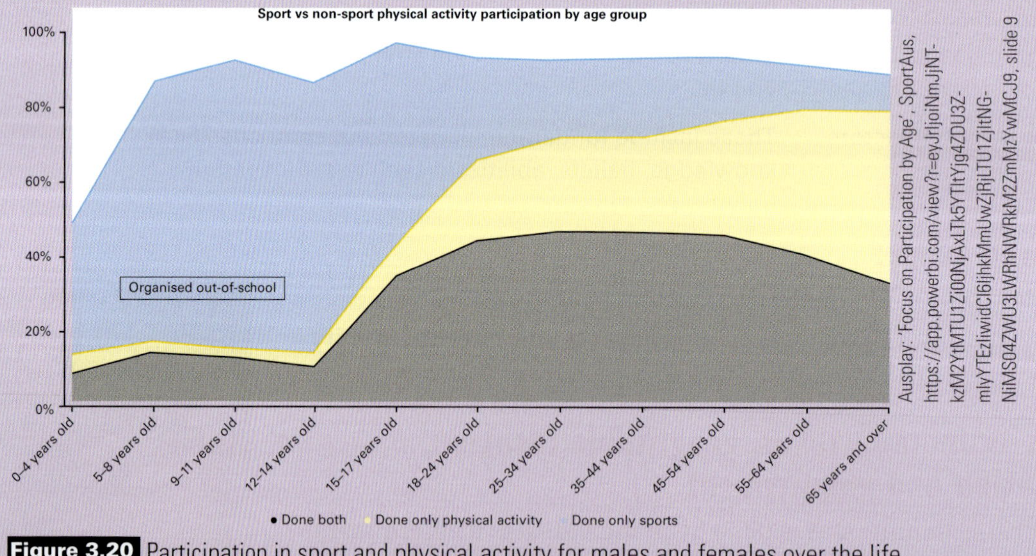

Ausplay: 'Focus on Participation by Age', SportAus, https://app.powerbi.com/view?r=eyJrIjoiNmJjNT-kzM2YtMTU1Zi00NjAxLTk5YTItYjg4ZDU3Z-mIyYTEzIiwidCI6IjhkMmUwZjRjLTU1ZjItNG-NiMS04ZWU3LWRhNWRkM2ZmMzYwMCJ9, slide 9

Figure 3.20 Participation in sport and physical activity for males and females over the life course (2020)

Discuss

1. Use the information to compare the participation rates of males and females and of the different age groups.
2. Propose three reasons why you believe male participation in sport is higher than that of females.
3. Generalise a statement that summarises the participation in physical activity from 0 to 65 years of age.
4. What is the relationship between participation in sport and physical recreation and age?

9780170463102

5 The greatest difference between male and female participation rates is in the 15–17 age group. Suggest three reasons why this might be the case.
6 The 65+ group has the lowest rates of physical activity participation. What are the consequences of a lack of physical activity in the age 65+ age group for the individual, and for the community?
7 Identify a suitable form of physical activity for each of the age groups shown in the graph. Justify why you have chosen the activity and how it meets the needs of the age group.

Where you live is also a determinant of physical activity levels. In Australia, people living in a capital city are more likely to be involved in sport and physical recreational activities. This may in part be due to the greater access to facilities and recreational services in capital cities, compared with other areas in Australia. Participation rates vary for each of the states and territories, with the Australian Capital Territory having the highest participation rates (93.2 per cent) compared with all other states and territories (84 to 89 per cent).

There are a number of other personal factors that influence behaviour patterns in relation to physical activity. Knowledge of the health-related benefits of physical activity, personal beliefs and attitudes regarding physical activity, motivation and enjoyment all affect levels of participation. Another factor that influences participation in physical activity is personal skill level. For example, sports such as surfing, boogie boarding, canoeing and kayaking, water-skiing and kite surfing would not be safe options for a person who has never learnt to swim.

Shutterstock.com/EpicStockMedia

Figure 3.21 Lack of skill can limit participation in some recreational activities. For kite surfing, you need to be able to swim.

iStock.com/Spotmatik

Figure 3.22 Your family, parents and siblings can influence the types of physical activity that you do.

Social factors

The people surrounding an individual influence physical activity behaviour. Families and friends provide social support for an individual's involvement in physical activity. Parents can be role models for their children by being active themselves: they can organise family activities that have a physical activity focus, such as a bike ride or bushwalk. Parents also support their children to be physically active by paying for uniforms, equipment and memberships to sporting clubs, as well as driving them to training and games!

Friends and peers can also influence an individual's level of participation in physical activity by being encouraging. However, peer pressure can also be a negative influence on behaviour and choices. Chapter 4 explores peer pressure and how different strategies can be used in decision-making. Physical activity often results in new friendships with teammates; teams are also often formed among a group of friends.

Schools, workplaces and community groups can all provide a social environment in which to be active.

Figure 3.23 Parents help their children be active by transporting them to games, paying for equipment and cheering them on.

Newspix/Steven Crabtree

Figure 3.24 There are many different apps available to record, track and share your physical activity behaviours.

Alamy Stock Photo/Donisl

FAST FACT

Social media is prominent in the fitness industry. There are many apps that can record, track and then share your personal fitness achievements. Through social media, online communities can be formed that engage a far wider audience than more traditional organisations.

CASE STUDY

WEARABLE ACTIVITY TRACKERS INCREASE PHYSICAL ACTIVITY PARTICIPATION

Identify

In 2019, a group of researchers conducted a meta-analysis to find out whether wearable activity trackers had an impact on physical activity participation. A meta-analysis is when a researcher looks at all of the previous experiments and research on a subject and analyses the findings to come to conclusions about the subject.

Understand

The researchers looked through several databases and then analysed the impact of wearable trackers on both physical activity participation and sedentary behaviour. They found that the use of a wearable activity tracker resulted in an increase in daily step count, moderate and vigorous physical activity and energy expenditure. There was also a decrease in sedentary behaviour.

Discuss

1 Why would people using wearable technologies have higher levels of physical activity and less sedentary behaviour?
2 Why would it be a good idea to record daily and weekly step counts and physical activity intensities?
3 Do you believe it would be a good idea to include an activity tracker on a Year 7 booklist so all students in the school will eventually have one? Briefly discuss.

Cultural factors

Different cultures have vastly different attitudes and preferences when it comes to sport and physical activity. Earlier, this chapter explored the 'typical' Australian and how sport has shaped Australian culture. The diversity of cultures that make up Australia today means that opportunities to be active need to cater for the multicultural population. The high levels of participation in soccer in Australia have been partially attributed to our migrant population. With diversity comes new experiences and opportunities to engage with games, sports and activities from different cultures. Sport can be used to break down cultural barriers within the community and provide minority groups with social support to participate in traditional activities. Bocce, yoga, badminton, martial arts and table tennis are just a few sports that have been introduced to Australian sporting culture through the migrant population.

Getty Images/Michael Regan

Figure 3.25 Australian women's doubles badminton team Leanne Choo (right) and Renuga Veeran (left) at the 2012 London Olympic Games.

BARRIERS TO PARTICIPATION

The same factors that make us more likely to participate in physical activity can also act as barriers. When asked, many people give the following reasons for not being physically active:

- lack of time
- dislike of exercise
- feeling too tired
- lack of company
- lack of money.

Any one or more of these reasons can become a barrier for an individual at different times throughout their lives. For example, a student may find that they have the time to be active but not the financial means, whereas a parent of young children may have the means but not the energy. The key to making physical activity part of a daily routine is finding the right activity that suits your needs.

Worksheet 3.6

Cost

Cost is often a barrier to participation in sport and physical activity. The cost of some activities means that a person cannot always participate in the sport or activity of their choice. There are many costs associated with some activities, including membership fees and cost of equipment, uniform and coaching. Many sporting clubs and associations have strategies for reducing the cost, especially for young people. For example, the 'Boots for Kids' program takes donated football boots and distributes them to remote Indigenous communities.

Fairfax Syndication/Pat Scala

Figure 3.26 At the Gunbalanya School in Arnhem Land, Junior Dirdi relaxes after a long afternoon kicking the footy

CASE STUDY  BOOTS FOR KIDS

Identify

The Boots for Kids program is an initiative to improve connectedness and engagement within Indigenous communities. However, there are many more beneficiaries than the kids who receive the boots to wear. Read further and try to identify who all of the 'winners' are as a result of participating in this program.

Understand

The Boots for Kids program kicked off on Wednesday, with the help of Olympic champion Cathy Freeman and Hawthorn football star Shaun Burgoyne.

People with used football boots are being encouraged to drop them in collection boxes at Coles stores, from where they will be cleaned and donated to children in remote Indigenous communities in the Northern Territory and Western Australia.

Indigenous AFL star Burgoyne said his experience in Indigenous communities had taught him that this type of program could have a strong impact.

'I think it's great because a lot of people out there have used footy boots that they don't use,' he said. 'They might just throw them away, but with this you can come down to Coles, chuck them in the box and they'll get shipped out to people in communities that really need them.

'It's just giving something back. I'll be going home to get all my spare boots and dropping them off tomorrow.'

Freeman added: 'I think it will have an impact on Indigenous families and kids in those far-flung regions of Australia, because they'll realise that giants like Coles are open-hearted and wanting to make a difference in their lives.'

Giving something to kids and keeping them connected to community and education is behind *The Age's Boots for Kids campaign.*

The footy boots will go to Indigenous kids in remote communities in Arnhem Land, the Pilbara and the red centre.

Linfox trucks will transport them to remote communities.

Discuss

1 In your own words, describe what impact the Boots for Kids has on the community. Identify at least three benefits from the initiative.
2 Research another initiative aiming to make sport more accessible in Australia. For example, the National Aboriginal Sporting Chance Academy (NASCA).

Weblink
NASCA

INVESTIGATION  COST OF SPORT

Purpose

Some sports have relatively low costs and others are much higher. This investigation will allow you to compare the costs associated with various common sports and physical activities.

9780170463102

Method

Perform a cost analysis for the activities in a table format like the example provided.

1 First, identify the requirements of the sport under each of the headings. Once you have a list of all of the clothing, shoes, equipment, fees and other costs, estimate an amount for each and calculate the total. Assume 40 weeks of play in a year.

Scaffold
Cost of sport

Sport/activity	Clothing	Equipment	Membership	Coaching	Other	Total
Tennis	Tennis skirt/ shorts, T-shirt/top	Tennis shoes, racquet, balls	Club fees	30 min/ week (individual)	Ball money	
Cost	$100	$300	$150/year	$30/week	$5/ week	$1950/ year
Walking						
Cost						
Gymnastics						
Cost						
Bushwalking						
Cost						
Skiing						
Cost						
Football (soccer, rugby, AFL)						
Cost						

Discussion

1 Compare the annual costs for the different activities listed. Which activities have the highest costs?

2 Elaborate how cost might be a barrier for participation in certain activities and provide two examples to support your opinion.

3 Develop two strategies for any of the activities listed to help overcome this barrier.

Access and location

Where you live can sometimes be a barrier to participation in physical activity. Imagine if you lived in Alice Springs but wanted to take up surfing, or if you lived in far north Queensland and wanted to snow ski! Just getting to a location to be able to participate can be a huge undertaking.

Figure 3.27 Some gyms are open 24 hours a day. What other initiatives can you think of that are aimed at increasing access to sport and recreational activities within your community?

Sometimes you have the facilities in your local area but there is no public transport, or the centre is not open at suitable times.

People with disabilities face additional challenges to accessing sporting and recreational facilities. Some places have designed their facilities to allow access for people with disabilities, but many have not. This makes it even more difficult for those with additional needs to access services and facilities that would benefit their health and wellbeing.

FAST FACT

There is an increasing demand in the fitness industry to offer personalised services to clients. One solution to this is for gyms to offer 24-hour access. This would allow people to train when it suits them. This is beneficial to shift workers and those who would not otherwise be able to access gym facilities during 'regular' opening hours.

CASE STUDY ACCESS FOR ALL ABILITIES PROGRAM

Identify

The Access for All Abilities program aims to provide increased active participation opportunities for people of all abilities.

Understand

The program uses a community development approach, working with various communities, clubs and organisations to influence local-level planning and to increase community awareness of people with disabilities, promoting inclusive and accessible sport and recreation environments, activities and events.

Everyone benefits from Access for All Abilities; not only do you improve your health and wellbeing, but there are also opportunities to meet new people, make friends, learn new skills and, above all, have fun. Access for All Abilities may even kickstart you into a brand-new sporting career. There are sporting competitions to enjoy within your local communities that offer improved levels of inclusiveness and liveability as well as an improved sense of belonging to a community.

The Access for All Abilities program is principally funded by the Victorian Department of Health & Human Services through Sport and Recreation Victoria (SRV).

Discuss

1 Visit the Access for All Abilities website and identify three programs available in your local area.
2 Create a leaflet for your local community to advertise the programs and attract new members.

Weblink
Access for All Abilities

9780170463102

Alamy Stock Photo/martin berry

Figure 3.28 Many local councils are building recreational centres, which provide multiple indoor and outdoor sporting and recreational facilities, such as pools, gymnasiums, basketball courts, all-weather fields and courts for soccer/hockey/tennis, etc.

INVESTIGATION

HOW ACCESSIBLE IS YOUR LOCAL COMMUNITY RECREATIONAL FACILITY?

Purpose

To investigate how accessible a local recreation facility is for different population groups within the community.

Method

1. This investigation requires a visit to a local community recreational centre or facility. Students should contact the recreational facility and explain to management that they are conducting research for a school project and would like to conduct an audit on the types of activities offered, access and programs for various groups, and to generally gain more insight into what is offered at the centre/facility.

Discussion

1. List all of the activities offered by the recreation centre to members of the community.
2. Identify physical characteristics of the facility that provide access for as many people as possible, such as signs written in languages other than English, ramps for wheelchair access and visual aids for those with impaired hearing.
3. Are there specific programs for different groups within the community? For example, older adults, women, mums and bubs, unemployed, English as an additional language.
4. Propose changes to promote greater access for all members of the community. Identify three changes you could make to the physical environment and three changes you could make to the programs offered.

1. List three social, three physical and three emotional benefits associated with participation in regular physical activity.
2. Which two physical activities tend to be the most common across most of the lifespan? Why do you think these activities feature so prominently?
3. Describe how three personal factors influence participation in physical activity.

1. Use a program such as Wordle or Tagxedo to create your own visual representation of sport in Australia.
   a. List three sporting events or sporting identities that you recognise as being influential in defining Australia's sporting identity.
   b. Collate everyone's responses manually or using software such as Google Docs, then make your word cloud.
   c. Analyse the visual to find the most common responses, and discuss why the event or person stands out as contributing to Australia's sporting culture and identity.
2. Stereotypes can act as either enablers or barriers to people participating in different sporting activities. Construct a table that identifies population groups and typical sports that they have been associated with – e.g. Australians playing AFL or Malaysians playing badminton. As Australia has become more and more multicultural, do you believe these stereotypes will continue to exist? Briefly discuss.
3. Try to recall instances when your friends have either been a positive influence in you taking up a sport, or have provided barriers to you engaging in a particular sport or physical activity. Provide a brief account of either of these.
4. Look back over the influences on the decision to participate in physical activity and answer the following questions:
   a. What sport and/or physical activity do you do?
   b. Discuss two individual, two social and two cultural factors that influence your decision to participate or not in sport and physical activity.
   c. Who or what is the biggest influence on your decision to participate or not?
   d. How does your decision to participate or not to participate in physical activity or sport impact on your health and wellbeing?

1. There have been some memorable times in Australian sport.
   a. What stands out for you when you think of a significant sporting achievement or memorable moment in sport?
   b. What contributed to the outcome?
   c. When and where did this occur?
   d. Who was involved?
   e. Why is this event significant to you?
   f. How did this event help to shape Australia's sporting identity?

   Use a graphic organiser to sort and present your information.

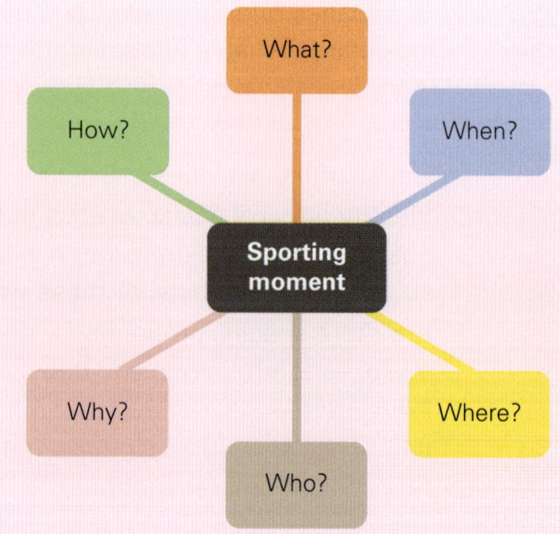

2. Conduct a web search for programs that are designed to increase access to physical activity for different groups

within our community – similar to the Access for All Abilities Program. Provide details about the program in terms of where it is run, who it is aimed at, how it increases participation options, any government funding and how it is advertised/promoted.

3 Create a mind map to show the relationship between a sport that you are interested in and all of the people who contribute to the successful running of that sport. Remember to include everyone from the groundkeepers to the car-park attendants!

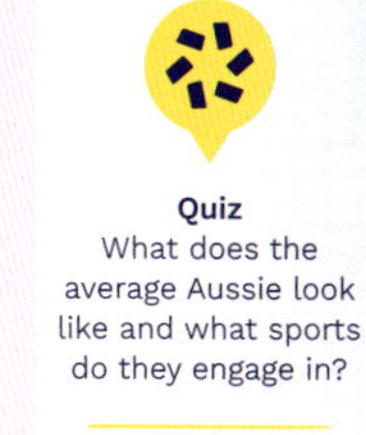

Quiz
What does the average Aussie look like and what sports do they engage in?

HOW CAN WE ENHANCE THE HEALTH AND WELLBEING OF THE COMMUNITY?

To improve the physical, social and emotional health of the community, government and non-government organisations are constantly developing programs to cater for the differing needs of various subpopulation groups within communities. A program may be implemented in a school, workplace or community setting.

Individual, social and cultural factors all influence physical activity behaviour. The physical environment and the policies surrounding physical activity are also important in changing the physical activity behaviour of people in the community. When a strategy addresses all of these factors, it is more likely to bring about a change in behaviour.

SCHOOL SETTINGS

Earlier in this chapter you saw how active kids are more likely to be active adults. To have greater success in increasing physical activity levels, a strategy needs to consider individual, social, environmental and policy factors.

Worksheets 3.7 and 3.8

- **Individual factors are likely to increase physical activity:** Offer activities that appeal to different ability levels, interests and age groups rather than a 'one size fits all' approach. For example, offer dance, games and sport and recreational activities such as ten-pin bowling in the physical education curriculum.
- **Social influences encourage physical activity:** Allow students to be active with friends, and use peer coaching, mentoring programs and 'house' systems to encourage social interaction and support networks for students.
- **Provide facilities to be active:** Allow students to borrow sporting equipment at lunchtime and recess. Provide change-room facilities, secure bike racks and shaded play spaces.
- **Implement policies to support physical activity:** Include the required amount of health and physical education in the curriculum, allow students to wear runners to school and have a policy on the expected level of participation in school sporting carnivals and events.

FACE TO FACE Be more active

Discuss with the person next to you how your school could increase the opportunities for students (and teachers) to be active during their breaks (recess and lunchtime). What could be offered to encourage students to be active before and after school? Use the provided information about school settings to prompt your discussion.

INVESTIGATION  RESEARCH ASSIGNMENT: STAFF AND STUDENT WELLBEING

Purpose

To develop a set of proposals to enhance the health and wellbeing of the staff and students in your school.

Method

Part 1

In groups of three or four, develop a proposal for a strategy to increase the physical activity levels of the staff, students or both in your school. Your proposal must address each of the factors that influence participation in physical activity.

Your proposal must include the following information (given as an example):

Title: Forest Hill Heights Joggers Club

Aim: To increase the physical activity levels of the Forest Hill Heights Secondary College staff and students

Details of the activity: Jog, run or walk one lap of the track (1.2 kilometres). A record is kept of the number of laps each person completes over the course of a term. At the end of each term, awards are given.

Who: All staff and students are encouraged to participate.

When: 8.15 a.m., three mornings per week

Where: Walking/bike track behind the school that runs along the creek and back up to the school gate

Facilities required: Staff or parent helpers to record completed laps, folder of class and staff lists

Justification: We think this strategy will work because you don't have to be very fit or especially good at sport to walk, and you can build up to jogging or running. You also don't need to buy any special equipment in order to participate. Everyone can join in with their friends and walk and talk at the same time. The school will need to allow students to wear their runners to school if they join the joggers club.

You must submit your proposal to your teacher and/or the school principal for approval.

Part 2

You now need to promote your strategy to your target audience (staff, students or both) in a multimodal campaign.

You will need to create a paper-based and digital marketing tool to promote your strategy. Obtain the required permissions, then distribute your advertising material appropriately.

Part 3

Implement your strategy within your school. Depending on your target group, this may be during lunchtime or PE classes, or after or before school.

Discussion

Reflect and evaluate by critiquing your strategy.

1 Critique the success of your strategy in enhancing the health and wellbeing of your target population.
2 What are the likely benefits of your strategy?
3 Do you think the strategy is sustainable? What barriers would exist to maintain the program within your school? How could you modify the program to overcome these barriers?

COMMUNITY SETTINGS

Within the community there are many different strategies designed to increase the health and wellbeing of individuals and the community as a whole. Open spaces such as parks, beaches and recreational facilities provide ample opportunities to be active. The natural environment, such as beaches, lakes and forests, can be difficult to get to easily, which is why the built environment can support and encourage physical activity in more convenient ways. Consider the bike paths, playgrounds and walking tracks in your community. Are they well maintained? Are they well lit? Is there shade provided? Are there drinking fountains and toilets in playgrounds and open spaces? Local councils need to consider planning and policy decisions that have an impact on the provision of indoor and outdoor spaces.

Dreamstime.com/Filedimage

Figure 3.29 Some community settings, such as Melbourne's Royal Botanic Gardens, encourage people to be active in open spaces.

WELLBEING CHECK IN

WALK AROUND THE BLOCK

This activity has been developed in collaboration with Dr David Bakker of MoodMission

Identify

Some of history's greatest minds got their best ideas on a good old-fashioned walk. Maybe you can, too.

Understand

Walking is a great form of exercise because it can help us think. Intensive exercise such as running diverts blood away from the brain towards the muscles, but walking allows our brains to keep functioning. As you walk, there are things to observe, sounds to notice and other small distractions that can inspire creative thinking and processing of feelings.

Practise

1 Organise to go for a short 10–15-minute walk. It might be around your suburban block or up and down the road.
2 Pick a time in the day that suits – avoid walking at night.
3 When you go on your walk, try not to take any distractions. That means your phone stays in your pocket and your earphones stay off. You're trying to give your mind some space to think.
4 Try to notice things in your surroundings as you walk. Can you see any trees? Other plants? Features in your environment?

Reflect

How did you feel when you got back? Any different to before? Did you have any ideas or make any plans while you were out?

The federal, state and local levels of government implement strategies for improving the health of the community. This list includes some examples of strategies that are aimed at improving health and wellbeing and increasing physical activity levels within the community:

- Stephanie Alexander Kitchen Garden National Program
- Healthy Spaces & Places
- Live Life Well (NSW initiative)
- Measure Up
- Swap It, Don't Stop It
- Active After-School Communities Program
- Get Up & Grow
- Active for Life
- Regional Sport Program.

COMMUNITY ACTION

Local governments often consult with different community groups to ensure that the facility or program being developed meets the needs of those who will be affected by its development. Communication between residents, facility users, businesses and government is imperative to the planning, design and management of a new facility or program. Feedback from these key stakeholders can come from surveys, letterbox drops and public meetings. This ensures that the outcome will be beneficial to all involved. Everyone, not just the government, is responsible for contributing to the health and wellbeing of their community.

> **FAST FACT**
> Since 2005, an average of two skate parks per week have been built in Australia!

Initiatives that have been instigated by young people that have had a positive influence on the health and wellbeing of their community include proposals and design input for skate parks.

WORKPLACE SETTINGS

Adults spend many of their waking hours at work. This makes workplaces an excellent setting to promote physical activity. There are many benefits to the employer, including increased morale and productivity and reduced absenteeism. Staff have a social support group in the people they work with, and employers can support workplace programs by allowing flexible work hours, subsidising the cost, providing and installing equipment and facilities, providing on-site programs and educating employees on the benefits of the program. There are several workplace programs promoting physical activity, such as:

- Global Corporate Challenge
- Healthy Workers
- Creating Healthy Workplaces.

Table 3.3 Workplace benefits of better health and wellbeing: comparing healthy and unhealthy Australian workers

Unhealthy	Healthy
18 days sick leave per year	2 days sick leave per year
Self-rated performance of 3.7 out of 10	Self-rated performance of 8.5 out of 10
49 effective hours worked (full time) per month	143 effective hours worked (full time) per month
High-fat diet	Healthy diet
Low energy levels and poor concentration	Fit, energetic and alert
Obese or overweight	Healthy body weight
Irregular sleep patterns	More attentive at work and better sleep patterns
Poor stress management techniques	Actively manage stress levels

Source: 'The health of Australia's workforce', Medibank, November 2005, p.7. https://www.medibank.com.au/Client/Documents/Pdfs/The_health_of_Australia%27s_workforce.pdf

The Heart Foundation has produced a *Healthy Workplace Guide* to help employers implement change within the workplace to support employees in leading healthy lifestyles. The guide provides examples of activities to promote nutrition and physical activity in the workplace. It classifies activities by whether they target people (social), the working environment (physical environment) or organisational policies (policy).

CASE STUDY
HEART FOUNDATION – HEALTHY ACTIVE BY DESIGN

Identify

The Heart Foundation has created a program known as "Health, Active by Design" which advocates for built environments that promote and support heart health, healthy eating, active living and physical activity. The program particularly focusses on getting more people active more often, through policies, programs and partnerships.

Understand

Cycling Brisbane is a behaviour change and promotion initiative directed by Brisbane City Council to encourage residents to ride more often. The initiative works closely with the council's bikeway planners to promote new and existing bikeway infrastructure, as part of the Better Bikeways for Brisbane program. The Cycling Brisbane program includes a mix of marketing campaigns, free workshops for people to improve their bike riding skills, an online route planner and other web-based resources. The free membership program provides exclusive discounts to businesses operating locally and a monthly newsletter to keep members updated on the latest bikeway projects and promotions. Launched in March 2014, the program now has over 18 000 members and growing. Information and resources are published on the website.

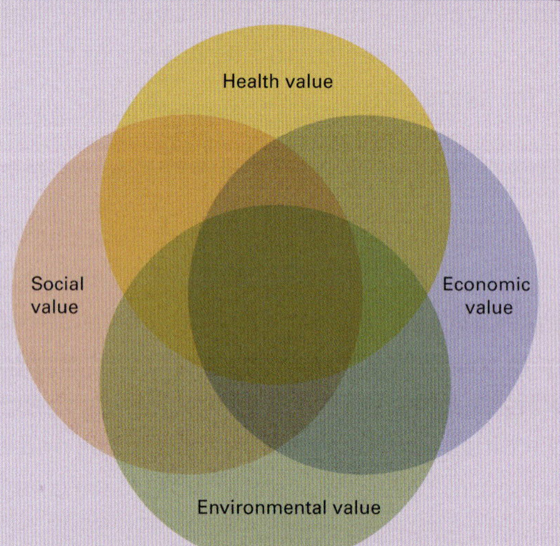

Figure 3.30 Various interrelated values that influence participation in physical activity

Discuss

Critique the Brisbane City Council's attempt to increase the amount of people regularly using cycling as a form of active transport:

1 Investigate the current number of members associated with the program – has this increased since the program was introduced?
2 Examine the website that accompanies this initiative and list three other forms of active transport that the Brisbane City Council encourages on it's purpose built paths.
3 Consider how your school could encourage and support more active transport to and from the school each week. Provide your opinion as to the top 3 barriers students may list as to why they would not consider active transport to/from school.

REVIEW

1 Create a table that shows at least two advantages and two disadvantages of being a member of a gymnasium that is open 24/7.
2 Other than improved health, list three benefits associated with having a workplace that promotes and supports its workforce in being active both in and outside workplace settings.

1 Reflect on your own local council and list three strategies it has used to promote increased physical activity amongst the local community.

2 Many schools are implementing bans on mobile phone use as a way of trying to get students more physically active during breaks, among other reasons. Try to think of any other policies schools could put in place to increase physical activity levels of both students and staff.

3 Make a list of local community venues (natural or built) that promote physical activity, recreation and sport. Additionally, indicate the types of activities that can be undertaken at each of these venues.

1 'I just don't have enough time' is often given as a reason why people aren't more physically active. Design a table that provides other reasons/excuses why people do not participate in physical activity in one column. In the opposite/next column, provide a strategy that might overcome the stated reason for not being active.

Reason for not participating in physical activity	Strategies to overcome reason
I don't have time	

2 Many people argue that the Olympic and Paralympic Games should be held together, rather than the Paralympics following on from the Olympic Games. Discuss the benefits associated with this merger and how it might lead to healthier communities.

3 Search for health and physical activity programs in your state or territory. They may be government or non-government programs.

a Try to find programs that cater to the following groups older adults; mums and bubs; ethnic minorities; unemployed; First Nations Peoples; people with disabilities.

b For each program, identify the setting and provide a brief description of the strategy in a table (as in the first example).

Scaffold
Health programs

Target population	Program name	Setting	Brief description
Kids	*Five ways to a healthy lifestyle*	*Community*	*1 Get active each day* *2 Choose water as a drink* *3 Eat more fruit and vegetables* *4 Switch off the TV or computer and get active* *5 Eat fewer snacks and select healthier alternatives*

Quiz
How can we enhance the health and wellbeing of the community?

WHAT NEEDS TO BE CONSIDERED WHEN MAKING HEALTHY CHOICES?

This chapter has explored the risks of being inactive, the influences on the choice to be active and the places and programs available for physical activity. What's right for you? Everyone should be active in some way every day, but the messages about physical activity are not always clear. The media and reality television programs, 'bootcamps' and personal trainers, gyms tailored just for women, programs that promise 'just 30 minutes a day', coaches, trainers, teachers, friends and family all provide conflicting information about the amount and type of physical activity individuals should be doing.

INVESTIGATION ⇨ FITNESS FOR SALE

Purpose

To evaluate and critique some physical activity and exercise options such as bootcamps, body-weight workout challenges and 'guaranteed-results' fitness equipment advertised on TV.

Method

Select one of the listed options: bootcamps, body-weight workout challenges or 'guaranteed-results' fitness equipment. Find out what you are required to do, what time commitment and financial costs are involved and any 'guarantees' or promises made.

iStock.com/CREATISTA

Figure 3.31 Bootcamps often bring groups of people together to improve on their fitness.

Getty Images/Lorado

Figure 3.32 Body-weight challenges such as the '12-week body transformation' promote quick, intense weight loss.

Scout Kozakiewicz

Figure 3.33 Fitness equipment that is 'guaranteed to work', often sold via TV 'infomercials'

Discussion

1. Compare the physical activity requirements of the activity you have investigated with the recommendations made in Australia's physical activity guidelines (both for children and young people, and for adults).
2. How reliable is the information provided by the company promoting the activity or product?
3. Do they make any claims about how you will benefit from the program? If any guarantees are offered, how would users be likely to make any claims against these?
4. Discuss any strategies used in an effort to make the activity/product seem user friendly. Do you think a commercial initiative (a program that is designed to make money for an individual or a company) can be impartial in providing information about physical activity? Discuss.

There are a number of reliable sources of information about including physical activity in your daily routines and making involvement in physical activity a lifelong commitment. You can always start with your parents, your school and your teachers. If they are unsure, they can direct you to a community-based provider of health information. There are many health services within communities that can be accessed. You may need to use one or more of these people to find the information you need.

Some reliable sources of health information particularly related to physical activity include:

- local doctor or general practitioner (GP)
- Australian Government Department of Health
- exercise physiologists (Exercise and Sports Science Australia, ESSA)
- qualified fitness professionals
- physiotherapists
- VicHealth
- The Heart Foundation.

Worksheet 3.9

CASE STUDY ⇨ HEALTHY LIFESTYLE

Identify

Analyse the scenario below to help Olivia.

Understand

Olivia is 17 and is feeling as though she is overweight and not fit and healthy. She knows she doesn't do enough physical activity, but she finds it hard to fit in any sport or exercise with school, homework, her part-time job and spending time with her boyfriend, Harry. Harry says she looks great and shouldn't worry about being healthy at her age. Olivia searches online for an average weight for a 17-year-old girl and finds that, according to the website, she is 5 kilograms heavier than she should be! Her mum says that Olivia is just like her, and tells Olivia that she will never be 'skinny' because she is 'big boned'. Olivia is confused and unsure of what to do; all she wants is to be fit and healthy.

Discuss

1. List three sources of information has Olivia accessed.
2. Identify which of these sources of information you believe to be reliable.
3. Provide three reliable sources of information Olivia could access to get some advice on a healthy weight range for her age.
4. Outline why these three sources are appropriate choices for Olivia.
5. Propose four ways in which Olivia could change her daily routine to increase her physical activity levels.
6. Discuss why it is important for Olivia to increase her physical activity levels.

IT'S YOUR CALL – TO BE OR NOT TO BE ACTIVE?

Worksheets 3.10 and 3.11

Quiz
What needs to be considered when making healthy choices?

There are many reasons to be physically active. This chapter has investigated the influences on decisions to be physically active and looked at different strategies in various settings that offer opportunities to be physically active. However, the decision is yours. It is the responsibility of the individual to look after their own health and to make the decision to participate in behaviours that will have a positive impact.

You might decide to get up half an hour earlier and go for a walk in the mornings before school. Finding ways to meet Australia's physical activity guidelines is an important step in reducing the amount of time spent being inactive. Put the phone down, turn off the computer and get active!

9780170463102

CHAPTER 3 REVIEW

1. What proportion of the Australian population do you believe fit the stereotypical image of an average Australian?
2. Body Mass Index is an estimate of the total amount of body fat an individual has. List two shortcomings that make BMI inaccurate in children older than five.
3. Recall what waist measurement for both men and women is an indicator of increased health risks.
4. The average Australian's BMI and waist circumference are getting higher, which increases the risk of cardiovascular disease. List at least three different types of cardiovascular diseases.
5. State two of the biggest causes of being overweight and obese.
6. The Australian Physical Activity and Sedentary Behaviour Guidelines have been updated with the Australian 24-hour Movement Guidelines for Children and Young People (5–17 Years). Provide an opinion as to why.
7. Compare the participation rates higher for recreational activities or for 'traditional' sports and list any differences.
8. Discuss why participation in some organised sports is increasing. List two of these sports.
9. List three factors that can affect participation in physical activity.
10. In your own words, what are perceived barriers that can limit an individual's or a group's participation in physical activity? What are perceived barriers?
11. Summarise how much physical activity should be incorporated into our daily routines to promote improved health and wellbeing.

DEVELOPMENT AND MANAGEMENT OF MENTAL HEALTH

4

IN THIS CHAPTER

You will look at the issues surrounding mental health and young people, looking at things like resilience, stress, peer pressure, body image, where and how to get help and how your school can promote mental health and wellness. Sometimes the stigma surrounding mental health issues and illnesses can put people off talking about mental health altogether. It is important that we talk about mental health to reduce the stigma that so often prevents people getting help.

Shutterstock.com/GoodStudio

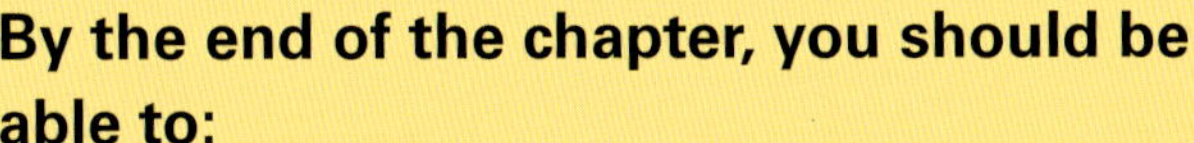

By the end of the chapter, you should be able to:

- identify how individuals have an impact on the identity of themselves and others
- identify the impact body image, stereotypes and resilience have on mental health and how they influence the way individuals think and behave
- identify the impact the media has on developing identities and mental health
- identify self-help strategies to support friends and family who are going through a challenging time
- identify strategies to de-stigmatise mental illness in the community
- plan and use positive health practices, behaviours, networks and resources in your community
- design, implement and evaluate strategies to enhance mental health and wellbeing in your community and design a mental health promotion campaign that connects to your community to enhance the wellbeing of students in your school.

HOW CAN I DEVELOP AND MAINTAIN GOOD MENTAL HEALTH?

Quiz
Pre-chapter

Before you start, take the pre-chapter quiz to find out how much you already know.

Good mental health is a state of wellbeing whereby a person is able to cope with stress and live productively in their community. When an individual's mental health is good, they feel positive about themselves and calm in stressful situations; they can feel confident and make rational decisions. They feel positive about life and deal with the daily ups and downs. They are also able to work towards goals and have respectful relationships with others.

Shutterstock.com/woocat

Figure 4.1 What do you think makes good mental health?

Mental health includes the following aspects:

- whether you like yourself
- your outlook on life
- how you respond to your feelings
- how you cope with stress
- your confidence in yourself
- your self-esteem
- your future vision of yourself.

Mental health is something everyone has, just as everyone has physical health. And just like physical health, there are things we can do to look after our mental health!

Mental health issues are things that affect the way individuals feel, think and behave. These can include depression and anxiety disorders. A mental illness is a diagnosable condition that has a deep effect on quality of life.

Mental illness can affect thoughts, feelings, actions and memory. A mental illness is usually longer lasting than mental health problems, and causes more distress and disruption to life. The following are some examples of mental illnesses:

- depression
- anxiety
- eating disorders
- psychosis
- self-harm
- excessive alcohol and other drug use
- **schizophrenia**
- **bipolar disorder**.

schizophrenia a mental illness that affects a person's capacity to think, act and feel; they may experience confused thinking, delusions and hallucinations

bipolar disorder previously known as manic depression; a mood disorder in which individuals experience extreme mood swings that affect their daily functioning

FAST FACT
Mental illness can occur at any age; however, anxiety disorders and depression are quite common in young people.

9780170463102

WELLBEING CHECK IN

Identify

A lot of mental health issues are maintained by vicious cycles. If you're aware of these cycles, you can take steps to prevent them spiralling.

This activity has been developed in collaboration with Dr David Bakker at MoodMission

Understand

We all have some cyclic patterns in our lives. For example, every time you do something that makes you feel good, you are more likely to do that thing again because you have received a reward. When you do it again, you may receive another reward, and this keeps going around and around, encouraging the behaviour more and more. While some cycles be small and relatively benign, others have much bigger, more negative consequences for our wellbeing. For example, if you feel sad and depressed, you may not be in the mood to get out and do enjoyable activities. When you don't do enjoyable activities, you have fewer positive emotions, which means you stay depressed. Most modern psychological treatments aim to help people break the vicious cycles that give rise to their depression, anxiety, or other mental health issues.

Practise

Which of these cycles do you think you might have a risk of falling into?

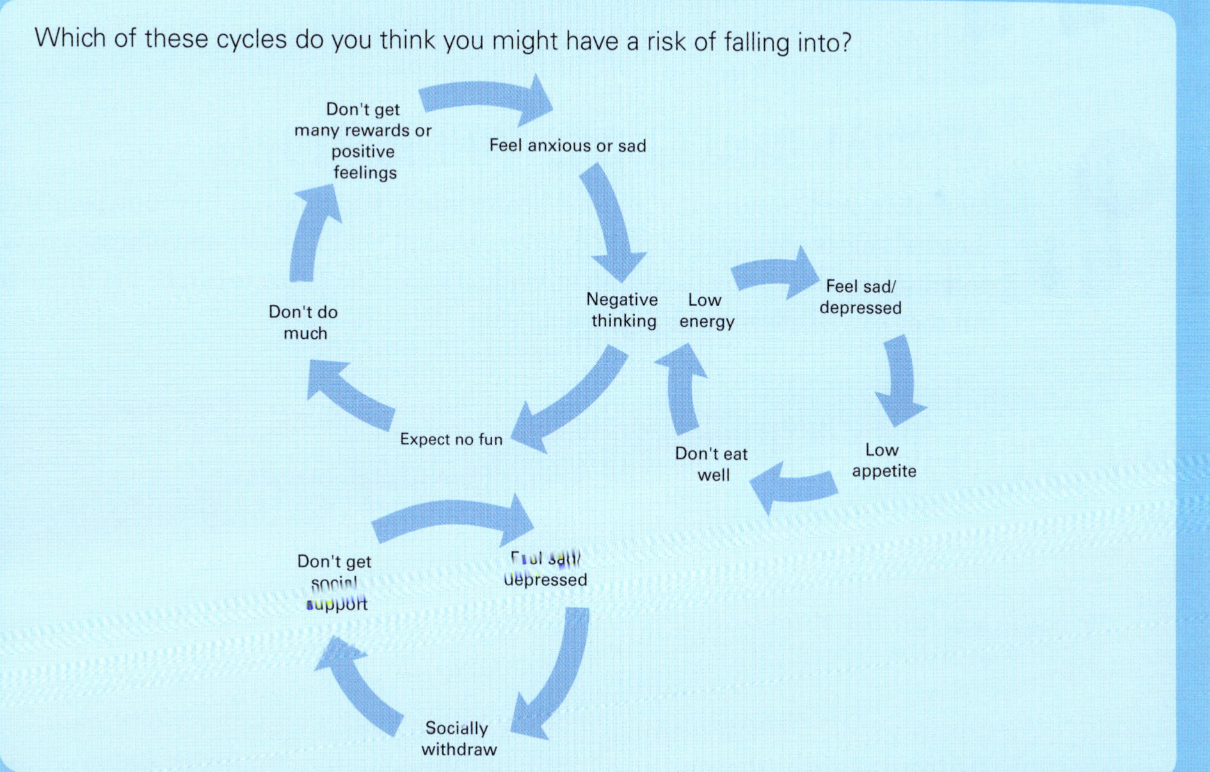

Reflect

Identifying your vicious cycles is important because then you can find ways of breaking them. What do you think might help to break the cycles you have chosen? Maybe one of the strategies or points discussed in this book?

Video
Mindfulness
Mindful movement

UP AND MOVING: Bursting the myths

Your teacher will put you into groups and give you each a balloon to blow up and tie. Discuss the following statements about mental health. Discuss the statement number that is written on your balloon.

Shutterstock.com/Martynova Anna

Figure 4.2 Get ready to burst the myths!

1 People can get over mental health illnesses, they just need to stop complaining and get a better attitude.
2 All young people try alcohol and other drugs, it's a rite of passage and isn't responsible for mental health issues.
3 Depression is one of the most common conditions in young people, it is something they can just get over.
4 Most people don't know what to say to people with mental health issues/problems, it's not something they feel comfortable with. However, they will easily empathise with someone who has a broken leg/cancer/headache, etc.
5 When someone has a mental health illness they will always need to be worried about it coming back.

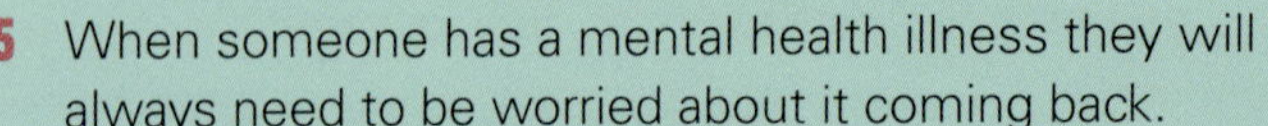

Burst your balloon and see what the slip of paper says on it – lets burst some myths! Please dispose of balloons responsibly. You could also use envelopes instead.

Worksheet 4.1

MENTAL HEALTH AND YOUNG PEOPLE

Australian studies show that mental health issues and illnesses in young people are on the rise. This is obviously of great concern. Mental health issues and illnesses have the potential to affect every part of an individual's life. The impacts can be life threatening, but they can be alleviated by accessing help.

INVESTIGATION: YOUNG PEOPLE AND MENTAL HEALTH IN AUSTRALIA

Purpose

To get an accurate idea about how many young people in Australia are affected by mental health illnesses.

Materials

Can we talk? Seven year mental health report – Mission Australia and Black Dog Institute

Weblink
Can we talk? Seven year youth mental health report

9780170463102

4

Method

1 Discuss how many Australians aged 15–19 report experiencing psychological distress. Consider why these numbers have risen since the last report.
2 Discuss how mental health illnesses affect a young person's life.
3 Justify whether there are differences in groups of young people. Consider gender and Indigenous status, for example. What are the differences and why are these differences evident?
4 Consider the main issues of concern for young people. Do these differ according to age, gender or Indigenous status? Why or why not?
5 Do most young people access help? If they do, where do they go to get help?
6 List recommendations that have been made for schools to help young people with mental health illnesses.

Discussion

As a small group or with the whole class, discuss the following:
1 What role should schools play in the identification and management of mental health illnesses?
2 What can schools do to prevent mental health issues and illnesses?
3 What can friends and family can do to prevent mental health issues and illnesses?
4 What can you can do to prevent mental health issues and illnesses?

FACE TO FACE **Debate – why are there so many mental health issues and illnesses now?**

Debate one of the following statements:
- 'There are more mental health problems and illnesses than ever before.'
- 'People have a better understanding of mental health and therefore more people are getting diagnosed.'

POSITIVE MENTAL HEALTH

If someone has positive mental health, they will generally be more resilient when problems present themselves. Resilience is necessary in life to 'bounce back' from the challenges that life brings. If someone's resilience is low, little setbacks or problems might really upset them and cause them to become stressed; they may even lead to mental health problems. Resilient people are more likely to:

- take care of themselves
- adapt to changes
- be connected to family, friends and school
- make safe and healthy decisions
- concentrate on the things that are good in life
- maintain a positive body image and attitude about themselves.

iStock.com/Ottilie_S

Figure 4.3 A smile a day …

FAST FACT

Resilience is the ability to bungee jump through the pitfalls of life. Even when hardship and adversity arise, it is as if the person has an elasticised rope around them that helps them to rebound when things get low and to maintain their shape as a person.

Surviving to thriving: promoting mental health in young people, A Fuller, ACER Press, Melbourne, 1998

UP AND MOVING Resilience

Find another student and answer the first question. Your teacher will give you 3–5 minutes to discuss each question. When you get the signal to stop, pair up with another student and discuss question 2. Continue doing this until you have completed the activity. You will have canvassed a lot of different ideas by talking to many students in your class.

1 Explain what Fuller is saying about resilience in the fast fact box above. Why does he use a bungee jumping analogy?
2 State how you would know if you had resilience.
3 How would you know if your friends had resilience?
4 Discuss if you believe people are born with resilience.
5 Create a list consisting of each of the following:
   - three characteristics a resilient person has
   - two examples of events that might test someone's resilience
   - three things someone can do to improve their resilience
   - three things a parent can do to improve their child's resilience
   - three things a teacher can do to improve a student's resilience
   - at least two items you would pack in a 'resilience toolkit'.
6 Represent Fuller's quote about resilience in a picture, collage or word cloud.

Figure 4.4 We ask for help when we can't solve a maths problem. Why do we find it hard to ask for help with a mental health problem?

ASKING FOR HELP

Asking for help with your mental health is a positive step that could prevent a lot of problems later on. It is easier to prevent something before it occurs than to have to treat it when you feel like things are getting out of your control. If you are feeling low, and can't work out why, asking someone you trust such as a trusted adult just to listen or offer advice could help you understand what's going on. It may seem like a small issue to you, but talking about it might stop or lessen your worry. If your issues are becoming too much, if they are lasting more than a week or two or if you are thinking about harming yourself, please seek help immediately. Things can and will often get better.

9780170463102

FAST FACT

Using Figure 4.5, write down six things that you could do if you were worried about a friend. Identify a way that you could approach your friend. Bringing up your concerns can sometimes be hard to do.

LOOK out for the signs
Want to know more about what to look for? Check out the signs and symptoms for anxiety and depression.

TALK about what's going on
Talk about what's going on ... check out some tips on how to talk about depression and anxiety.

LISTEN to your friend's experiences
Listening to your friend's experiences is really important. Sometimes you don't even need to say much.

SEEK help together
Learn about the different types of help available and where you or a friend can go to talk to a professional.

Adapted with permission from Beyond Blue

Figure 4.5 Beyond Blue suggests these four steps you can use to help a friend.

CASE STUDY

THE IMPACT OF COVID-19 ON SOCIAL ISOLATION AND LONELINESS IN AUSTRALIA

Identify

When COVID-19 hit Australia in 2020, there was a period of social isolation enforced to prevent the spread of the virus. This led to many Australians not being able to have physical contact with their family, friends, peers and colleagues.

Loneliness and social isolation in Australia

A recent snapshot from the Australian Institute of Health and Welfare collates data on Australian experiences of social isolation and loneliness since the start of the COVID-19 pandemic. It highlights the impact of the pandemic in exacerbating pre-existing risk factors for social isolation and loneliness. In this infographic, we share some of the findings.

Loneliness and social isolation: what is the difference?

'Loneliness is a subjective state of negative feelings about having a lower level of social contact than desired.'

'Social isolation is seen as the state of having minimal contact with others.'

Percentage of people reporting feeling lonely 'most or all of the time' during the past week by age group (April 2020 to July 2021)

14, 12, 10, 8, 6, 4, 2, 0
April 2020, August 2020, April 2021, August 2021
18–24, 25–34, 35–44, 45–54

Impacts of COVID-19 on loneliness

54% of respondents in a 2020 Australian survey reported feeling more lonely since the start of the pandemic

In April 2020, 49% of respondents in an Australian study said they had not met with others socially since the start of the pandemic

In October 2020, during their third consecutive month of a Stage 4 lockdown, Victorians were twice as [illegible] to those living in other Australian states

Factors identified to reduce the risk of loneliness

Having paid work | Caring for others | Engaging in volunteer work | Maintaining community memberships | Owning a pet | Being in a relationship | Living with family

Source: Australian Institute of Health and Welfare. (2021). *Social isolation and loneliness*. Accessed January 2022 via: https://www.aihw.gov.au/reports/australias-welfare/social-isolation-and-loneliness-covid-pandemic

wellbeing by Teacher
www.teachermagazine.com/wellbeing

Zoe Kaskamanidis, *Teacher* magazine. Reproduced by permission of the Australian Council for Educational Research.

Figure 4.6 Loneliness and social isolation in Australia

Discuss

1 Consider how social isolation and loneliness has an impact on mental health.
2 Discuss reasons why mental health might be impacted differently for everybody consider: school aged children and youth, university students, the elderly, people working for a living, people with caring responsibilities stay at home parents, medical staff, etc.
3 Discuss, if you feel comfortable, how you, or your family, dealt with any social isolation or loneliness you might have felt during COVID-19.
4 Propose two strategies for dealing with loneliness for each of the following people:
   - Ethel is in a retirement village and her children and grandchildren have not been able to visit her for the last 2 years.
   - Ray is single and lives alone. It is so difficult to meet a partner during COVID-19 lockdowns.
   - Georgie has been completing schooling at home, and their parents won't allow them to go out anywhere as they have an immuno-compromised baby at home.
   - Jet has missed Years 7 and 8 at school and now enters Year 9 without really connecting with any of his peers. He doesn't want to go to school.

GETTING HELP

Getting help with a mental health problem, for yourself or someone else, can be a difficult thing to do. It can be difficult to seek help for yourself, as you may not want to admit that you are not coping. Seeking help for someone else can also be hard, because you can't be sure exactly what they are going through or the type of help that they need. If you haven't had to find it before, you may not know who or where to go for help. In addition, there can be a **stigma** attached to mental health problems, which can also make seeking help more difficult. There are people who are specially trained in the area of mental health who can provide you information, a listening ear or counselling.

stigma
a mark of disgrace that sets a person apart from others

Weblink
Explore the webpages of these organisations

Resources

There are many organisations that can provide resources to assist in a wide variety of issues that involve mental health. Some are listed in Figure 4.7.

Kids Helpline is Australia's only free (even from a mobile), confidential 24/7 online and phone counselling service for young people aged 5 to 25.

Figure 4.7 Resources that can help you or a friend

UP AND MOVING Where to go for the right help

The websites of the organisations listed in Figure 4.7 are a good start when deciding where to go for help. Read the following scenarios, and in your group choose three. You must decide which resources you would recommend for each scenario.

Use the table in Worksheet 4.2 to decide which resources will work for your chosen scenarios. Then, write in the specific web address for the site you would recommend. Remember to identify the scenario you are working with at the top of the table.

Worksheet 4.2

Scaffold
Where to go

- Mia has stopped going to dance classes and always makes an excuse not to do PE. She used to love it! You ask Mia what is going on and she tells you that her dance instructor didn't give her a lead role because she couldn't fit into the costume.
- Ryan found out that his dad has been diagnosed with a mental illness. Ryan is distraught. He comes to you because he needs some information and doesn't know where to turn.
- You haven't seen your friend Sieu-Ting at school for almost two weeks. Her mum calls you one day and asks if you would come over. Your friend confesses to you that every time she starts to get ready for school, she has an anxiety attack and can't move. She just crawls back into bed until it is over. She seems really scared.
- Your younger sister tells you that someone is saying hateful things about her online. She is afraid to use the computer because every time she does there's another mean message. You suggest she tells her teacher, but she doesn't want to.
- Cooper is devastated because he has just broken up with his partner. No one at school knows that Cooper is gay, except for you. You can't seem to get him interested in anything and are worried about the exams coming up. You know Cooper wants to do well so he can go to university next year to study law.
- Faiz has been acting really weirdly; he doesn't spend time with his friends anymore, not even you. You are really worried because he doesn't seem to be looking after himself. He's physically 'there' in class, but his head is somewhere else. The teachers are getting annoyed with him because he just ignores them. When you asked him about it, he got really angry and told you to go away.

1. Summarise what you could do once you have found some resources that could help the situation.
2. State what you might say in each case.
3. Propose what you would do if your friend or sibling has one of the following responses:
   - gets angry
   - gets emotional
   - does nothing
   - asks you go with them to access more help.
4. What would you do if your friend or sibling doesn't want any help but you are still worried about them? Where would **you** go for help?

1 Consider what it means to have resilience.
2 Identify the types of traits resilient people often have.
3 Justify whether you are resilient or not.
4 List three different resources and their websites or phone numbers you could use if you're worried about your mental health.

1 Devise a ranking system from most resilient to least resilient. Discuss what items were on your ranking system.
2 Consider older people – teachers, parents, caregivers, or grandparents, for example. Does age make you more or less resilient? Discuss your answers.
3 Consider children aged 0–5 years old. Does their youth make them more or less resilient? Discuss your answers.
4 Watch the eight-minute video clip by Dr Brian Walker from CSIRO Sustainable Ecosystems, Australia: 'The best explanation to resilience'.
   a As a class, discuss your thoughts about his explanation of resilience.
   b Based on the video clip, do you agree or disagree with this explanation of resilience?
   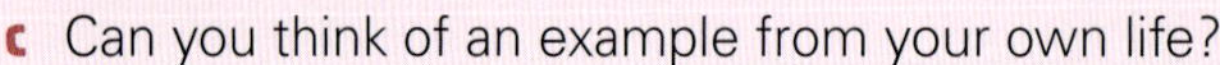
   c Can you think of an example from your own life?
   d Share your ideas with the class.

Weblink
Resilience

1 Investigate what your peers think we should be doing in our everyday life to build our resilience. List the top five responses.
2 Using your list, write/perform a 30-second TV advertisement for building resilience in adolescents.
   a Interview your peers again and ask what types of messages appeal to them.
   b Think about the best way to get all that information out quickly (song/rap/poem/jingle/spoken word/dance, etc.).
   c Record your ad and discuss this activity with your peers. What worked? What didn't? How might you do things the same/differently if you did this activity again?
3 Choose one of the resources from Figure 4.7 and create a profile on them. Identify:
   a who they are
   b what they do
   c when they could be helpful to you.

HOW DO I DEVELOP MY OWN IDENTITY?

Worksheet 4.3

Everyone is different in their own way; everyone has their own identity. Your identity is made up of the way you see yourself, and your characteristics, attributes and skills. The past influences your current view of yourself, and your current view of yourself

9780170463102

influences your behaviour in the present. You also set goals in the present, which will affect your future view of yourself.

Everybody is different, and that should be celebrated. Can you imagine a world where everybody looked the same, everybody could do the same things and everybody liked the same foods? What would school be like if this were the case?

Other factors also have an impact on your identity, right throughout your life. These include, but not limited to, family, friends, peers, school, culture, religion, laws, groups you belong to and the media. Some of these factors have more impact than others; some have an impact for a short time, others for a long period of time.

As an example, family has more of an impact on younger children – they need their family's help and often think and behave the way their family does. As children get older, peers and the media take on a more important role, as adolescents experiment with their independence. What makes you choose the clothes you wear? When you make a decision, do you consider the opinion of others?

Worksheet 4.4

Self-esteem, peer pressure, stereotypes, the media and body image all have varying impacts on identity. It all depends on how strong your identity is to begin with. If you know who you are and what your identity is, you are less likely to be pressured to do something you don't want to.

Getty Images/PM Images

Figure 4.8 Be true to who you are; you don't have to follow the crowd.

Worksheet 4.5

SELF-ESTEEM

Weblink
Watch the headspace video clip of teenagers talking about their ideas about self-esteem.

Self-esteem is how people feel about themselves, which has an impact on their identity. To have healthy self-esteem, people must respect and love themselves. As the saying goes, 'you can't truly love someone else until you truly love yourself'!

WELLBEING CHECK IN: WOULD YOU TALK TO A FRIEND THAT WAY?

This activity has been developed in collaboration with Dr David Bakker of MoodMission

Identify

We talk to ourselves in our own heads all the time, but are we treating ourselves fairly?

Understand

Most people find themselves engaging in 'self-talk' all the time. This is you talking to yourself in your own head. Sometimes our self-talk can get nasty, and we can be really hard on ourselves. This is called negative self-talk, and might include thoughts like, 'You idiot!' or 'Why are you so lazy, you've got to work harder!' Negative self-talk can make us feel pretty bad about ourselves and can lead to darker, depressive thoughts, as well as low self-esteem.

Practise

1. What are some negative things you've said to yourself before? Write down at least five examples.
2. Now read them back. Would you ever say these things to a friend? Chances are you wouldn't, unless you're wanting to start a fight. So why say them to yourself?
3. Now write down at least five things you could say to a friend who was in the same situation you were when you spoke to yourself that way. This might include things like 'That sounds really hard,' or 'You're trying your best'.
4. The next time you catch yourself doing negative self-talk, try to remember to be a friend to yourself.

Reflect

Did you feel funny about being kind to yourself? Some people feel this way, especially if they think they should be hard on themselves in order to stay motivated or be held accountable. But it's important to remember that this actually isn't the case, and kindness is much more motivating than cruelty.

FAMILY AND FRIENDS

Our family and friends can have an impact on our developing identity. Our family, from birth, have had an influence on the development of our values and morals. They set up an expected way of behaving and work towards achieving their hopes and goals for you. Our friends can introduce us to different ways of being, as their families have different influences than ours.

Sometimes what our families want or expect can be in opposition to what our friends want us to do or be. This can be very difficult to negotiate as often it ends up in confrontation with one side. Sometimes we want to be able to do what our friends do, but we know that goes against what our parents/carers want us to do. Often this is because our parents/carers want to keep us safe and don't want us taking risks. It is important for you, if you are ever unsure, to weigh up both (or more) sides and if you are still unsure, seek out help. Check out the face to face activity below for some examples and consider how you would tackle them.

FACE TO FACE The Hot Seat

The following scenarios are examples of when you have to make a decision when there are competing expectations on you. Get into a group of 3 and have the young person sit down, have the parent and the friend stand on each side of the young person. The friend and the parent will take turns trying to convince the young person to do what they want them to do. Try to come up with at least 5 reasons each as to why the young person should do what you want them to do. At the end the young person has to vote for who was most convincing and why.

Skye is stuck and doesn't know what to do. There is a party this weekend and they know there will be alcohol. Skye's parents have said they are not to go to any parties if alcohol was available. Skye's friends all want Skye to come to the party, it's their first one in Year 10 and a lot of the senior students will be there.

JB is at a party and is about to get a lift home. JB's ride has been smoking dope all night and although he says he is OK to drive JB doesn't want to risk it.

JB's mum said that if ever he gets in a situation where he needs help to call her – no matter the time. It is 2:25 a.m. and he really doesn't want to tell her why he needs a ride home. JB's friend is trying to convince JB to get in the car so that they can all go to another party. JB doesn't know what to do because he wants to go to the other party but knows the risks of drug driving.

Bree got permission to ask a few friends over for the night, they are having a great time until one friend suggests that they all watch some porn. Bree doesn't really want to, but she doesn't want to look like a prude. Her parents have set a lock on all devices so that what they can watch is limited. Bree knows the password and could access anything online.

Onyx feels like they cannot be themselves. Onyx identifies as gender diverse. At school Onyx has seen other students come out as gay or have transitioned. Onyx has always been made to really comfortable at school, especially with a great group of friends who tell Onyx just to be themselves. At home is a really different story though. Onyx has tried to engage the parents in discussion about gender, but they want to hear nothing about it. Dad said, "You are born a man – You are a man – no questions asked, now start acting like a man".

Jake is getting more and more upset every day. Jake has meet this really awesome girl and has asked her to go out. She said yes! The problem is that Jake's mum won't allow Jake to see this girl, unless they are under her roof and in her sight. Jake's girlfriend is getting really sick of not being able to see him, as they don't go to the same school.

Questions:

1 Discuss how it feels having to make a choice.
2 Justify who was more convincing.
3 Discuss what options you have when there are two competing opinions that you value.
4 Discuss how these scenarios could impact your mental health.
5 Discuss how you could help a friend in this situation.
6 Identify why parents might have a different opinion to you.
7 Discuss strategies that help when talking to parents.

PEER PRESSURE

Traditionally families, school, religion or culture teach individuals right from wrong, and this can have an impact on their values, morals and behaviour. Other outlets, such as music, media, friends or peers, may cause a person to question their values and consider other options. For some, adolescence is a defining time, where things are experimented with and decisions made about what feels right. Unfortunately, sometimes individuals are not confident about their own decisions, or they want to fit into a group, so they allow others to make these decisions for them.

iStock.com/sturti

Figure 4.9 Resisting peer pressure can be hard.

FACE TO FACE Think, pair, share: peer pressure

1 Has there ever been a time when you felt pressured to do something?
   » Did you do what you wanted to do? How did that feel?

   OR

   » Did you do what the other person or group wanted you to do? How did that feel?

2 Have you ever been a part of a group that has pressured someone else into doing something they might not have wanted to? How did that feel?

UP AND MOVING Scenarios

In small groups, consider what you would do in each of the following situations, then discuss or role-play them. In each scenario, consider the following factors:

- who is being pressured
- what type of pressure is being used
- how the person being pressured might be feeling.

1 You are working in a small group and have to present your findings to the class. Charlie is trying to convince the group to make Talisha present the group's work. Talisha has a serious stutter and gets very embarrassed when she speaks in front of the class. The rest of the group is pressuring her to take the role.

2 You are at a party. You promised your parents there would be no alcohol at the party, which is the reason you were allowed to go. There is alcohol at the party; you have decided not to drink, but all your friends are trying to make you.

3 You want to get back at a kid who has embarrassed you and you have decided to get your revenge online. You make some nasty comments on social media and ask all your friends to do so as well.

Extend this by doing a role play with these scenarios, where an individual or group are trying to convince someone to do something they don't want to do. At the end you could rewind and take the role play back to when a decision was made, this time make a different decision and see what the outcome will be this time.

Worksheet 4.6

Who is affected by peer pressure and how are they affected? It can be hard enough to make decisions by yourself, let alone in front of others, especially when emotions are involved. Have confidence in yourself and the decisions that you make. Everyone makes mistakes, so don't be hard on yourself. Learn from them, and the next time you make a decision it will be easier.

THE IMPACT OF THE MEDIA

More than a decade ago, the Chair of the Ministerial Media Code of Conduct on Body Image stated:

> 'We live in an age where we are surrounded by an image of an ideal body shape, communicated by a constantly evolving media. Much of what we see is a fusion of reality and fantasy, with many of the images presented to us having been altered or enhanced in some way.'

Source: L Sanger, Chair of the Ministerial Media Code of Conduct on Body Image Working Group, Final Report July 2007

9780170463102

Figure 4.10 Constant pressure from the media influences the way individuals think.

Do you think this is still the case? Why/why not?

The media plays a role in setting the expectations of what a male or female should look like. Young people are often heavily influenced by these images, as they can be very persuasive in promoting the benefits of looking and behaving a certain way. For example, media images can send the message that if you look this way you will be happy, popular, accepted, have friends, be loved, have fun, be able to afford clothes, have freedom and so on. Unfortunately, the claims they make rarely, if ever, lead to the promised benefits. Instead of understanding that the media promotes unrealistic expectations, people often end up blaming themselves for failing to achieve. This then impacts on their feelings of self-worth, and therefore their mental health and wellness.

FACE TO FACE A media parody

Watch Jesse Rosten's video clip, *Fotoshop by Adobé*, which is a parody of popular media campaigns for beauty.

1. How are popular advertising techniques used in this parody of a serious advertisement?
2. Discuss in pairs why photoshopping occurs and what impact you think this has on young people. What does it say about society?
3. The media plays a role in shaping identities in areas of gender, culture, religion, ability, race and sexuality. Think about what you see in the media every day: does it reflect Australia as you know it?

Weblink
Fotoshop by Adobé.

INVESTIGATION AN ALIEN INVESTIGATION

Purpose

What is presented in the media does not always reflect what we see in our everyday lives. This investigation aims to look at these differences and explore why they might exist.

Materials

You will need to watch some TV shows featuring ads, and have a look at community places featuring advertising, e.g. shopping centres, movie theatres, bus/train/tram stops, social media, etc.

Method

For the first section you will be working by yourself, the second you will be talking to peers in class and the third working on your own again.

1 Pretend you are an alien who has just landed on planet Earth. All you have seen so far are representations of humans in the media. What would your impression be of humans, based on what you see in the media?
   - Describe a typical human woman.
   - Describe a typical human man.
   - Describe the diversity in the human race.
2 Now that you have seen some actual humans, you are confused. These people don't look anything like the people you've seen in the media. You want to know what is going on. You find some individual humans to talk to. Ask them about their experience with the media, using the following questions to get started:
   - How does the media influence your views on your body?
   - Is there an expectation on Earth to look a certain way?
   - How important is it for you to look a certain way?
   - Does the media have any impact on who you are?
3 Write a letter back to your home planet describing what you have found out.

Shutterstock/Tomacco

Figure 4.11 If aliens tried to act like humans, how would they behave?

You are no longer pretending to be aliens! Get into groups so that you can discuss what you have learned.

1 Summarise the main issues you were writing about in your letters.
2 Identify if anyone in the group has ever had an experience where they have felt 'different'. How do they describe the feeling, and what did they do about it?
3 Would you say that Australia is very diverse? Why or why not?
4 Consider if there is any such thing as a typical man or typical woman. Explain your answer.

FACE TO FACE Social media

Weblink
Social media vs reality.

Worksheet 4.7

Watch the video about social media vs reality, then discuss the questions in small groups.

1 Why is getting 'likes' so important to these people?
2 Why do they lie online to make their lives look better/perfect?
3 How do you think that it makes them feel if they don't get the 'likes' they want?
4 How do you think their mental health is?

Social media is improving the ways that we can connect with friends, families and other people or groups that we are interested in knowing about. It has been especially helpful during worldwide lockdowns for keeping in touch and seeing people online.

However there are many pressures, particularly on young people, with respect to the ways that posting occurs and is expected. It is easy to think that everything you see online is real. Often people post embellished examples of their lives (see Face to face 'Social media' above). Posting on social media can have both positive and negative effects. If you post something and it gets a lot of attention or likes, it might make you feel good and so you keep doing it. If you get negative messages or are bullied, it can have a negative impact on your mental health when you might think that everybody has a better life than you. This can lead to you feeling sad, mad, jealous or feeling like you are not good enough. Some people even resort to faking their own lives online to help get over these feelings. If you are having any of these feelings, it can be a good idea to take a break from the online world.

Sometimes it is easier to see negative things online when it is about someone else, e.g. seeing bullying happen. Often when it is about ourselves, we can think that they might be right. If you see bullying happening online, and you feel safe to do so, call it out. Sometimes a kind comment can alert the writer to really think about their behaviour. If the bullying gets worse or continues you have every right to report and block. Report to the app you are using initially. You might want to take screenshots as proof of the bullying. You can also report to the esafety Commissioner who can help young people negotiate the law.

If you ever need help or someone to talk to there are specific help lines for young people that you can access for free and confidential counselling, information and help.

Kids Helpline and Headspace are great resources where you can also have an online chat 24/7 every day of the year.

Weblink
eSafety for diverse groups

Diversity

Australia is home to a diverse range of people, which makes for an interesting and exciting society. The media doesn't always reflect this diversity, and this can be an isolating experience for those who don't see themselves represented. This can lead to them feeling like 'there is no one like me' or 'I have to change to fit in'. How would this impact an individual's mental health?

STEREOTYPES

From the moment of birth an individual can be cast into a stereotype. The first question the parents or guardians may be asked get asked is, 'Is it a boy or a girl?' This often determines how the baby is thought about from the outset. Friends and family may use this information to decide what colour clothes to buy the baby or what type of toys to give them. Imagine being told what type of clothes to wear or what type of toys to play with just because you were born a certain gender! Unfortunately, these types of stereotypes can follow babies through childhood, even into adulthood.

Getty Images/Constance Bannister

Figure 4.12 Stereotyping based on gender has changed over time.

Stereotypes are preconceived perceptions of the way a person or group should look, act, believe and so on. Stereotyping someone is hurtful because it dismisses all the qualities that they possess outside the stereotype. Stereotyping can lead to bullying or, worse, people living their lives filled with hate for another group or individual just because they 'fit' a particular stereotype. Can you think of some stereotypes you have heard before? There are so many! Think of how many stereotypes you would fit into. Think about how many parts of you a person would miss out on if they only saw you in one way.

Figure 4.13 Harrison France, photographer of these 29 photos, stated, 'Although stereotypes will probably never disappear, I think it is important to highlight the fact that people aren't always what they seem and that labels can be offensive and are, most likely, incorrect.'

CASE STUDY ⇨ WHERE DO I FIT?

Identify

Stereotypes influence the way we, and everyone else around us, act, even if we might not mean to let them.

Understand

I'm Sasha and I'm 15. I live with my mum and three brothers; two older and one younger. My mum is a beautician and LOVES all things 'girly'!! I'm the opposite, I'm more like my brothers. I wear pants or shorts – I can't remember the last time I wore a dress or skirt. I love footy and hanging out with my male friends; I don't have a lot in common with my female friends.

I've been asked to be bridesmaid in my auntie's wedding and I DON'T want to wear a dress. Why do I have to? It's just not me! I'm really torn because I love my auntie, but she is not respecting my wishes. I get so much pressure to be more girly as it is, not just from Mum, but Dad, school and all the media out there – it is pressure! I get called butch and asked if I am a lesbian all the time. I am different from all the other girls. I don't understand why I have to act like them. It is getting harder and harder to be myself and I feel like I'm going to be swallowed into this image of what a girl should be. So what do I do? Wear the dress, don't be in the wedding party, or stand firm and insist I get to wear pants?

Discuss

1 Weddings can be very traditional and there are specific 'ways' that people are often expected to dress. A lot of money is spent on weddings and some brides and grooms have very fixed ideas of how they want the photos to look. Make a list of all the things you have heard that you should/shouldn't wear to a wedding. For example, some people believe you shouldn't wear black because it is what you wear to a funeral. Does Sasha have the right to ask to wear something different?
2 Clarify what Sasha means when she says her mum likes all things 'girly'.
3 Consider why society might want Sasha to 'act more like a girl'. What does 'acting like a girl' mean?
4 List three types of pressures Sasha might be experiencing.
5 Why do you think people might ask Sasha if she is gay?
6 Identify any situations where stereotypes can influence the way that you act, dress, behave or think.
7 If you were Sasha, what would your decision be? Justify your position.

UP AND MOVING Boys won't be boys

Watch the TedTalk about what we teach boys, then complete the following activity.

1 Move into small groups of single-sex peers (if you are in a co-ed school). On a piece of paper, brainstorm a list of 'The best things about being a male'. Then do the same thing for the statement: 'The worst things about being a male'.
2 Your teacher will help summarise the two brainstorms, noting in particular the differences between the answers given by males and females (if co-ed).
3 Get into pairs (of different genders if at a co-ed school) and come up with three expectations of men that could lead to a mental health issue. Explain why this might be the case.
4 Identify what kinds of conversations Ben (from the TedTalk) says boys should be having.
   a How will these conversations help prevent mental health issues and illnesses?
   b Where could these conversations be taking place?
   c Who should be responsible for having these conversations?

Weblink
Boys won't be boys, boys will be what we teach them to be

FACE TO FACE Advertising debates

In your class, debate the following topics:

- 'The representation of women over time in advertising has had a negative effect on women.'
- 'The representation of men over time in advertising has had a negative effect on men.'
- 'The representation of women over time in advertising has had a negative effect on men.'
- 'The representation of men over time in advertising has had a negative effect on women.'

Is there ever a time when a stereotype is useful or positive?

1. Consider how stereotypes influence our mental health.
2. Discuss how the media reinforces stereotypes and therefore also has an impact on our mental health.

1. Even though Australia is seen to be a diverse nation, we still have stereotyping. Propose three reasons why this might be so.
2. Hypothesise where stereotypes come from.

Your identity is exclusive to you.

1. Create a list of all the ways that you feel pressured to behave or present yourself in a certain way. Consider parents or caregivers, other family members, friends, teachers, online audience, etc.
2. Discuss how you handle all these competing pressures that want to tell you how to behave or present yourselves.

HOW CAN I DEVELOP AND MAINTAIN GOOD BODY IMAGE?

Video
Body image: What role do stereotypes play in the formation of negative body image?

How can social media promote positive body image? Watch the video and start the discussion.

Body image is the way a person sees, thinks and feels about themselves and the way that they think others see them. Body image has several aspects – affective, perceptual, cognitive and behavioural.

- **Affective:** how you feel about your body; do you like the way that it looks?
- **Perceptual:** how you see your body; this may not be the same as how others see you.
- **Cognitive:** how you think or feel about your body, perhaps even a belief that you should be different from the way you are.

- **Behavioural:** the actions you take, or don't take, because you are not happy with your body.

The Butterfly Foundation

Figure 4.14 How do you feel about your body?

Forced choice

Consider the statement 'It is healthier to be thin than fat'. Your teacher will design a continuum in the classroom like this:

Strongly agree ←——————→ Strongly disagree

1 Regarding this statement, where do you stand on the continuum, from strongly agree to strongly disagree?
2 Discuss with the class the reasons for your response.
Your teacher might have more statements for you to consider.

FAST FACTS

1 Only one in five women are happy with their body weight!

2 About 45 per cent of men are unhappy with their bodies, compared with 15 per cent 25 years ago!

Source: www.betterhealth.vic.gov.au © Copyright State of Victoria 2020

CASE STUDY ⇨ THE BUTTERFLY FOUNDATION

Identify

The Butterfly Foundation, in partnership with Instagram, have developed a program called The Whole Me to celebrate body image in Australia.

Understand

The Butterfly Foundation is looking at body image both on and offline, and helping young people feel better about their bodies. Have a look at the site and see what you think about the resources they have developed: the facts and stats; the parents and teen guides; and the video series.

Weblink
The Whole Me

Discuss

1 Consider whether this is an effective way to tackle both body image and media portrayals online.
2 Discuss why this is different compared to other body image campaigns you have seen before.
3 Evaluate whether it should be used by schools to help tackle body image issues and, in turn, help young people with their mental health.

Worksheet 4.8

Negative body image

Negative body image doesn't just appear; it can take a short or a long time to build up. The way people think of themselves has a lot to do with how they feel about themselves. If their mental health is good, they are not as affected by negative events. They accept that these things happen and they get on with life. If their mental health is poor, these negative events can have a much greater impact and really bring them down. Negative body image can affect self-acceptance and self-esteem, which then affects whether we choose healthy behaviours.

These are some factors that contribute to a negative body image:

- being teased about appearance in childhood and adolescence (too thin, too weak or too fat)
- growing up with dieting parents, or one who was unhappy with their body shape
- a cultural tendency to judge people by their appearance
- peer pressure among teenage girls to be slim, go on diets and compare themselves with others; peer pressure among teenage boys to be tough and strong
- media and advertising promoting idealised images
- a tendency in the media to push fad diets and weight loss or muscle building programs
- well-meaning public health campaigns that urge people to lose weight
- emphasis on sports players and celebrities as role models for young people.

Source: www.betterhealth.vic.gov.au © Copyright State of Victoria 2020

9780170463102

CASE STUDY  RESET

Identify

The Butterfly Foundation has released a body image program just for boys called RESET, aimed at tackling negative body image issues and eating disorders in boys.

Understand

Headline statistics relating to males' body concerns and eating disorders include:

- Over-exercising and an extreme pursuit for muscle growth are often perceived as healthy behaviours for males.
- 90 per cent of adolescent boys report that they exercise primarily to gain muscle.
- Two-thirds of adolescent boys report making specific changes to their diet to gain muscle.
- 25 per cent of people experiencing Anorexia Nervosa and Bulimia Nervosa are male.
- Almost an equal number of males and females experience binge eating disorder.
- Eating disorders have one of the highest mortality rates of all psychiatric disorders and suicide rates are 20 per cent higher in eating disorder patients than in the general population.

Source: 'Media release – Launch of Australia's first digital body image program for boys', The Butterfly Foundation, 29 October 2018

Discuss

Explore the RESET website before answering the following questions.

1. Reflect on how male stereotypes have an impact on mental health, why do you think it is necessary to target boys specifically.
2. Discuss why boys might over-exercise and put their bodies at risk just to have defined muscle.
3. Discuss why boys might not talk about body image or access help easily.
4. Justify why this program is called RESET. What are they trying to reset?
5. Design a banner to hang in your school that sends a positive message about body image for all students.

Weblink
Explore RESET

Body image can be thought of on a continuum, from a positive or healthy body image to a negative or unhealthy body image. In the unhealthy range, there is a condition called **body dissatisfaction**, which has an impact on the way that someone thinks, feels and behaves. There is a high correlation between body dissatisfaction and anxiety, depression and eating disorders (such as anorexia, bulimia and binge-eating disorder). Body dissatisfaction affects both males and females.

body dissatisfaction a mental health problem where someone has negative thoughts or feelings about their own body

The media feeds a negative body image by promoting a climate of 'quick fixes' – particularly in the areas of health and fitness. The diet industry makes millions of dollars by promising a quick fix to weight loss. When a person's body image is low, they may believe the promises that the diet industry makes, and may invest in products they think will make them thinner or more attractive.

What does the word **diet** mean to you? Often people think it means restricting the amount of calories you eat to lose weight. This is not the true meaning of the word! A diet is what someone normally eats. This usually stays fairly consistent over time, and has to do with what you like, what is accessible and any dietary needs

diet types of food that a person, animal or community usually eats

Weblink
Watch the Dove Real Beauty Sketches video clip then consider: why do you think that the sketches are different; and what is the message these women are sending?

Getty Images/Influx Productions

Figure 4.15 Food has an impact on mental health!

you might have. This was explored in Chapter 2.

A child or teenager's diet is often similar to their parents'; this can change when they start cooking for the family or move away and cook for themselves. The word 'diet' actually has a lot of power, as it is the first recommendation made when someone wants to lose weight. How can we change the negative connotations of this simple word?

FACE TO FACE What the world eats

Do an internet search for 'Hungry Planet: What the world eats'. This is a photographic project documenting the diets of various families around the world.

In pairs, discuss your family diets:

1 Does your family's diet look the same as the photo of the Australian family's diet?
2 What impact does your diet have on your identity as an Australian or an individual?
3 What differences or similarities did you notice about the typical diets of the people from the other countries pictured?
4 Do these photos reinforce stereotypes of other countries around the world?

Weblink
Hungry Planet: What the world eats

REVIEW

1 Discuss how stereotypes influence young people.
2 List the types of stereotypes that can have a powerful influence on someone's body image.
3 Consider why body dissatisfaction even a 'thing'. Aren't we all taught to love our body and what it can do, regardless of height, size, shape, etc.?

REFLECT

1 Justify whether you think that young people today are receiving positive or negative messages about their bodies. Think about all forms of media.
2 Can you think about a TV show and an advertising campaign that portray diversity in people as a positive thing, and don't conform to stereotypes?
   a Discuss why you think they can get away with doing this. Consider the audience, the place it is shown, what it is selling (in the advertising campaign), etc.

EXTEND

1 Consider the following statement: 'Around the world, people are born with a body type that fits their environment.'
   a Discuss what this statement mean to you.
   b Choose two countries and explain why this happens – consider climate, geographical location, food supply available, economics, resources, etc.

9780170463102

c Consider what might happen if someone moves to live in a different country. Do they have to adapt their body to fit?

d People from all over the world live in Australia.
- Consider how stereotypes impact the way that these people acclimatise to their new home.
- Discuss whether you think people need to conform. Should they have to? What things might they need to conform to, to make life easier? What things might they not want to change and why?

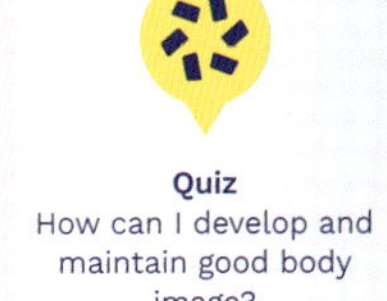

Quiz
How can I develop and maintain good body image?

HOW CAN I MANAGE STRESS?

Stress is a normal part of life and does not always need to be seen as a negative. Often stress is a powerful and positive motivator to get things done efficiently and in a timely manner. It helps keep an individual alert and ready for anything. Stress plays an important role in keeping individuals safe. It keeps their bodies alert in times of trouble so they are able to react quickly if needed.

There are many things that cause stress – family and friends, schoolwork, your expectations of yourself (or of others). These are all stressors. How much stress a person feels has a lot to do with their attitude to the situation. For example, one exam might cause you a lot of stress because you don't think you have done enough study, another might not even worry you because you know the content back to front. Sometimes the way you look at things is as important as what you do in a situation.

FAST FACTS

The problems from stress usually happen when stress is regular and doesn't let up. The chemicals the body releases can build up and cause changes that damage your physical and mental health.

Figure 4.16 Your attitude will impact your results.

FACE TO FACE Stress-contributing factors

In small groups, list all the things that cause you stress, then categorise these as either 'very stressful', 'moderately stressful' or 'a little bit stressful'.

Now look at your 'very stressful' list and answer the following questions:

1. Consider if this event is likely to occur.
2. State how much control you have over this event occurring.
3. Explain how much conflict will arise if this event occurs.
4. Propose how much your life will change if this event occurs.

Use the same questions for the 'moderately stressful' and 'a little bit stressful' categories.

5. Identify any similarities within your categories. For example, does your 'a little bit stressful' category have events that you can predict and have control over, with little or no conflict or change? Can you explain why this might be? Why do these four factors have such an impact on stress levels?

In most cases, everyone's body responds to stress in the same physical way. However, people vary in their emotional and behavioural responses. You and your friend might be in the same stressful situation, but only one of you shows any symptoms of stress (such as sweating, nausea or an increased heart rate). The way you interpret and respond to a situation determines how stressful it is for you. A previous bad experience may cause similar situations to be highly stressful.

WELLBEING CHECK IN

WHAT MAKES ME FIGHT OR FLEE?

Identify

Sometimes fear is useful and other times it just gets in the way. We usually call unhelpful fears about the future 'anxieties'.

This activity has been developed in collaboration with Dr David Bakker of MoodMission

Understand

The fight or flight response is where our body prepares us to deal with a threat by either running away from it or fighting it. This can be thought of as the body's alarm system. When we feel our heart start racing, our breathing get faster and shallower, and our muscles tense up, we know that our body has detected a threat. But the threat isn't always worth the response. For example, does it help to fight or run away if we need to get up in front of the class and do a talk?

Practise

1. Think about the last time you felt the signs of the fight or flight response.
2. Rate how strong the feeling was out of 10, with 10 being the most afraid you've ever been and 1 being totally calm.
3. Now rate how much danger you think you were actually in, with 10 being at risk of dying and 1 being totally safe.

4 Do you think you felt more fear than was justified? Maybe you didn't, and maybe you did.
5 The next time your body's alarm goes off, take half a second to notice what sort of situation you're in. Are you actually safe from harm?

Reflect

If your body is giving you false alarms, what do you think you could do to teach it to respond differently?

FACE TO FACE Responding to situations

1 In pairs, rank the following situations from 1 to 21, with 1 being the situation that would have the most impact on your life, and 21 the situation that would have the least.
 » You get into a school sports team.
 » Your favourite teacher isn't at school today.
 » You are fighting with a friend.
 » You are asked on a date.
 » Your house gets robbed.
 » You are teased.
 » Your internet is not working at home.
 » Your parent caught you looking at something inappropriate online.
 » Your assignment is now due on Monday, not tomorrow.
 » You break a violin string right before a performance.
 » You are in a car crash.
 » You are feeling depressed.
 » You haven't eaten all day.
 » You failed a maths exam.
 » It has been raining all day.
 » Your mum and dad have separated.
 » Your pet died.
 » You got a new pet.
 » You made a new friend.
 » You lost your voice.
 » You caught the flu.

2 Identify two possible emotional responses for the four situations that you feel would have the most impact on your life.
3 Determine whether the emotional responses are appropriate or not, and discuss why. Consider the following questions in your discussion:
 » What impact might this event have on your life?
 » How would your response to this event affect those close to you (e.g. family, friends, teachers)?

Scaffold
Impacts on my life

» Identify any emotional responses that are inappropriate and why.
» Describe what an extreme emotional response might look like.
» How might an extreme emotional response affect the situation or your relationships with others?
» What would happen if someone did not respond emotionally at all to the situation? Why might this be a problem?

4 As a class, identify the possible outcomes of these different responses.
5 Consider why all lists or responses may not be the same.

Stress in Years 11 and 12

The years of schooling often build up to choosing the 'right' things to do in your final two years, whether that is choosing the 'right' subjects to pursue or the 'right' training centre or work options to attend. There is so much pressure put on students, and possibly their parents or guardians to help, to make the 'right' decision. But there isn't one right decision, there might be several. This can be reassuring, but it can also make it stressful to pick.

When we put a lot of pressure on ourselves to achieve and we don't quite live up to that, this can also cause us stress, as we feel like we have let ourselves and others down. Some people know exactly what they want to do when they finish school. This might make it easier to make decisions, but if they fall short and are unable to start on their intended path straight away, it can be very disappointing. Other people don't have any ideas about what to do next, so their result might help direct them into a pathway.

Worksheet 4.9

When you do your best it opens doors to multiple opportunities, some you may not have even thought about. Stress is a big component of any exam period. Think about what has helped you in the past when studying for and taking exams. See if you can set up a pattern that ensures you have done what you need to with time to spare before the exam; his should help you feel better prepared.

It's important to keep things in perspective. Something might come up that impacts your attempts at your exams. If this happens, it is good practice to look after yourself and go and talk to someone who you trust. See if there is anything that can be done to help the situation. Some examples of situations that might occur that are beyond your control are:

- You become really sick during the study period.
- You get a concussion playing a fun Year 11 vs Year 12 footy match at school.
- Your parents split up.
- Your grandparent passes away.

Dealing with stress

The way that someone responds to a stressful situation can actually either decrease or increase their stress! If a person's mental health is good, they can often think logically through the situation and trust themselves to make good decisions. Can you think of a time where you responded to a situation in a way that calmed you or others down and made the situation less stressful? There are so many ways to deal with stressful situations. Sometimes all it takes is a deep breath in and out to relax and give the brain a moment to contemplate the situation.

INVESTIGATION ⇨ DEALING WITH ANXIETY

Purpose

To discover methods to calm an anxious mind

Materials

Using your computer to search, look for anxiety management strategies on the following sites: Beyond Blue, Better Health Channel, headspace, ReachOut Australia.

Weblink
Beyond Blue Better Health Channel headspace ReachOut Australia

Method

1 In groups, come up with the most common strategies these sites promote.
2 Write up a timetable for a week showing how someone with anxiety could fit in daily strategies to help themselves.

Discussion

1 Discuss if the websites' strategies overlap.
2 Consider how they promote the strategies that they do – what is it about these strategies that works?
3 How easy/hard was it to schedule daily strategies into a timetable?
4 State two pieces of advice you would give someone to help them keep up the strategies that help.

FACE TO FACE Ways to deal with stress

1 In pairs, choose one of the following scenarios (or your teacher might allocate you one).
   - You have a major French assignment due soon, which involves presenting in front of the class in French! Your teacher has said you must complete this assignment to pass the subject, but you cannot stand talking in front of people, especially your peers. (you, teacher)
   - Your mum recently married a man who has three other children, and they all moved in over the weekend. You now have to share a room with a 12-year-old! On top of this, you are having a hard time with your stepfather, who thinks you should be spending a lot more time studying than you are. (you, mum)
   - Your grandpa, who you get along really well with, has recently been diagnosed with cancer. He isn't expected to live for very long and has decided that he doesn't want anyone to see him like this, so is refusing visitors. (you, grandpa)

» You spent ages on your art piece this term. Your teacher had set a difficult task that had to be delivered online. You think you will get a really good mark for this project because it was creative and looked fantastic. You have just gone to upload your finished project you can't seem to find it on your computer! (you, teacher)
» Your best friend just started seeing someone new. You are really happy for them, but things have already changed, and it's only been a few days! You used to catch up with your friend before school and at lunchtime, and then walk home after school together. You have only caught a glimpse of your friend all day and end up walking home alone. You have other friends at school, but no one like your best friend. You are feeling left out and alone. (you, best friend)

2 In pairs:
   - brainstorm all the possible ways to deal with the stressful situation
   - think about which ways of dealing with stress are healthy and which are unhealthy
   - think about what criteria you used for deciding whether your solution was healthy or unhealthy.
3 In brackets after each scenario there are two people listed. You are to each take on one of these roles and develop a conversation to express your thoughts, opinions and beliefs.
   - Was it easy or hard to express your thoughts, opinions and beliefs? Why?
   - Did your partner acknowledge your feelings in the discussion?
   - Did you feel you were expected to behave a certain way? What impact did this have on your discussion? If you behaved differently than expected, would it change the result?
   - What did each person want from the discussion? Did they get it in the end?
4 Come back to the group and share your experience with the class.

Even after you have talked to someone about the thing that is causing stress, you may still have questions about the situation. Or you may have been told that it is something you just have to deal with. Sometimes you may need to access help or support. Finding support systems is an important step in adolescence, as you build the resources that you will use in adulthood.

FACE TO FACE Support

Everyone has times in their lives when they need support. There might be a person, place or thing that can help you in a situation when you might need advice or are not coping.

1 Brainstorm examples of support when dealing with stressful situations. For example, a counsellor might be able to support you by helping organise your thoughts and understanding how you feel about a situation, and this might help you make better decisions.

9780170463102

2 Discuss the following questions:
- How does support assist you in times of stress?
- How do you know when you need support in times of stress?
- What would happen if you didn't have access to support in times of stress?

MENTAL HEALTH AND STIGMA

Almost 75 per cent of people with a mental illness have experienced stigma. When a person is labelled by their illness (e.g. 'a diabetic'), they are then seen as part of a stereotyped group. Stereotyping can create prejudice, which leads to negative attitudes and discrimination. Experiencing or assuming there is stigma can have an impact on whether someone will access help and what sort of help they access. They may be more likely to access anonymous help, such as the internet, even though it may not provide the most accurate information or the best support.

Negative attitudes don't just come from people; there are services that discriminate against people, too. Don't accept anything less than respectful treatment from any service you come into contact with.

Figure 4.17 We all need support

CASE STUDY ⇨ ADDRESSING THE STIGMA

Identify

You are a council member who works in a low socio-economic area with great cultural diversity. There are high levels of government-funded housing, migrants and mental health issues in the area. The local primary and high schools are close, and there are always a lot of young people around. The local church attracts a lot of people on Sundays; however, they don't stick around afterwards.

Understand

You have had complaints from the staff who work at the only community house in your area. They are concerned they are going to lose their jobs because no one ever comes in to use the facility. There are two regular staff members, a psychologist who comes once a month and a social worker who comes twice a year. The volunteers who previously came every day are no longer seen. The two staff members think the programs they plan are fantastic; they came up with the ideas on their own! The programs include sewing (you bring your own machine), trips to the local swimming pool, high-impact aerobics, photography courses and computer workshops on software such as Adobe Photoshop, FrontPage and Garage Band. They even offer pizza and ribs or pie nights once a week. You go and visit the community house and this is what you see.

Figure 4.18 Newtown Community House: this is what you see.

Discuss

1 You are to report back to the council about this community house. The council wants to see the house being used, otherwise they will shut it down and sell it off. You have six months to do something about it. What are your recommendations for addressing the availability of mental health services in the community?

Consider the following points:

- Is there still stigma attached to seeking mental health services?
- How does the house look – is it welcoming?
- Staff attitudes – are the programs they are developing appropriate and accessible to all members of the community? Is the food they are providing enticing for all members of the community?
- Community attitudes – is there community input and involvement?

Stigma and discrimination occur in society. There are many groups of people who experience these. Can you think of some of these groups? Stigma and discrimination need to be called out as they can lead to mental health issues, and discrimination is against the law! The Mental Health Commission in Western Australia suggests a few things that can be done:

- Learn the facts about mental health and illness and share them.
- Learn about people's personal experiences.
- Speak up when family, friends or even the media make misleading statements or promote negative stereotypes.
- Offer support, just as you would with someone who has a physical illness.
- Treat people with respect, regardless of their mental health.
- Don't discriminate.
- Speak about your experiences, if you feel safe to do so.

FAST FACTS

1 People with mental health illnesses experience stigma and discrimination that may delay or stop them from accessing help.

2 Stigma and discrimination can come from the general population or organisations, or even friends and family.

3 Stigma and discrimination can occur in any environment.

4 Being an up-stander when you witness discrimination can be a way of calling it out.

CASE STUDY

DISCRIMINATION AND DEPRESSION

Identify

Almost 90 per cent of people with mental health issues say their lives are negatively affected by stigma and discrimination.

Understand

Mischa has lived with depression for most of her life. Usually she has it under control and lives a happy and fulfilling life. At the end of Year 10 Mischa's mother got a job interstate and the family had to relocate. It wasn't ideal, but the family pulled together and made it work. Mischa visited the school to enrol and choose her subjects. She wants to take psychology, because living with a mental health problem has made her want to help others as a career. But after seeing Mischa's file, the careers counsellor at Mischa's new school said that she will not be able to take psychology because, in his experience, taking such a class could affect Mischa's mental health.

Figure 4.19 What do I do?

Discuss

In pairs, discuss the following questions

1 Redraw Figure 4.19 and fill in or make a list of the possible responses to each of the 'I' statements Mischa might have in this situation. For example: 'I know that I am capable of dealing with this subject without it affecting my mental health.'
2 For each of the points you have just identified, consider how this might have an impact on others in Mischa's life. Consider her family, the school community and the community at large.
3 How will the counsellor's decision affect Mischa's life?
4 Is this careers counsellor discriminating against Mischa? Is their behaviour ethical?
5 Does Mischa have any power in this situation?
6 What would you suggest as an alternative response to her desire to study psychology?
7 If you were the school principal and Mischa's mother had just told you what the careers counsellor said to Mischa, what are three things you could do to improve this situation at the school?

1 Discuss what stress is and what types of things cause you stress.
2 Consider why we need stress.
3 Develop a body map outlining what our stress responses are and where they occur in the body. (You could draw an outline of a body on paper or outline someone's actual body on butchers paper.)

1 Propose a time where a positively stressful situation might turn into a negative. For example, you might win a money prize, but then all your 'friends' are after you for money!
2 Discuss a time where you felt extremely stressed.
   a What was it due to?
   b How did you feel?
   c What did you do to overcome the feelings of stress?

1 Make a list of all the things you do to lessen and overcome stress.
2 Now ask your peers what they do so that you have a list of even more ideas. If everyone's answers are very similar, you might need to do some research to find out how others deal with stress.
3 Develop and write an advice column in the school newsletter for all the Year 12 students getting ready for their exams. Offer them some insight into how they can prepare for and undertake exams with as little stress as possible.

HOW DOES MY COMMUNITY PROMOTE MENTAL HEALTH?

Health promotion aims to provide individuals and groups with strategies to improve their health through education and awareness. It allows these same people the opportunity to become more aware of their situation and at the same time give them more control over their health and wellbeing.

THE WORLD HEALTH ORGANIZATION

The World Health Organization (WHO) has identified 10 key action areas for health promotion:

1. Build healthy public policy.
2. Create supportive environments.
3. Strengthen community action.
4. Develop personal skills.
5. Reorient health services towards primary health care.
6. Promote social responsibility for health.
7. Increase investments for health development to address social inequities leading to poor health.
8. Consolidate and expand partnerships for health.
9. Strengthen communities and increase community capacity to empower the individual.
10. Secure an infrastructure for health promotion.

Can you think of any health promotion strategies your state or territory uses to improve health? Do these strategies include mental health? A unique and very effective way to look at health is in a holistic manner. This means that all dimensions of health (social, emotional, physical, cognitive and spiritual) are seen to make up the whole; if one dimension is missing, the rest may suffer. For many First Nations Peoples, dimensions such as extended family, community, environment and land play a significant role in health. They consider these dimensions to sustain health, and believe that each aspect needs to be looked after.

Scaffold
Explore some examples of state specific Health Promotion sites

The WHO has also set out a list of Sustainable Development Goals (SDG) that all United Nations member countries adopt. These are seen to be the most important goals to allow access for all. They also address important issues such as climate change and environmental sustainability.

Not every country has the same access to these areas of development. Australia is lucky, as it is able to work towards most of these goals. Each country should take each sustainable development goal and work through the 10 key actions, listed above, for health promotion. Each country will use a different health promotion strategy as they have different ways of doing things. For example, 'Life below water' may not be as pertinent to you if you are a landlocked country; however, you need to ensure that seafood does not become so expensive that your country cannot afford to import it.

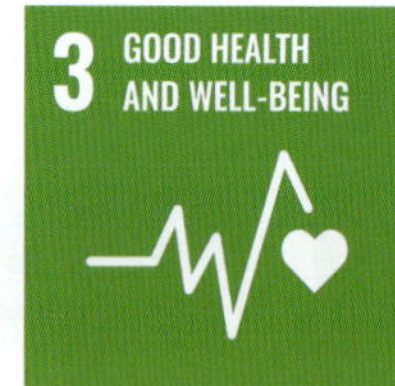

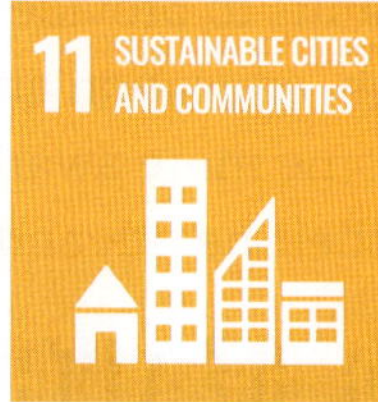

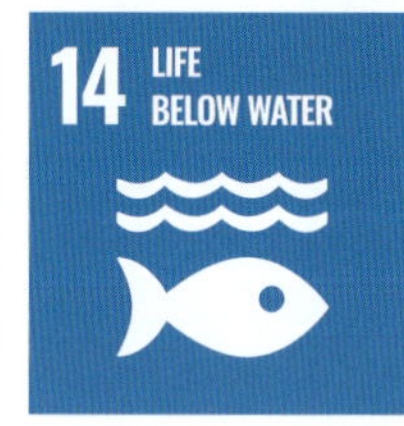

Figure 4.20 The 17 Sustainable Development Goals

FACE TO FACE How is Australia going?

Worksheet 4.10

With a partner, discuss how you think Australia is doing with each of these goals. Do you have any evidence for your evaluations? Can you think of other countries in the world that might have a different outcome for each goal?

CASE STUDY

SHAME TO DIE OF SHAME

Identify

The following image is an artwork by Wiradjuri artist Jordana Angus, which appears on a Queensland mobile BreastScreen van. The artwork is titled *Screening for Health, Not Screaming for Help.*

Figure 4.21 Detail of *Screening for Health, Not Screaming for Help* by Jordana Angus

Understand

This BreastScreen mammography van travels across Queensland to the border of the Northern Territory and across to the Torres Strait Islands, providing outreach mammography screening to all women. The aim is to encourage Aboriginal and Torres Strait Islander women, in particular, to have regular breast screening.

The artwork encourages women to be screened, rather than ending up having to scream for help. The painting uses the hibiscus flower to portray the breast in an indirect fashion. This is important because it makes coming for screening acceptable, rather than a shameful or clinical experience. The diverse colour palette of this painting reflects the need for the mobile service to be welcoming to all women in the areas visited. Yellow and blue represent the islands and green symbolises the rainforest. Pink and purple are used as they are the branding colours of the BreastScreen Queensland program, and pink is also the international colour for breast cancer awareness.

Weblink
BreastScreen

Discuss

1 How does the strategy address the UN health promotion action areas?
2 How could you measure the success of a health promotion strategy such as this?

HEALTH BEHAVIOURS IN OUR COMMUNITY

In all communities there are examples of good and poor health choices that people make. Can you name some? Good and poor health choices both have an impact, not only on the individual, but also the community in which they live. Can you identify a consequence for one of the poor health behaviours you have listed?

Worksheet 4.11

FACE TO FACE Choosing health behaviours – is it a right?

1 Identify social, cultural and economic factors that influence the following health decisions:
   - people who smoke
   - people who don't seek help for mental health issues
   - people who bully.
2 Does someone have the right to choose these behaviours? Why or why not?
3 Does someone have the right to discriminate against someone who chooses these behaviours? (For example, the boss refuses to give an employee overtime because the boss thinks they can't manage their mental health issues.)
4 These three behaviours have physical, social, emotional, spiritual and economic consequences. What are these? (For example, the Australian healthcare system is under stress because of illnesses caused by smoking; taxpayers' money goes, in part, to fund this system.)
5 Your teacher will organise a line debate, one side for and one side against the statement: 'Governments should be able to mandate against bullying.'

Racism

People with good mental health are usually more accepting of others' choices, unless they affect their own wellness. They may feel the need to stand up for themselves, and sometimes others, to stay safe. Racism is an issue that can affect mental health.

INVESTIGATION ⇨ WHAT IS THE IMPACT OF RACISM ON MENTAL HEALTH?

Purpose

To understand how racism has an impact on mental health.

Materials

Access the RacismNoWay! poster for schools

Weblink
RacismNoWay

Method

The middle of the poster lists all the effects of racism in schools.

1. Each member of your group will take on one of the subheadings: Educational outcomes, Individual happiness and self-confidence, School climate, Cultural identity, School/community relations and Student behaviours.
2. Make a list of all the ways that each dot point could have an impact on someone's mental health, e.g. poor school attendance – this will put the student behind in classwork and they won't know what or how to do what the rest of the class is doing and may feel stupid; they will also miss out on social time with friends and might feel isolated and lonely.
3. Have a look at how many of the impacts you have all identified and see if they cross over. Are they similar or different?

Discussion

1. Come back as a class.
   a. On the whiteboard, list five things someone could do if they felt they were a victim of racism. How could they access help based on their perceived level of safety?
   b. Choose one dot point from each indicator and make a statement about what you as a class can do to prevent racism in your school.

Weblink
Check out **RacismNoWay** an Anti racism education website for Australian schools

The Australian Human Rights Commission has developed a campaign to tackle racism in Australia. Their campaign, 'What you say matters', deals with the issue of racism in a way that connects with people. It also connects people to their communities.

9780170463102

UP AND MOVING What can be done?

Go to the website of the Human Rights Commission and look at the video clip, 'Racism. It Stops With Me'. Listen to the rap by brothablack: 'What you say matters'.

Can you think of other situations you have witnessed or heard of in school? Can you write a verse for any of those situations?

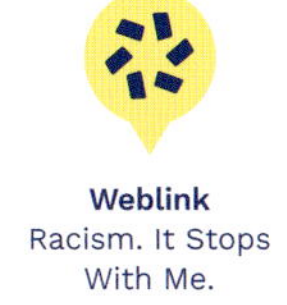

Weblink
Racism. It Stops With Me.

FACE TO FACE Create your message

Racism is not the only issue that can affect mental health.

1 In groups of two or three, research one of the following topics, or your teacher might allocate one to you. Use the resources listed on page 144 as a starting point for your research.

- Separation or divorce of parents/carers
- Cyberbullying
- Getting along with family
- Peer pressure
- Exams and studying
- Economic issues
- Relationship break-up
- Drugs and alcohol
- Illness or injury
- Bullying
- Friends
- Grief and loss
- Self-esteem
- Stress
- Conflict
- Disability
- Legal issues

2 After you have researched your issue, develop a campaign rap, song, poem or video that includes, if possible, all the following factors:

- what the issue is
- who the issue affects
- why it might be a problem
- how it affects mental health
- what feelings the issue produces
- when to seek help
- where to go for help
- what coping strategies could be used.

3 Think of a slogan for your rap/poem/video.

4 Discuss with the class how you might use these items to promote mental health and develop connections to the school community.

CASE STUDY ⇨ VICTORIA LAUNCHES NEW ANTI-RACISM CAMPAIGN

Identify

Victorian. And proud of it. shows the way forward to ensure we remain cohesive, and that every Victorian has the opportunity to be part of our state's success. *Victorian. And proud of it.* is also about sharing the stories, and celebrating the everyday examples of contribution and belonging that real Victorians from all walks of life make.

Source: © Copyright State Government of Victoria

Weblink
Victorian. And proud of it.

Understand

Racist tirades on public transport, bigotry in schools, and discrimination in the rental market face a crackdown as part of a state-wide push to counter the rise of the far right in Victoria.

Amid concerns the community has reached a tipping point, the Andrews government will today launch a new campaign to reframe the debate on multiculturalism in the wake of events such as Brexit, the rise of Donald Trump and the growth of One Nation …

The campaign begins on Sunday night with a television ad featuring Victorians from diverse backgrounds, and an underlying message of shared values and common goals.

'No matter where we're from, we all do our bit, making Victoria the best place on earth – so let's be proud of it,' says the voiceover.

A $1 million anti-racism action plan has also been developed with a focus on anti-Semitism and Islamophobia. The plan includes:

- New measures to crack down on commuters who racially abuse people on public transport.
- A broad-ranging review to stamp out discrimination and 'unconscious bias' against some people who seek approval to rent a home.
- A revamp of the Respectful Relationships school curriculum to teach students about prejudice and discrimination.
- 'Bystander awareness training' to help people respond accordingly when they experience or witness racism.
- Tougher complaint mechanisms to ensure the reporting of racism is taken seriously by authorities.

Source: 'Victoria launches anti-racism campaign to help counter rise of far right' by Farrah Tomazin, *The Age*, 18 February 2017.

Discuss

1 Why do you think that this campaign was developed?
2 What are they asking us to do?
3 Do you think this is realistic?

Knowing where to access support systems is an important skill. It makes getting help much easier, especially if you are familiar with the person, the service and/or the process. Knowing where support services are may also make seeking help much faster.

Young people spend a great deal of time at school, and all types of services that help young people maintain their wellbeing should be available. Do you think your school promotes mental health for its students?

Figure 4.22 A local community

UP AND MOVING My local community

In a small group, research the mental health services in your local area. Your teacher will advise you on how to access the required information (e.g. this could be a visit, online research or an interview).

1 Critique the appropriateness and effectiveness of the mental health services available for young people in your local community.
2 Evaluate how accessible the services are for young people, and propose changes to promote greater inclusiveness.
3 Determine which service(s) would be the best for each of the following people:
   - a young person with a mental illness
   - a young person in a crisis
   - a young person who needs advice about a parent with a mental health problem.
4 Investigate ways to store and share the contact information of these services with other young people.
5 Present your information to the class in a creative and innovative manner (e.g. a wix, prezi, role-play, campaign).

iStock.com/sturti

Figure 4.23 How does your school support mental health?

Dreamstime.com/Rawpixelimages

Figure 4.24 What facilities does your school have to help students, teachers and the community?

UP AND MOVING How does my school do?

1 Discuss what your school does to enhance the mental health of the school community (students, teachers, staff, parents, community).

2 Investigate any examples of mental health initiativesin your school. Here are some places to look:
- the office – student wellbeing or engagement policy
- the student wellbeing office or chaplaincy or first-aid office – posters, brochures, DVDs, websites
- around the school – posters, announcements, notices
- health teachers – past mental health initiatives (e.g. wellness week, health expo, relaxation workshops, health education curriculum)
- the canteen – healthy food to feed healthy minds
- the office or website – newsletters for parents and the community.

3 After you have completed this audit of your school's mental health strategies, as a class produce a list of ideas the school could consider to improve or enhance its mental health strategy.

Worksheets 4.12 and 4.13

Mental health promotion in my school

Design an afternoon where your year level can celebrate the diversity in your school. This activity enables you to be creative and use everything you have learnt in this chapter to promote belonging in your school as a mental health initiative. Try to include the whole school (students, teachers, staff, parents and community).

1 Discuss what health promotion is and how it impacts our communities.

2 Explain how racism is a community health issue.

1 Propose what topics/issues that you see needs to be promoted, either in your community or at your school.

2 Investigate what you would need to learn about the topic to be able to put a campaign together.

3 Discuss who you would target in your campaign strategy. What sorts of approaches might appeal to this group?

1 Debate whether good mental health leads to better grades in school.
   a Why or why not?
   b Who says? What evidence can you find to support this?

2 Discuss the types of programs, activities and student-led or community-involved initiatives that are schools putting in place to improve mental health? What impact do you think these types of initiatives will have?

Quiz
How can I develop and maintain good mental health?

9780170463102

CHAPTER 4 REVIEW

1 Discuss the differences between good mental health, mental health issues and mental illness.
2 List four benefits of good mental health.
3 Create a list of the negative impacts of mental health problems on Australia's economic health.
4 Propose five factors that shape your identity. Discuss how these have an impact on your mental health in both positive and negative ways.
5 Give an example of how a person might have an impact on someone else's identity.
6 Discuss why stereotyping someone can be so harmful.
7 Discuss how the media can have an impact on mental health.
8 Discuss what steps have been taken in Australia to oppose the media's impact on young people's mental health.
9 Justify why finding and/or accessing help for mental health issues can be difficult.
10 List the steps Beyond Blue recommends you take when helping a friend.
11 Discuss how your attitude has an impact on your stress response.
12 Discuss how emotions affect a stressful situation. Consider extreme emotional responses through to no emotional responses.
13 List five healthy ways to deal with a stressful situation. Provide an explanation of why each of these are healthy ways to deal with the situation.
14 Discuss whether there is a stigma attached to mental health.
15 Diversity in how cultures think about mental health are normal. Discuss how First Nation Peoples define and therefore treat people with mental health issues.
16 Why is mental health promotion an important community strategy?
17 Why is it important to be aware of mental health services, even if you don't have a mental health problem or illness?
18 Justify how improving mental health will improve lives.

SEXUALITY, GENDER AND RESPECTFUL RELATIONSHIPS

5

IN THIS CHAPTER

You will gain an understanding of the differences between sex, sexuality and gender and how these can be stereotyped to make young people extremely confused about what it means for them. Talking about the differences can alleviate a lot of myths and misconceptions. You will learn about qualities that are essential to positive, affirming and effective relationships. You will discover how relationships change and develop, particularly as you move through adolescence, and how these connections help to form an individual's personal identity. You will look at the issue of gender-based violence, which is having a horrific impact on the world. You will understand what this is and how we all can be active bystanders and not allow it to continue. The nature of sexual activity is explored, as well as the implications of beginning more intimate relationships. You will look at consent and the many complications of gaining and getting consent. Along with this comes decisions that need to be made, such as safer sex methods to prevent unwanted pregnancies and STIs. Easily accessible support is available – some support organisations are listed at the end of the chapter.

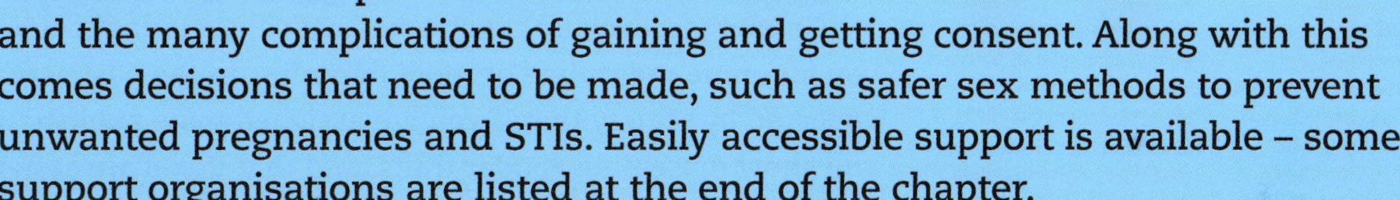

Shutterstock.com/Nicetoseeya

By the end of the chapter, you should be able to:

- understand the diverse nature of sexuality
- describe the qualities essential to forming and maintaining positive relationships
- define the factors that have helped to shape personal identity
- value and respect diversity in relationships
- understand the state of play regarding gender-based violence in Australia
- develop skills that enhance a respectful relationship
- define consent and develop an understanding of consensual relationships
- examine the potential consequences of developing more intimate relationships, such as pregnancy and sexually transmissible infections
- outline a variety of support services within the community.

WHAT'S THE DIFFERENCE BETWEEN SEX, GENDER AND SEXUALITY?

Quiz
Pre-chapter

Before you start, take the pre-chapter quiz to find out how much you already know.

sex whether a baby is born with male or female body parts

intersex a baby born with variations of any of the sexual characteristics (e.g. hormones, genitals, chromosomes)

Weblink
Explaining intersex.

Sex is the word that refers to the biological body parts a person has. At birth, you are assigned male or female sex. If you are assigned male anatomy at birth, you usually have male sexual reproductive organs. If you are assigned female anatomy at birth, you usually have female sexual reproductive organs. If you are born with any variations of these sexual characteristics (e.g. hormones, genitals, chromosomes) you are **intersex**. These variations can cause sexual ambiguity. This does not mean there is anything wrong with the baby. However, in the past, in these cases, doctors would convince the parents to decide what sex they wanted their baby to be and the doctor would perform surgery to make the baby either male or female. As a result, many children grew up in a body of the 'wrong' sex. These days, human rights have prevailed and children are more often allowed to wait and decide which gender they identify with. There is nothing wrong with a child being born intersex, so there is no urgent need to do anything. The child will grow just like any other, and as their identity and confidence develops, they will hopefully be able to articulate their individual needs.

gender refers to the socially constructed categories assigned to individuals on the basis of their apparent sex at birth. While other genders are recognised in some cultures, in Western society, people are expected to conform to one of two gender roles matching their apparent sex; for example, male = man/masculine and female = woman/feminine

norms standards that are seen as normal or acceptable

cisgender a term used to describe people whose gender corresponds to the sex they were assigned at birth

transgender umbrella term used to refer to people whose assigned sex at birth does not match their internal gender identity, regardless of whether their internal gender identity is outside the gender binary or within it. Transgender/trans or gender diverse people may identify as non-binary, that is: they may not identify exclusively as either gender; they may identify as both genders; they may identify as neither gender; they may move around freely in between the gender binary; or they may reject the idea of gender altogether

nonbinary a term used to describe gender identity that does not conform to traditional gender norms and may be expressed as other than woman or man, including gender neutral and androgynous

binary concepts binary relates to either/or; with a binary concept, there are only two options, not a scale of options

Gender is a socially constructed concept relating to whether you are more masculine or more feminine. In the past, it was expected by society that if you are assigned male at birth, you will show masculine gender traits, and that if you are assigned female at birth, you will show feminine gender traits. The construction of gender happens early on in a baby's life. This notion of gender '**norms**' is often reinforced by things like giving stereotypically male or female gifts, such as a truck for a boy and a doll for a girl, or choosing the colour blue for boys and pink for girls. These gender stereotypes can be reinforced in many ways, and can continue into adulthood, e.g. how a person is expected to behave and dress, the employment and education subjects they choose, etc.

We all have our own gender identity. This is determined by what we feel about ourselves, not based on any stereotype. If our sex and gender identity line up – for example, if a person is assigned female and their gender identity is feminine – they may be **cisgender**. If an individual's sex does not match their own gender identification, they may be **transgender** or **nonbinary**, which means their sex and gender do not line up or are not **binary concepts**. It can be very difficult for an individual to tell people that they do not fit their assigned sex. The amount of support they have from family, friends, schools and others will have an impact on their experience. Some families understand what their child has been going through and accept them no matter what. Others will enforce their assigned sex and continue to call them by their assigned name and dress them in sex-assigned clothes. This can be very damaging and confusing for the child as they try to navigate the conflict between their inner selves and the lack of support from their loved ones. The child's mental health can be put at risk, and they can be isolated from accessing help.

An individual's **sexuality** forms part of their identity and is constantly evolving as they mature. Sexuality is central to a person throughout their life. Sexuality is your feelings, thoughts, attraction to and behaviours towards another person. Sexuality is diverse and fluid. This means that everyone's definition of it can be different, and that it is not something we are 'stuck' with for life. You might have heard that you are either **heterosexual** or **homosexual**, but while some people do comfortably identify as one or the other, it doesn't have to be that way. Sexuality is also not a binary concept, it is more like a continuum, which you may move around on depending on the person you are with. If you are heterosexual, this doesn't mean that you will only have dreams or fantasies about someone of the opposite sex. You may have a sexual thought or even perform a sexual act on someone of the same sex, but not consider yourself homosexual. Your sexual identity is YOUR identity. Sexuality can be experienced and expressed in thoughts, beliefs, attitudes, values, behaviours, practices, fantasies, desires, roles and relationships. It is okay, and normal, to have questions about sexuality.

You might have heard of people who were in a heterosexual marriage and then, after the relationship ended for whatever reason, they later entered into a homosexual relationship. Does this mean they were always gay, and the marriage was a lie? Not at all! An individual can be attracted to whoever they like. LGBTQIA+ is an acronym used to refer to people who may be **lesbian**, **gay**, **bisexual**, transgender, intersex, **queer** or **asexual**/**aromantic**.

A person's sexual orientation refers to who they are attracted to. Some people have no questions about their sexuality, others are not sure. Either is okay. During puberty we all go through a same-sex attracted phase, it is a part of adolescent development. This can be confusing. Sometimes young people know for sure who they are attracted to and their thoughts, beliefs, attitudes, values, practices, fantasies, desires, roles and relationships all line up with their sexual behaviour, i.e. if they are a male and are attracted to other males, they have intimate or sexual contact with only males. It is okay to question your sexuality – not everyone knows straight away, and it is fluid, meaning it can change, which can also be confusing!

Alamy Stock Photo/Jeffrey Mayer

Figure 5.1 Ellen Page, who is now known as Elliot Page, has openly come out as transgender.

Weblink
Watch the video to find out more about transgender people.

sexuality sexual feelings that may be expressed through thoughts, emotions, sexual behaviour or sexual orientation

heterosexual an individual who is sexually and/or romantically attracted to the opposite gender

homosexual an individual who is sexually and/or romantically attracted to the same gender

lesbian an individual who identifies as a woman and is sexually and/or romantically attracted to other people who identify as women

gay an individual who identifies as a man and is sexually and/or romantically attracted to other people who identify as men. The term gay can also be used in relation to women who are sexually and romantically attracted to other women

bisexual an individual who is sexually and/or romantically attracted to people of the same gender and people of another gender. Bisexuality does not necessarily assume there are only two genders

queer a term used to describe a range of sexual orientations and gender identities, often used as an umbrella term to describe the full range of LGBTIQA+ identities

asexual a sexual orientation that reflects little to no sexual attraction, either within or outside relationships. People who identify as asexual can still experience romantic attraction across the sexuality continuum. While asexual people do not experience sexual attraction, this does not necessarily imply a lack of libido or sex drive

aromantic refers to individuals who do not experience romantic attraction. Aromantic individuals may or may not identify as asexual

INFLUENCES ON SEXUALITY

Physical, emotional and social experiences all form part of your sexuality and affect how it develops. The physical side includes the changes occurring both inside and outside of your body as you mature. The emotional side is expressed in many of your thoughts and feelings. Discussing some of these thoughts and feelings with people that you are close to and trust may help you cope.

Many different people and factors affect the social side of your sexuality, including family, friends, peers, religion, culture and the media.

Shutterstock.com/Monkey Business Images

Figure 5.2 Your family, including parents and siblings, can be one of the greatest influences on your sexuality.

Family

Perhaps one of the greatest influences on your sexuality is your family. The ways in which your parents or carers relate to one another will often influence your behaviours towards others. If your parents or carers are affectionate with one another, you may be comfortable expressing your feelings openly. Your parents or carers may also place boundaries on your behaviour in terms of freedom and independence, such as rules for dating.

WELLBEING CHECK IN

UNDERSTANDING EMOTIONS

This activity has been developed in collaboration with Dr David Bakker of MoodMission

Identify

What do emotions do for you?

Sometimes emotions are seen as annoying or wrong, but it's really important to acknowledge, understand and validate what emotions do for us.

Understand

There are three main purposes for emotions:

1. To motivate us and prepare us to do certain things
2. To communicate to others
3. To communicate to ourselves

For example, when you feel fear, it 1) prepares you to avoid a threat, 2) tells other people that you don't want to do something and can also inform them of the threat and 3) communicates to yourself that you are in danger.

We know that sometimes emotions aren't useful or justified. For example, if you feel scared or anxious around your friends, it could be an instance of social anxiety, which you then want to figure out ways of changing. Regardless of whether an emotion is useful or irrational, it still helps to understand what your body and brain is trying to achieve by triggering that emotion.

9780170463102

Practise

Think of the last time you felt a strong emotion. It helps to pick an example that relates to a situation that directly involved you, e.g. not a sad movie you watched. Now note down:

1 What was the emotion?
2 What was it motivating you to do?
3 What was it communicating to others?
4 What was it communicating to you?

Tip

Sometimes one part of an emotion is so strong that it makes it difficult to notice other parts. That's perfectly normal. If that is your experience, make a note of it in the table below. The more you practise observing your emotions the more you will start to notice.

Emotion	Specific function	Body responses	Thoughts	Behaviours
Anger	Lets us know when we've been wronged. e.g. not being allowed to go to my friend's party even though my parents originally promised that I could	e.g. hands clench, muscles tense, face gets hot	e.g. 'I've been treated unfairly', 'this shouldn't be happening'	e.g. make aggressive gestures, throw something, yell at someone, walk out
Anxiety/ Fear	Focuses us on escaping from danger			
Happiness	Focuses us on continuing to engage with things that are pleasurable and meaningful			
Sadness	Focuses us on what we value and communicates that we need support			
Love	Signals that something is going right			

Reflection

Did you act with your emotion's wishes? Or did you avoid doing something the emotion was motivating you to do? Sometimes we find ourselves doing something contradictory to our emotions, e.g. sadness communicates to other people that we need support, but lots of people don't show their sadness to others. What might be some of the reasons for this?

Peers

peer group group of people of approximately the same age and interests

How your **peer group** behaves, speaks, thinks and dresses may affect how you express your sexuality. Your school and education may also have an impact. If you attend a co-educational school, or have siblings of a different sex, you may be able to develop more effective relationships with the opposite sex than someone who attends a single-sex school, for example. The sex of your teachers may have an impact, if you see them as role models, and so might the subjects you have the opportunity to select from.

Shutterstock.com/Beatriz Vera

Figure 5.3 Just like family, your peer group can influence how you identify and express your sexuality.

Culture

Your culture may have its own expected codes of conduct. In most traditional First Nations communities, boys are initiated into manhood in a cycle of ceremonies where they are taught their traditional songs and shown the dances that are associated with them.

The men who live in Vanuatu, a small island nation in the middle of the South Pacific, take part in a yearly harvest ritual called land diving. Villagers build wooden towers reaching heights of more than 30 metres. The men tie a vine to a platform on the tower and tie the other end around their ankles, then dive from the platform headfirst. The goal of the jump is to land just close enough for the diver's shoulders to touch the ground. Any miscalculation on the length of the vine means either serious injury or death. This ceremony dates back nearly 15 centuries and serves two purposes. It is performed as a sacrifice to their gods to ensure a bountiful yam crop, and it also serves as a rite of passage to initiate the tribe's boys into manhood. Boys as young as five years old will take part in the ritual, which is often preceded by circumcision. The boys start out jumping low, but will work their way up as they get older. The higher a man goes, the more manly he is considered by the tribe.

Alamy Stock Photo/Penny Tweedie

Figure 5.4 Boys learning a traditional ceremony in the Central Desert region

Weblink
Land diving in Vanuatu.

Alamy Stock Photo/Hemis

Figure 5.5 Land diving in Vanuatu

Religion

Your religion may influence your sexuality in terms of sexual standards and norms. For example, in many faiths, sexual relationships are permitted only between a husband and wife, so **chastity** before marriage is considered a positive and essential part of life.

chastity not engaging in sexual relationships

Media

The media and advertising have an influence over what is perceived as 'normal'. For example, they can show advertisements that demonstrate sexual attractiveness as glamorous and sophisticated. Unfortunately, they often portray unrealistic images that many young people feel they should emulate in order to be like their favourite celebrities.

Worksheet 5.1

The media can also portray positive images of acceptance of those who choose to live outside society's standards and sexual norms.

FACE TO FACE Media influences

1 In pairs, analyse how this advertisement may influence adolescent behaviour.
2 What message is it representing?

Getty Images/David Cooper

Figure 5.6 The media can influence what you think about sexual attractiveness.

SEXUAL ORIENTATION AND GENDER IDENTITY

Worksheet 5.2

Today's society recognises many different kinds of relationships, not just heterosexual relationships. People today are aware that many people have a sexual orientation that does not fit this heterosexual perspective. Same-sex couples living together in the same household have only been included in the Australian Bureau of Statistics Census of Population and Housing since 1996, demonstrating the changes in societal acceptance of same-sex couples as a type of family in the national population. There has been a steady increase in self-identification of both male and female same-sex couples since 1996.

Video
Sexual orientation and gender identity. Watch the video and start the discussion.

Weblink
Watch the National Institute for Challenging Homophobia (NICHE) video about Tony's experiences as an intersex person, and then write a synopsis of the challenges Tony faced growing up as an intersex person.

Young people questioning their sexual or gender identity can often feel isolated, and may be more at risk of experiencing mental health problems. This is often associated with community attitudes rather than their sexual or gender identity. They may also face difficulties linked to disclosure, or 'coming out'. It can help to think of this as 'inviting someone in' rather than 'coming out'. You are in charge of who knows and what they know. It is important that everyone has someone to talk to, someone they trust. You also have a right to expect privacy, you do not have to tell anyone.

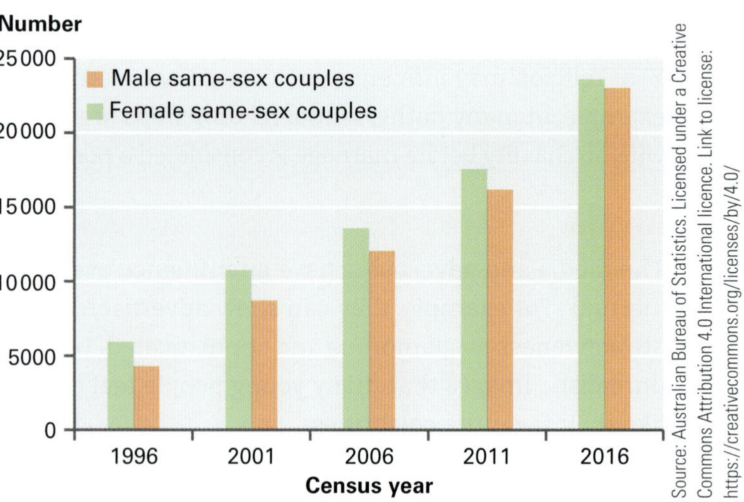

Figure 5.7 Number of self-identified same-sex couples in Australia, 1996–2016

Just as it is important to acknowledge someone by their name, it is also very important to acknowledge their correct pronouns. This can be as easy as asking them what pronoun they would like to be called, e.g. him, her, they, them. Doing this can have an incredible impact on how safe someone feels, as it tells them their classmates are respecting them for who they are, not what they are.

There are support groups who can help young people who are feeling confused about or who want to celebrate their sexual identity or orientation; one is Minus18. Other organisations, such as Parents and Friends of Lesbians and Gays (PFLAG), offer support and information to families and friends of LGBTQIA+ people.

CASE STUDY

MINUS18

Identify

Minus18 is a community-based, non-profit organisation that supports young people of diverse genders, sexes and sexualities, and their families and friends. Its aim is to create acceptance, supporting young people to build resilience and achieve their potential.

Discuss

1 Review this website and evaluate the services Minus18 offers young people.
2 Discuss how the organisation promotes inclusiveness within the community.
3 Investigate campaigns or activities they run or support to help inclusion of LGBTIQA+ people in the community.
4 Discuss why they run workshops. Who are they targeted at? What are these workshops about?
5 Create a PMI diagram, discuss the Plus/Minus/Interesting features of the website.

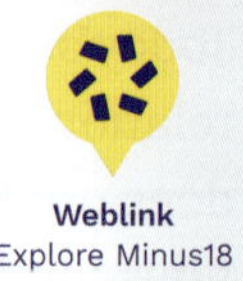

Weblink
Explore Minus18

9780170463102

STEREOTYPES ABOUT SEXUALITY

In society, there are images and ideas of how different genders are supposed to act. These are portrayed through media advertising, gender expectations in terms of a career or how each gender is expected to act in certain situations. These are called stereotypes. A stereotype is a widely held belief about a particular thing or person based on their characteristics, which may not always be true. These stereotypes can cause confusion for young people who are trying to understand their sexuality. The media portrayal of what a man or a woman should be often ignores the continuum of behaviours between hyper-masculine and hyper-feminine. For example, a girl may have short hair and dislike wearing dresses, but this does not mean that you can tell her sexuality; she is merely expressing her personal identity.

Getty Images/Ethan Miller

Getty Images

Getty Images/Kevork Djansezian

Figure 5.8 Celebrities such as Miley Cyrus (left), Ellen DeGeneres and Portia de Rossi, and Billy Porter may demonstrate alternative perspectives to traditional stereotypes.

FACE TO FACE Quick quiz

Discuss the following statements. Which ones represent stereotypes?

- Only males should ask people out and pay for dates.
- It is the female's responsibility to perform household tasks such as cleaning and washing.
- Males should never play with dolls.
- Males should not show emotion and cry.
- Females can manage a career and a family.

Worksheet 5.3

FACE TO FACE Stereotyping

A man and his son are driving in a car one day when they are involved in an accident. The man is killed instantly. The boy is knocked unconscious, but he is still alive. He is rushed to hospital and will need immediate surgery. The doctor enters the emergency room, looks at the boy, and says …

'I cannot operate on this boy, he is my son.'

So, the question is, how is this possible? The answer is simple:

The doctor is the boy's mother.

1 Did you automatically assume the doctor was male?
2 What does this say about your own perceptions of gender roles?

1 What is the difference between sex and gender?
2 Discuss how your family has an impact on your personal identity.
3 Discuss how your family has an impact on your sexual identity.

1 List three different gender stereotypes.
2 Consider how these stereotypes might make growing up a confusing/difficult time for the following people who are:
   a gender diverse
   b transgender
   c intersex.
3 Consider how these groups of people might be treated by others who believe the stereotypes.
4 Consider what you could do that might change this experience for these groups.

1 Discuss why you think some people have a hard time talking openly about sex, gender and sexuality. Consider people's morals, values, religion, cultural beliefs, political beliefs, etc.
2 Discuss where you think the old idea that everybody should be heterosexual comes from. Is there any historical significance?
3 Explain to someone why binary concepts can be detrimental to building close-knit communities.

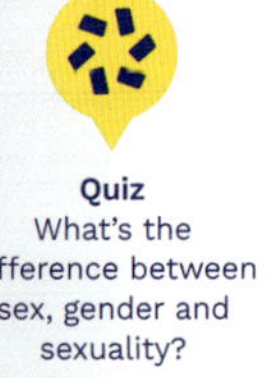

Quiz
What's the difference between sex, gender and sexuality?

HOW DO I DEVELOP RELATIONSHIPS?

Relationships are an important part of everyone's life as they fulfil social, emotional, intellectual, physical and spiritual needs. Think about how many relationships you have in your life – with family, friends, partners, school, groups, family friends and others. Life without relationships would be really difficult. (See also Chapters 4 and 6.)

CHARACTERISTICS OF POSITIVE RELATIONSHIPS

Building positive and respectful relationships is one of the most important aspects of human life. From the moment babies are born, they establish bonds with those around them – first with their parents, then with friends throughout school, and then intimate partner relationships later in life. People who are in loving, supportive relationships are more likely to feel healthier, happier and more satisfied with their life, reducing the chances of experiencing mental health issues.

Many different relationships are formed during a lifetime in order to satisfy a variety of needs. In an ideal world, a family provides basic human needs such as food, love, affection and safety. Friends give a sense of belonging. School, work or sport relationships can give a sense of achievement or accomplishment of goals.

Shutterstock.com/Rawpixel.com

Figure 5.9 Friends can give you a sense of belonging, while school relationships can give you a sense of achievement.

FACE TO FACE Mind map

In a mind map like the segment shown here, list all of the relationships you currently have in your life. Next to each relationship, provide two or three qualities or aspects that you gain from being involved in that relationship.

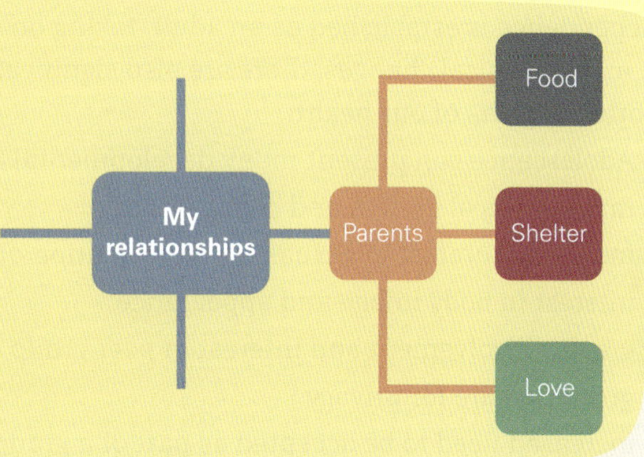

FACTORS AFFECTING RELATIONSHIPS

Family

The relationship formed with family is very important in satisfying needs and developing a values system, particularly for young people. Your family should support you, provide guidance and reinforce right from wrong. However, not all families are always supportive. As individuals move through adolescence, their relationships start to change. As adolescence is a time when teens are seeking independence from their family, they will establish strong friendships during this time and may even develop sexual relationships. With this greater freedom and independence comes more responsibility to look after yourself and your body.

FACE TO FACE Families

A family is defined by the ABS as two or more persons, one of whom is at least 15 years of age, who are related by blood, marriage (registered or de facto), adoption, step or fostering, and who are usually resident in the same household.

Source: Australian Bureau of Statistics, T2901.0 - Census of Population and Housing: Census Dictionary, 2016, ABS Website, accessed 04 June 2022. Licensed under Creative Commons Attribution 4.0 International licence

This includes couples with children, couples without children, newly married couples, same-sex couples, single parents with children and siblings who live together. In pairs, discuss the following questions:

1. Do your families fit into this definition?
2. Do all types of families fit into this definition?
3. Can you write a better definition?

Adolescence

Adolescence can be defined as the period of rapid change between childhood and adulthood. It begins with the changes associated with puberty and ends as independence is established as an adult, taking on roles and responsibilities. As well as the many physical changes, there are also significant developments during this time in all other aspects of our health.

Adolescence can present many developmental challenges, such as:

- rapid physical growth and change
- increased interest in sexuality and formation of sexual identity and preference
- interest in body image and appearance
- greater involvement and interest in peer group rather than family
- greater desire for privacy
- increased need to be accepted as part of a group
- expansion of cognitive and intellectual capabilities.

Each adolescent grows at a different rate from any other and has different experiences – no two people are the same! Between the ages of 10 and 14, people who have ovaries are often taller and more physically developed than people who have testes, as they often go through puberty earlier. Some people with testes develop musculature and deep voices long before their peers.

There is a predictable pattern of physical changes that occur in the body during puberty, but you never know exactly when these will occur. The emotional and social changes will also happen, but will be less predictable. For example, social groups may change, and sexual attraction may begin early for some and not others. You will never know when it is going to happen until it happens to you!

Worksheet 5.4

Interests also start to vary during adolescence. You may find that your friends suddenly have a crush on the latest band, start to dress differently or involve themselves in a new hobby. They may form opinions and values that are different from yours, influenced by their background, culture, religion, family, community and the media. These influences help to establish a person's values, attitudes and beliefs. Younger children rely on family to guide them and satisfy their needs. As children move towards adolescence and young adulthood, the peer group becomes more important and may influence decision-making.

INVESTIGATION ⇨ CHALLENGES OF ADOLESCENCE

Purpose

To research and discuss the challenges that young people face growing up.

Materials

You will need a search engine on a computer/tablet

Method

Form teams of three or four. Each team takes on the role of a group of writers for a major TV series who have been given the task of developing a storyline about the pressures of adolescence. You must prioritise the three major challenges that adolescents face. Consider the following questions in your answer.

1. Your group will be given one of the following adolescents: a Year 12 student, a gay male, a person with a disability, a person who speaks English as a second language, a person living in a rural area, a refugee, a transgender female, a national athlete.
2. Consider the biggest challenges this person faces throughout adolescence.
3. Consider the decisions they will need to make to negotiate their way through these challenges.
4. What are the consequences of these decisions?

Discussion

1. Discuss whether the key challenges were the same for everyone. Discuss the challenges that may be faced by different sexes, genders, sexualities, abilities, cultures, living localities, etc.
2. Explore the many factors may influence these challenges. What if an adolescent experienced one of the following situations:
   - lived with a parent who had a mental health issue?
   - lived in poverty?
   - lived with a family who moved every two years with the military?
3. Make a list of other diverse backgrounds and experiences that may influence the transition through adolescence.

Communication

Communication is essential to developing positive and respectful relationships during adolescence. Each type of relationship formed can be usually enhanced by developing strong and effective communication skills, and improved through practice and reflection.

Effective communication skills involve:

- displaying active listening by paying close attention to the speaker
- maintaining eye contact
- showing interest in what the speaker is saying
- asking questions for clarification
- displaying non-verbal communication cues such as facial expressions that are consistent with the verbal information.

Being empathetic to and understanding of differing points of view helps foster positive relationships.

WELLBEING CHECK IN

REFLECTIVE LISTENING

This activity has been developed in collaboration with Dr David Bakker of MoodMission

Identify

Do you have a friend who you feel really gets you? Here's how to be that friend for someone else.

Understand

It's important to feel heard and understood when we talk to people. However, sometimes it's not obvious that someone understands, and it can feel like they just don't get it. Psychologists, counsellors and other people whose job it is to talk to and listen to people are trained in something called reflective listening. This basically involves reflecting back what the person has said to you to show them that you have heard. Practising reflective listening can strengthen relationships by a) helping you remember what the other person has said, and b) making the other person feel heard.

Practise

1. Have a conversation with a friend.
2. When your friend says something, reflect it back to them in a slightly different way. For example, if your friend says, 'I'm feeling pretty stressed about the test this afternoon,' say back to them, 'Yeah, I can see this afternoon's test is stressing you out'.
3. You can then go on to ask a question or talk about something related, e.g. 'What about it is stressing you out?' or 'I'm feeling pretty stressed about the test too.'
4. Try practising it with different friends or family members.

Reflect

Do you think your good friends hear you and understand you? They may not do this reflective listening all the time, but they may say other things that show that they listen and care. What are some of these other things? And how could you do them more for your friends?

Conflict resolution

Not all relationships are positive. Inevitably, in any relationship, disagreements will occur and conflicts may arise. Conflict may not necessarily be bad for the relationship; it may assist with consideration of the other person's perspective and help to gain a greater understanding of their point of view. When conflict arises, it is important to be able to **negotiate** and resolve disagreements through conflict resolution and a willingness to **compromise**. Mediation may involve a third party who can listen to both points of view and then give some perspectives on how both sides can make some concessions in order to reach an agreement.

negotiate to discuss in order to reach an agreement; discussion may involve some bargaining and reviewing needs and wants

compromise a way for two parties to reach agreement by each making concessions

assertive to clearly state one's needs and wishes with respect both for oneself and the other people involved in the communication

If a person takes an **assertive** approach to communication, they will usually be more successful in getting their point across without hurting the other person's feelings. In any relationship, a give-and-take approach is most effective. (See more about assertiveness on page 211 in this chapter and on pages 251–4 in Chapter 6.)

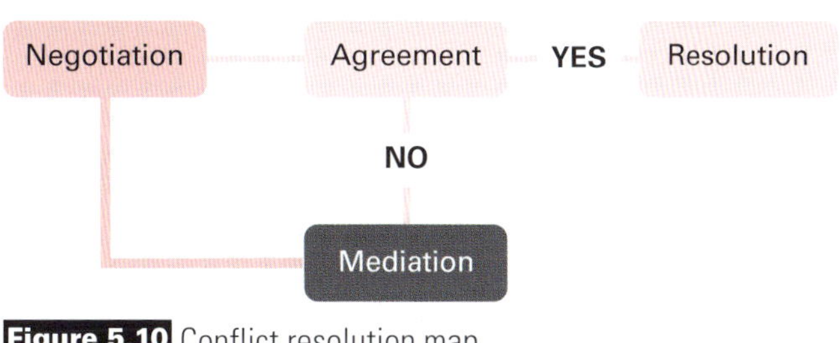

Figure 5.10 Conflict resolution map

PERSONAL IDENTITY

From birth, an individual develops a sense of who they are from the relationships they form with family, friends and the community. These connections help to build a personal identity, shaped by the person's environment and background. An individual's personal identity is made up of their past, their present and their future. The development of our identity is never finished and is not stuck, it is not something we ever achieve. It is constantly developing.

Alamy Stock Photo/YAY Media AS

Figure 5.11 Your personal identity is shaped by your environment and background.

Table 5.1 Influences on personal identity

Past influences	Present views	Future goals
Cultural heritage	Religion	School
Upbringing	Education	Career
Parents' interactions	Values/beliefs	Relationships

UP AND MOVING — Personal identity

1 As a class, brainstorm where identity comes from. Specific examples may include gender, cultural background or religious affiliation.
2 Next, create your own unique Personal 'I' Chart – Who am I? – by drawing your outline and brainstorming all the unique qualities that make up YOU. Then try to identify where these qualities come from. Your identity is made up of all of these things, it is who YOU are.
3 Share the two parts of your identity you are most proud of with another class member.
4 Design a badge or T-shirt that makes a statement about your identity.

Worksheet 5.5

CASE STUDY ⇨ CONFLICT OF VALUES

Identify

Values are attributes that a person considers to be good, appropriate and important to guide and determine how they live their lives. In each of the scenarios below, assess how different values have created the conflict.

Understand

1 Eric is gay and has just come out to his mother. He wanted to live his best life, and couldn't hide anymore. He thought by telling his mum, she could soften the blow for his dad, who doesn't like gay people. Eric was shocked when his mum started to get angry with him; he realised that she wasn't as open-minded as he'd thought. Eric wanted to run away from home. He knew things would just get worse when his dad got home.

2 Ahn's parents or guardians have moved to Australia to ensure their children get access to a good education and to set them up for life. They have gone without financially for so many years so their kids could go to the best private schools. Ahn has been talking, with the careers counsellor at school and has decided that next year he wants to go to TAFE and learn how to be a chef. He was so excited to tell his parents when he got home, but he didn't anticipate their horror, or their anger at having 'wasted' their money on sending him to a private school so he could be a lawyer or doctor.

Shutterstock.com/BearFotos

Figure 5.12 Eric was shocked when his mum started to get angry with him; he realised that she wasn't as open-minded as he'd thought.

3 Sarah has known her parents/guardians don't think much of school and where it can take you, but Sarah has a great relationship with one of her teachers, who has mapped out a way that she can achieve her goal of being a teacher. She is so excited that this could actually happen. But when she tells her parents her plans, they get angry with her, saying she knew she would have to leave school and get a job to help support the family (Sarah has four younger siblings). Her parents want her working full time, as neither of them are able to work and living on a government pension is hard.

Discuss

1 For each conflict, describe the different values that have led to the conflict.
2 Evaluate the impact that culture, religion or beliefs can have on family relationships during adolescence.

Culture

In addition to dealing with the developmental aspects of adolescence, Australian young people who come from a different cultural background may need to deal with their family's expectations about maintaining the values, customs and traditional roles and responsibilities of their heritage. This may lead to an identity crisis and conflict in their relationships.

Australia is a multicultural nation, and many of the people you meet and form relationships with will have a different background from yours. Developing an understanding about different religions and cultures helps us to understand and respect people from different cultures. Greater understanding and acceptance of the fact that everyone is different helps build tolerance and empathy. Ultimately, everyone is a person and deserves to be treated with respect, regardless of their ethnicity, religion, race or sexual preference.

Shutterstock.com/GraphicGeoff

Figure 5.13 We are all people, regardless of our ethnicity, religion or race.

UP AND MOVING: Personal, Social and Cultural Identities

In your groups you will be given dice to determine a character for you to do this activity with. You will roll the dice 3 times. You can only use each number once. If you do roll the same number, you will need to roll again.

- The first time you will roll will determine your character's Personal identity. For example, if you roll a 3 your character's personal identity is Introvert.
- The second time you roll will determine your Social identity and
- The third time will determine your Cultural identity.

	Personal	Social	Cultural
1	Speaks 5 languages	Goes to nightclubs/bars	First Nations
2	Non-binary	High school teacher	Greek
3	Introvert	Frequents the dog park	Sudanese
4	Athletic	Member of a sporting club	Muslim
5	Transgender	Goes on social cruises	Indian
6	Influencer	Accesses social media	Aussie

When you have the three identity components, discuss the following questions:

1. What stereotypes are there in society about your character?
2. How do these stereotypes impact our identity?
3. What would you like the world to know about your character's identity?

Pair up with another group with a different character. Imagine that the two characters got into a relationship. Discuss the following questions:

1. What would you need to talk about to help your partner understand your identity?
2. How could you provide support and be understanding of your differences?
3. What would a respectful relationship look like in this relationship?

1 Discuss the impact your family have on your relationships with friends, partners and other family members.
2 Discuss how your peer groups have an impact on your relationships with friends, partners and family.
3 Discuss how expectations about relationships from family and friends impact on your identity.

1 Discuss how your peers have an impact on your identity.
2 Discuss how you felt talking about the challenges for adolescents.
3 Justify whether these challenges the same the world over. Give some examples from other countries.

1 Discuss what would happen if the values of your friends and family are different to yours.
2 You have noticed that your friend isn't hanging out with you as much as they used to. You try texting them, but don't get a reply. You are getting frustrated because you know your friend has started going out with someone and is spending more time with them. In a few days you see that they are Facebook official and realise you've been dumped! You go over to their place to confront them.
   a What do you say?
   b How will you get your point across when they just want to spend all their time with their new partner?
   c How do you leave the conversation? What have you both decided?

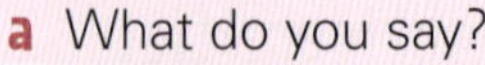

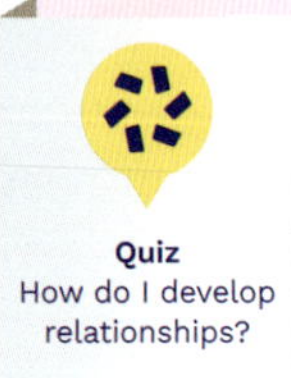

Quiz
How do I develop relationships?

HOW CAN I BE RESPECTFUL IN RELATIONSHIPS?

Shutterstock.com/CartoonDesignerFX

Figure 5.14 Communication gets the job done.

Relationships require work to help them grow; you need to contribute to a healthy relationship in order to make it stronger. Strong relationships can withstand a lot, but if you aren't willing to be an equal member of a relationship, then you may find it is not working. Respectful relationships are those that have a healthy give-and-take, where the people in the relationship show traits such as loyalty, empathy, understanding, acceptance, trust, confidentiality and equity. Relationships help a person build their identity and are important in developing self-esteem.

FORMING RELATIONSHIPS

Adolescence is a time when you become sexually aware of yourself and others. You may experience new feelings and become sexually attracted to other people. You may start to think about dating, sexual intimacy and making a commitment to one person in a relationship. Often there is peer pressure to begin dating or to go out in groups of partner couples. There are also benefits to being single and feeling confident about this.

FACE TO FACE Developing Relationships Online

Worksheet 5.6

Scaffold
Single vs dating

Stella is 15 and has noticed recently that her friends seem to be doing more things without her. She's feeling left out – just last weekend they all went into the city without inviting her. She hasn't told her guardians about what is going on, even though they ask after some of her old friends and what they are up to.

Stella is spending much more time by herself in her room, and especially online on several social sites, trying to connect with new friends. She was excited when a good-looking 16-year-old boy, Tom, liked some of her messages, then DM'd her. They struck up a conversation, and Stella started to confide in Tom about what's been happening with her friends. Tom was very sympathetic, saying he couldn't understand how a cool girl like her could not be liked, that he's never met anyone as special as her, and he really notices how mature she is. In fact, he suggests that maybe her friends are excluding her because they are jealous, and just not as grown up. Stella was flattered that someone as good-looking as Tom would fancy and really get her, and agreed to meet up 2 weeks later. They keep chatting online, mainly really deep chats late into the night. However, when she meets up, to her surprise, Tom is a 45-year-old man.

Scenarios developed by Elizabeth Clancy and Bianca Klettke at Deakin University (2022), from research and with thanks to peer-aged focus group contributions.

Discuss:

- How did Stella not know he was a much older man?
- How can you be sure you know who you are talking to?
- What would you do if you were Stella and this man introduced himself as Tom?

UP AND MOVING The perfect partner

As a class, brainstorm all the qualities that you would look for in a partner. Rank the qualities on the board from most important in a friendship/relationship, to least (but still important). Write each quality on a card.

As a class, use these cards as bricks and build a class wall with them, placing the ones that are most important on the bottom of the wall, and gradually moving up to the top layer. Then discuss the following questions:

1. Why are some qualities more important than others?
2. Consider what would happen if one of the qualities on the top of the wall (one of the lesser important qualities) was lost. For example, if 'they share your vision' was lost in a relationship, what would this mean for your friendship/relationship.
3. Consider what would happen if one of the qualities on the bottom of the wall (one of the more important qualities) was lost. For example, if 'trust' was lost in a relationship, what would this mean for your friendship/relationship? What would it do to your relationship wall.
4. Is it realistic to expect a partner to have all the qualities you are looking for?
5. Is it possible for you to have all the qualities someone else is looking for?

Intimacy

People who care about each other are said to have formed an intimate relationship. This does not necessarily mean they are sexually active. There can be intimate friendships where two people feel comfortable telling each other their innermost thoughts and feelings, and accept each other for all that they are as individuals. This is a **platonic** friendship. Individuals who are in caring, loving relationships are more likely to develop intimacy where they feel they can be appreciated for being themselves without fear of being ridiculed or hurt.

platonic a relationship with no sexual or romantic contact

WELLBEING CHECK IN → WRITING A GRATITUDE LETTER

This activity has been developed in collaboration with Dr David Bakker of MoodMission

Identify

We all have people in our life who we are grateful for in some way. But we don't always thank them.

Understand

Expressing thanks is a powerful prosocial behaviour. It brings people together and strengthens relationships. But sometimes things go unsaid or unexpressed, and the people who make positive impacts on our life go unthanked. Writing a letter of gratitude can make both the writer and the receiver feel good.

Practise

1 Think of someone who you are grateful for but haven't necessarily shown a lot of thanks. They might be a family member, friend, teacher, sports coach, bus driver, etc.
2 Once you've thought of your recipient, write them a short letter expressing your thanks. It's usually best to hand-write this, as it's more personal that way, but you could also type it and print it out if preferred.
3 If you can, deliver your letter in person. You might feel a little shy or embarrassed about doing this, but trust us, you will get a nice reaction from your letter's recipient.
4 When you deliver your letter, you may want to ask your recipient to read it while you're there.

Reflect

Did you enjoy this experience? If you did, try making this a habit, writing a few letters a year. You might even get some letters in return.

RIGHTS AND RESPONSIBILITIES

When establishing new relationships, it is important to have a clear idea of both your and the other person's expectations. Every relationship has its own rights and responsibilities. In 1948, the General Assembly of the United Nations adopted and proclaimed the 'Universal Declaration of Human Rights'. Examples of these rights include the following statements:

- **We are all free and equal:** We all have our own thoughts and ideas. We should all be treated in the same way.
- **No one should be discriminated against:** Everybody has rights that need to be acknowledged, no matter what their gender, race, religion or sexual preference is.
- **We are all equal before the law:** The law is the same for everyone. It must treat us all fairly.

FACE TO FACE Rights and responsibilities charter

If you had to design a charter of rights and responsibilities for relationships, what would you include? The relationships could be with your family, friend or with a partner. Complete your charter in a table similar to the example provided or use the scaffold.

Rights	Responsibilities
Be respected	Behave respectfully and respect the rights of others

1 List one right and one responsibility you feel are most important in building relationships.
2 Share and discuss your choice with a partner. Now you should have two rights and responsibilities in your charter.
3 Discuss these choices in a group of three, then justify your choices to the class. Develop a class charter of rights and responsibilities for relationships.
4 How might the charter vary if it was applied to a platonic friendship compared with a more intimate relationship with a partner?

Scaffold
Rights and responsibilities

ROLE OF POWER

The *Oxford Dictionary* defines power as 'the ability or capacity to do something or act in a particular way; to direct or influence the behaviour of others; or the physical strength and force exerted by something or someone' (Source: Lexico.com). Power can be used positively; for example, enforcing the school rules in order to keep students safe. It can also be used negatively, such as when one person may pressure the other to engage in sexual activity.

Everybody deserves to be in a respectful relationship, but sometimes relationships are not respectful. It can be difficult to know if a relationship is respectful, especially in its early stages. Actions that we initially think are loving might be possessive or controlling. For example:

- *He wants to know where I am all the time.* Why does he need to know this? Is he keeping tabs on you and who you are with?
- *She wants to spend all her time with me.* Why does she want to do this? What about her friends? It's not a good idea to dump your friends for a new partner.

- *They won't let me pay for anything.* This might sound great, especially if you don't have any money, but are they controlling what you are doing as well, and making all the choices? Do they expect something from you because they paid?

Sometimes the relationships that you have been exposed to in the past can have an impact on how you behave in future relationships. For example, if you see your father being abusive to your mother, you might think that is the way males treat females. If you see your father/guardian acting in a loving, caring manner, on the other hand, you might copy or expect this in your relationships.

If you felt you were in a relationship that was not respectful, what would you do? It is important to be able to talk to your partner. If they won't talk or don't want to talk about it, you might need to think about whether they are right for you. You might also be feeling under pressure to be in a relationship because all your friends are, and you want to feel loved as well. But it is not worth getting into a relationship that is not respectful, ever.

Why won't you talk to me?

Call me!

Are you with Simon?

Where are you?

Pick up your phone now

Why aren't you answering my calls?

Shutterstock.com/Bogdan Sonjachnyj

Figure 5.15 Is this love?

FACE TO FACE Maintaining respectful relationships

Working in groups, read the scenarios provided. In each scenario you find yourself in a difficult position. You have a relationship with this person; how do you keep it respectful?

- Your friend Ranjit has been going out with his partner almost as long as you have with yours. He tells you that he is going to have sex with his partner for the first time this weekend. You have heard through some friends that his partner is going to break up with him on Monday.
- Your sister Bec is going to have some friends around while your parents guardians are away. She is supposed to be looking after you and your younger brother. You have heard her talking to her friends about bringing alcohol and drugs to the party.
- Your parent or guardian has given you an ultimatum: 'stay at school or get out of the house!'. You want to leave home because of the abuse, but you are afraid that if you're not there your younger brother will suffer like you have.
- You start to tell your teacher about some things your friend is saying and doing. She interrupts you and tells you about mandatory reporting. This means that if you say something and she believes someone is at risk of harm, she will have to tell someone else. You realise she thinks you are talking about yourself. You want to get help for your friend as you think it might be really serious.

Discuss the following questions about each situation:

1. What do you do right now?
2. What choices do you have in this situation?

3 What part does empathy play in these conversations? Explain how someone in one of the scenarios might show empathy.
4 Who in the scenarios has power? How can someone who doesn't have much power themselves access power?
5 Maintaining respectful relationships can be hard, especially in difficult situations like these. What steps would you take to ensure the relationship stays respectful?
6 Where would you go for help in dealing with each of these situations?

Abuse in a relationship

When one person exerts control and there is a lack of respect, there can be a power imbalance that might lead to an unhealthy or even abusive relationship. Someone who loves and cares for another would not participate in abusive behaviour. Table 5.2 shows signs of abusive behaviour.

Table 5.2 Signs of an abusive relationship

Types of abuse	Examples
Verbal	insults, taunts, name calling, teasing
Psychological/emotional	threats, controlling behaviour, jealousy, blaming, **gaslighting**
Physical	hitting, grabbing, shoving, pushing
Social	exclusion from group social activities, ignoring your presence, speaking about you while you are nearby
Sexual	any unwanted and uninvited physical advances or sexual activity
Cultural, religious or spiritual	not allowing someone to practise their cultural, religious or spiritual traditions or beliefs
Financial	controlling all the money, expecting someone to always pay, keeping family in constant debt

gaslighting psychological manipulation of a person usually over an extended period of time that causes the victim to question the validity of their own thoughts, perception of reality, or memories and typically leads to confusion, loss of confidence and self-esteem, uncertainty of one's emotional or mental stability, and a dependency on the perpetrator.

Source: Merriam-Webster, https://www.merriam-webster.com/dictionary/gaslighting

Video
Gender stereotypes: What are gender stereotypes and what impact can they have on us?
How do gender stereotypes contribute to gender-based violence?
Watch the video and start the discussion.

Gender-based violence

'Gender-based violence', 'violence against women and children', 'domestic violence' and 'family violence' are all names for the frightening societal issue that is currently rife around the world; in Australia almost one woman per week is being killed by their current or former intimate partner. The federal government has acknowledged that gender-based violence is an area of concern and has devised 'The National Plan to Reduce Violence against Women and their Children 2010–2022' to uphold the rights of women across the nation.

Gender-based violence statistics

Since the age of 15:

3 in 10 women and 4 in 10 men
have experienced **physical violence** by a partner, other known person or a stranger.

1 in 5 women and 1 in 20 men
have experienced **sexual violence** by a partner, other known person or a stranger.

The most common location for violence to occur

For women:
her home or the perpetrator's home

For men:
in a place of entertainment or recreation

Perpetrators

1 in 3 Australians experienced violence by a male perpetrator.

1 in 10 Australians experienced violence by a female perpetrator.

1 in 4
women have experienced violence by an intimate partner. This could be someone they live with, have previously lived with, or have never lived with.

1 in 6
women have experienced physical violence from a partner.

1 in 4
women have experienced emotional abuse from a partner.

Based on Australian Bureau of Statistics data

Figure 5.16 Gender-based violence statistics

Violence and gender

There is a gendered nature to violence in Australia, and around the world. The statistics show us that 95 per cent of victims of violence, whether they are male or female, experience this violence from a male perpetrator. There are some very significant ways that male violence is different.

Males are more at risk of violence:

- from another male
- at the hands of a stranger
- in a public place
- as a one-off incident.

Women are more at risk of violence:

- from a male
- at the hands of a current/former partner
- occurring in a private place (like the home)
- on an ongoing basis.

Gendered norms are everywhere in our communities, and they can have a real impact on the way that people are respected, or not. Men, stereotypically, are meant to be more aggressive, strong, loud. They are expected to be CEOs and CFOs of major businesses, judges, surgeons, etc. Women, on the other hand, are stereotypically cast as the opposite – feminine, caring, calm, passive, weak, quiet, and are expected to take on caring roles, such as teacher, nurse, etc. Even today, women are not paid as much as men, and few sit at the head of companies compared to men. This gap has an impact on how males and females feel about themselves, and therefore how they conduct themselves. Can you think of any gender stereotypes that keep males and females in certain roles? Do you think that gender stereotypes will still exist in 50 years? Why/why not?

Worksheet 5.7

FACE TO FACE 'Like a girl'

Have a look at the clip 'Like a girl'. In pairs, analyse the advert.

1 What is your reaction?
2 How would you feel if you were a girl and someone said 'don't run like a girl'?
3 How would you feel if you were a boy and someone said 'don't run like a girl'?
4 Why do some people say things like this to put others down?

Weblink
Like a Girl

To do the work required to eliminate gender-based violence, we first need to identify the things that promote gender-based violence. We can then address these to diminish the violent effects.

UP AND MOVING Public service announcement (PSA)

Design a new product or service that helps young people to develop or maintain respectful relationships. Develop a 30-second public service announcement (audio or video). There are resources on the internet that you can use as examples, such as 'Just Spray it PSA'.

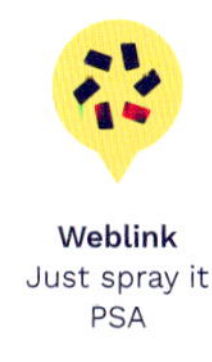
Weblink
Just spray it PSA

Weblink
Check out the facts and figures on gender-based violence from Our Watch.

FACE TO FACE Gender-based violence

Explore the Our Watch figures on its website by following the weblink. Then, in small groups discuss the following:

1 They say that this is a preventable problem. What do they mean by this?
2 Are you surprised by any of the statistics? Why/why not?
3 Consider how gender-based violence impacts you, even if you don't know of anyone who has been affected.

Worksheet 5.8

Discrimination and harassment

All relationships come with rights and responsibilities to ensure each person feels equal and respected in the partnership. This leads to positive relationships, where there is an atmosphere of open communication, trust and honesty.

Any relationship that demeans or intimidates someone hinders the establishment of these qualities of effective relationships.

Weblink
Change the story: A shared framework for the primary prevention of violence against women in Australia

Discrimination

Discrimination can be defined as treating people unjustly or being prejudicial in your treatment of people, based on their race, age, gender, sexual preference or because they have a disability. Fear and ignorance are often the most common causes of discrimination – some people fear things that are unknown or different to them. Discrimination occurs when someone is treated less favourably based on a particular personal characteristic.

Harassment

Harassment is any type of behaviour that is intimidating, humiliating or threatening. Harassment may be in the form of comments, stalking, texts, emails, letters or notes. These are some of the forms harassment can take:

- personal harassment – making fun of someone based on their appearance
- racial harassment – making negative comments about someone's ethnic origins
- sexual orientation harassment – treating someone differently based on their actual or perceived sexual preference in a partner.

Sexual harassment involves any unwelcome sexual behaviour that offends or intimidates someone. It is flirtation or sexual contact that is not consensual and mutual. It may involve staring, inappropriate touching, jokes of a sexual nature, sending explicit material via text message or email, or intrusive questioning about someone's personal life.

Discrimination and harassment are also discussed in Chapter 6, pages 248–9.

Worksheet 5.9

Homophobia

Adolescence is a time when young people are questioning who they are, what they stand for and how they would like to live their lives. It can be a time when the aspects of sexuality are explored, including sexual preference and gender identity. Not everyone is the same, and this may cause some people to fear or dislike those who walk a different path to them. If this fear or dislike is in regard to people who are same-sex attracted, it is termed 'homophobia'. Homophobia is a type of discrimination.

FAST FACTS

Research regarding SSAGQ (same-sex-attracted and gender-questioning young people) has shown that:

- 57 percent of SSAGQ young people had experienced verbal abuse
- 15 percent had experienced physical abuse
- 23 percent had experienced sexual assault or harassment because of their sexual or gender identity
- 60 percent felt uncomfortable or unsafe at school.

Writing Themselves In 4 – The Health and Wellbeing of LGBTQA+ Young People in Australia, Adam O. Hill et al., La Trobe University, February 2021

Most of this discriminatory activity takes place in schools. No one should feel unsafe within a school environment. The whole school community, including teachers, students, parents, guardians and staff, should be involved in supporting conduct that is non-discriminatory and inclusive of all its members. It is the law. This may involve using appropriate language to describe sexual orientation, not participating in bullying, either as a bully or a silent bystander, and teaching curriculum content that affirms diversity and equal opportunity.

There are many LGBTIQA+ festivals and events across the country, where people can celebrate gender and sexual diversity with others. The Australian Pride Network provides a list of celebratory events in all states and territories: Sydney Gay and Lesbian Madi Gras, WA PrideFEST, Brisbane Pride Festival, Melbourne Midsumma Festival, NT Parrtijima, ACT Fair Day, TasPride Festival and SA Feast Festival as examples. These events give LGBTIQA+ voices a way to express themselves in safe and supportive environments.

Weblink
The Australian Pride Network: LGBTIQ+ celebrations around the country

Shot by Suzanne, shotbysuzanne.com.au

Figure 5.17 Everyone in the community can help to reduce discrimination and harassment.

1 Define what a respectful relationship is.
2 Identify whose responsibility is it to ensure a relationship is respectful.
3 List 10 behaviours that you would expect in a respectful relationship with someone. How might you negotiate these at the beginning of a relationship?

1. Consider why you think there is such a problem with gender-based violence around the world.
2. Consider whether you have seen any changes over the last few years that you now recognise have been helping Australians deal with gender-based violence. Consider the news, media, advertisements, respectful relationships in school, fundraising, etc.
3. If you were asked to solve the problem, make a list of the first five things you would do.

In Parliament the Australian Prime Minister gave an apology speech to Brittany Higgins and many other women who were not able to speak up. Brittany alleged that she was raped in 2019 in the office of her then boss. When a review of sexual misconduct in parliament was held they 'heard from 1,723 individuals and 33 organisations'.

The Prime Minister said:

'I am sorry, we are sorry. I am sorry to Ms Higgins for the terrible things that took place here. The place that should have been a place for safety, that turned out to be a nightmare'.

1. Discuss what a place of safety means in a workplace.
2. Discuss what the prime minister means by a 'nightmare' and what are the implications for her working life going forward.
3. Discuss why you think that so many women were unable to come forward to tell their stories in public.
4. Discuss the implications for Brittany being a spokesperson and the bravery that took.

Weblink
Full article about the Prime Minister's apology speech to Brittany Higgins.

HOW CAN I MAKE DECISIONS ABOUT SEXUAL RELATIONSHIPS?

Shutterstock.com/Rtimages

Figure 5.18 Decisions about intimate relationships can be difficult.

One of the most common decisions in relation to intimate relationships is whether you wish to participate in sexual activity with your partner.

The complex decision of whether or not to have a sexual relationship needs to be made rationally. You need to analyse several factors:

- ask yourself if you are ready for a sexual relationship
- consider the alternatives available to you
- gather as much information as possible on each alternative
- carefully contemplate the possible outcomes of each alternative
- make a valued and well-thought-out decision.

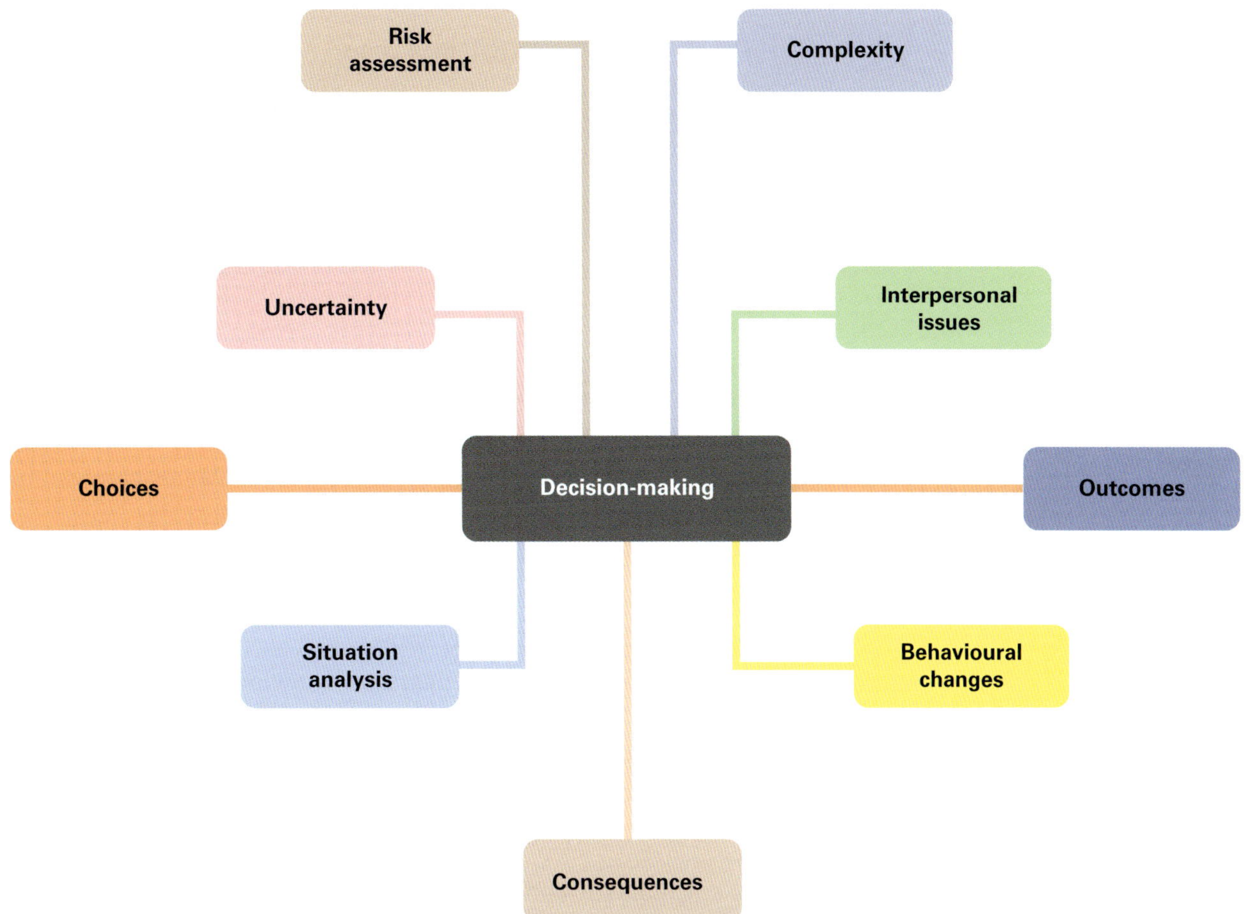

Figure 5.19 Decision-making needs to consider a number of different factors.

This may take time. You should never feel forced into making a rash or impulsive decision. You need to acknowledge your own feelings, values and attitudes in the process. Also acknowledge your body's responses; butterflies in your tummy may be a sign you are nervous about something good, but it may also mean that you are doing something you are not ready for. Listen to your body, it will tell you a lot! Becoming sexually active before you are ready could hurt you physically, socially and emotionally. Remember, not all intimate relationships have to involve sexual activity, and engaging in sexual activity does not prove that you care for that person!

FAST FACT

The National Survey of Australian Secondary Students and Sexual Health in 2018 found that the majority of students have experienced some form of sexual activity by Year 12. This includes deep kissing, being touched or touching other's genitals, giving or receiving oral sex and penetrative sex. Although the media perception is that almost all young people are involved in sexual intercourse, just under half (approximately 45 per cent) of Year 12 students and approximately 66 per cent of Year 10 students have not experienced sexual intercourse.

FACE TO FACE Teen sexual behaviour

1 Where do we get ideas about how many young people have had sex in our communities?
2 Discuss how this compares to the statistic above for Year 10s and Year 12s.
3 What pressures do young people face to have sexual intercourse early?
4 Why do you think this is the case?

Worksheet 5.10

Weblink
Go to Youth law for any questions you have on the Law.

Video
Consent: How can we ensure enthusiastic consent when engaging in sexual activity with a partner? Watch the video and start the discussion.

Worksheet 5.11

Weblink
Watch the cup of tea analogy to giving consent. What are your thoughts?

Worksheets 5.12 and 5.13

When deciding on the right time to begin a sexual relationship, there are a number of factors to be considered:

- **Personal values:** decisions and actions that conflict with personal values can lead to feelings of discomfort, uneasiness and guilt.
- **Readiness:** emotional and social maturity need to be on par with physical maturity.
- **Communication:** thoughts and feelings should be discussed openly with each other. There should also be the option to say 'no' at any time.
- **Risks of sexual behaviour:** there must be awareness and knowledge of risks of pregnancy and contraction of a sexually transmissible infection (STI).
- **Consent:** both people in the relationship **must say 'yes' to each activity, each time**.

CONSENT

Consent (often referred to as enthusiastic consent) **MUST** be gained from all people involved in any sexual activity, otherwise the sexual activity is against the law. Make sure you understand the laws relating to consent. The legal age for consensual sex varies from 16 to 17 years of age across Australian states and territories. Along with age restrictions, there are also other laws that you must abide by, such as:

- They must be able to give free consent – they are not being bribed or forced to say yes.
- They must be conscious to be able to say yes – they are not asleep or unconscious.
- They must not be affected by drugs or alcohol – if they are affected, they cannot give consent.
- They must fully understand what is about to happen to them and who is participating.
- If they do or say nothing, they have not given consent.

Consent is also something that can be taken away at any time. If you give consent to participating in any sexual activity with another person, you can say stop at any time during that activity, and your partner needs to stop. This is the law. If you get consent, it is important that you ask if what you are doing is okay. Any new activity requires new consent. You also need to gain consent every time – just because someone gave consent in the past does not mean you can assume consent in the future.

UP AND MOVING Is this Consent?

Read each scenario and decide whether the person has given consent for the sexual activity:

- Kris has been drinking and she thinks her long-term partner wants to have. Kris is not quite with it, so she goes to bed. She wakes up a few hours later and feels like she has had sex but doesn't remember any of it.
- Dani has autism spectrum disorder and takes the bus to school every day. She is the first to be picked up and the last to be dropped off. The driver has been talking to her each day and they are becoming good friends. She told him she has been wanting to buy air pods for a while now. He suggests that at the end of the ride they have oral sex and he will buy her the air pods. She agrees.
- Ollie is feeling so much pressure from his friends to get a girlfriend, they all have one and tease him because he doesn't. He doesn't feel ready. At a party his friends had set him up and shoved him and a girl who was willing to kiss him in a cupboard. She kissed him before he even realised what was happening.

Discuss:

- Was consent given?
- Was the person in a position to give consent?
- Is the sexual activity legal?
- How could you rewrite the scenario to include a statement of consent?

Being able to say 'no'

Assertiveness is the ability to speak up for what you think and feel in a respectful way. It involves expressing your feelings clearly and in a non-aggressive manner. Try to use 'I' statements, such as these examples:

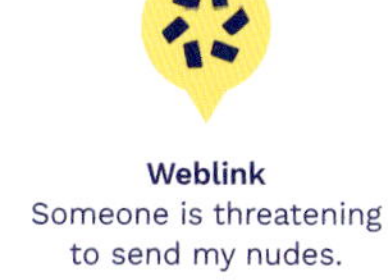

Weblink
Someone is threatening to send my nudes.

- Right now, I'm feeling …
- I don't like it when …

It's okay to say 'no' to someone pressuring you into an activity if it is an act or behaviour that makes you feel uncomfortable. There are ways of defusing this – you could respond with some humour, make another suggestion or walk away from the situation. Staying true to your values and standards is an important part of emotional maturity.

In terms of a sexual relationship, no means no! This is easy to say, but when the relationship is important to you, it is not always easy to say no. You need to develop the skills to do so.

Remember, it is your body, and you have the right to refuse sexual contact at any time.

Alamy Stock Photo/Blulz60

Figure 5.20 You have the right to say 'no' to any act or behaviour that makes you feel uncomfortable.

Newspix/David Sproule

Figure 5.21 Drinking alcohol inhibits your ability to make informed decisions.

Drinking

Drinking alcohol increases the risk of sexual harm because your inhibitions may become lowered, and you are less likely to take control in unwanted situations. Your ability to protect yourself against unplanned, unprotected or forced sexual activity is reduced. Safe sexual practices such as the use of a condom are also less likely to be followed.

If someone is affected by alcohol or drugs, they cannot clearly communicate and decide whether they want to engage in sexual activity, and thus they can't give 'informed consent'. Anyone who has sex with a person who is affected in this way is committing sexual assault, possibly rape, which can result in criminal charges.

10:35
Tuesday, August 23

Sexting is the act of sending or receiving sexually suggestive or explicit messages, videos or photos.

Swipe up to unlock

Figure 5.22 Sexting may be in the form of words or images.

Sexting

Sexting is the sending of 'sexy' texts. This might be in the form of words or images on any form of mobile or online device, such as a mobile phone, Skype and social networking sites such as Facebook, Twitter, tumblr, flickr, YouTube, Instagram or Snapchat. It is illegal to take, send or receive sexual images of anyone under the age of 18. Once you send an image or text to someone, you have lost control of it and it could easily end up in the wrong hands. Even if you have set the image to disappear after a few seconds, people may take screenshots and pass the image on. Deleting material from your Facebook page or phone does not necessarily mean it is no longer accessible. It becomes part of your digital footprint and may have consequences in the future.

CASE STUDY ONLINE IMAGES

Identify

Examples of image-based abuse include:

- A current or ex-partner shares an intimate image on social media without your consent.
- You send your intimate image to a friend or acquaintance and that person posts it online without your consent.
- A work colleague Photoshops an image of you making it look like it is an explicit image of you and shares it broadly via email.
- A violent or abusive partner posts explicit images online to further hurt or humiliate you.
- A person threatens to share an intimate image if you do not give them money, perform a sexual act or provide more images.

Weblink Quick guide on image based abuse.

Understand

5

1 Pressured by mates to share...

Sam is hanging out with his mates after school. They're all checking their socials. Sam's phone pings, and he smiles – it's a message from his girlfriend Zala. It's their 3-month anniversary and things are going really well - they message every day, more when they can't meet up, sometimes just memes or jokes. There's also a few hot pics that Zala sent him a week ago.

Sam's looking forward to meeting up with her on the weekend at a party. Zala's bought a new dress and sends Sam a few shots of herself – she looks awesome and he's excited. Sam's mates see him smiling at the phone and start teasing him when they realise it's Zala – they want to see photos of her. Sam shows them these new shots and they start asking about whether there are 'other' types of photos. Sam hasn't told anyone about the sexy pics Zala sent him. Just then, Max grabs his phone and starts to scroll finding those sexy pics. He jumps up and starts showing the pics to the group. Sam's not sure what to do. The shots look amazing and Zala's super-hot, so he feels proud to be her boyfriend, but he knows she wouldn't want the others to see those pics. Max pushes him to share the shots with the group, given how hot she looks, and Sam's heart sinks.

2 Receiving unwanted images.

Luke is 13 and just started Year 7. He's worried about a lot of things: most of his friends went to a different high school so he doesn't know anyone here. How will he fit in, get to know people, make friends? Along with all the new changes happening, he is now catching the bus to and from school, so his parents have gotten him a phone. Luke's excited, he's finally got social media, like the other kids at school, and can keep in touch with his old mates from primary school.

By late February, he knows his way around, has got a group of mates and feels like he is starting to fit in. But, one afternoon on the bus home, he gets a text from an unknown number. Assuming it is a new friend from school, Luke clicks on the notification, but is confused when he realises it's a nude of a random girl, whose face he can't make out. Luke's uncomfortable and doesn't know how to react. Should he tell his mates? Show them the picture? Respond?

3 Received an unwanted image, shared it with her friends...

Ella is in Year 9. She's got some really good friends who do everything together, especially her best friend Tiana. A few girls in the group have boyfriends, a couple who Ella knows from school, and some others she's seen around at parties. Tiana has been dating Josh for 2 months, after they met at a friend's 15th. Ella only knows Josh well, enough to say 'hi' at parties, but only knows what Tiana tells her.

After playing netball on Saturday with Tiana, she sits down to catch up on some homework. Her phone lights up with a Snapchat notification, from Josh. Ella's surprised, as they haven't really talked outside hanging out with Tiana and her friends. She opens the snap, to see that it is a dick pic. Ella is shocked, and takes a photo of the conversation on another device, sending it to Tiana straight away, hoping to warn her friend about Josh.

As she tries to finish off her homework, Ella starts to get worried. Tiana always replies straight away to her messages, and it's been a full hour since she sent the photo. She's also nervous about catching up tomorrow, since a huge group of friends including Josh and Tiana are all going to the beach. What if Josh has gotten to Tiana first? What if he's twisted the story? What if her best friend thinks she's done something wrong? What will all her other friends say?

Developed by Elizabeth Clancy and Bianca Klettke at Deakin University (2022), from research and with thanks to peer-aged focus group contributions.

Discussion

Discuss in your groups:

1 What do you do right away?
2 Consider whether this is image-based abuse.
3 Propose a strategy to deal with the situation in the short term
4 Consider whether you would go to anyone for help with the situation and who that would be.

Weblink
Righting the wrong.

Access the following website called Righting the wrong
https://www.esafety.gov.au/key-issues/image-based-abuse/quick-guides/righting-the-wrong

1 Develop a list of 10 things you could say to a friend that has shared an intimate image they didn't have permission to share.
2 Develop a list of 10 things you would say to a friend who is the victim of image-based abuse.
3 Propose three reasons why someone might send an intimate picture/recording of someone else without permission.
4 Investigate the laws for image-based abuse.
5 Investigate what your school rules and consequences for image-based abuse are.

1 Make a list of all the things you need to consider prior to starting an intimate or sexual relationship with someone.
2 Define what consent is and discuss how do you make sure you have it.
3 Discuss what you would do if someone is threatening to post nude photos you have sent them online.

1 Currently, many young people are being coerced to send nude pictures of themselves. Discuss why you think this is.
2 Currently, most young people who start having sexual intercourse are using condoms. Discuss why you think this is.
3 Come up with three arguments for and three arguments against the following statement: You cannot give consent to sexual activity unless you have a blood alcohol level (BAC) of below 0.05.

Grace Tame was sexually assaulted as a child and again by a teacher at her high school. She was the Australian of the year in 2021. Watch from 2.20-4:42 of her speech to the National Press Club.

Weblink
Grace Tame's National Press Club speech.

1 Discuss why someone would 'turn a blind eye' to sexual assault that was taking place.
2 Discuss why it is so difficult for someone to 'publicly come forward with their stories'.
3 Discuss what the impact of survivours such as Grace Tame, Brittany Higgins, Chanel Contos and the #MeToo movement have had on how we talk about respectful relationships in our communities.

HOW CAN I HAVE SEX SAFELY?

There is no such thing as 100 per cent SAFE sex, unless you abstain entirely from any sexual behaviour. So it is important to practise SAFER methods to try to avoid things like sexually transmissible infections (STIs) and unwanted pregnancies.

SEXUALLY TRANSMISSIBLE INFECTIONS

If you decide to become sexually active, you are putting yourself at risk of contracting a sexually transmissible infection (STI), particularly if you practise unsafe sex. An STI is an infection that can be passed from one person to another through engaging in vaginal, anal or oral sexual intercourse. Most STIs are carried through body fluids, although some can be contracted through skin-to-skin genital contact. STIs are classified as: bacterial (such as chlamydia, gonorrhea and syphilis), viral (such as HIV genital herpes, genital worts and hepatitis B) or parasitic (such as pubic lice and scabies).

iStock.com/AmmentorpDK

Figure 5.23 You cannot tell just by looking at someone if they have a sexually transmissible infection.

Symptoms of an STI may include the following:

- pain when urinating
- unusual discharge from the penis or vagina
- sores, lumps or blisters around the penis or vagina that may be itchy
- unusual bleeding from the vagina or bleeding after sexual intercourse.

Most STIs have no obvious symptoms. Someone with an STI may look perfectly healthy – you cannot tell if someone has an STI just by looking at them. Using a condom helps to reduce the spread of STIs, however, they are not 100 per cent effective against all STIs. The most effective way to prevent an STI is to practise **abstinence**.

abstinence not participating in any form of sexual activity

If you are involved in a sexual relationship and you have had unsafe sex or think you may have an STI, you should seek medical help from your doctor, medical professional or a family planning clinic. Table 5.3 will give you some information about the types of STIs that exist. In the next part we will discuss two of these – chlamydia and HIV.

Chlamydia

One particular area of concern among young people is the increase in chlamydia infections. Chlamydia rates have quadrupled in the past decade, mainly among teenagers.

Table 5.3 STIs – what you need to know

Group	STIs	How transmitted	Prevention	Symptoms	Treatment
Parasitic	Pubic lice Scabies	Skin-to-skin contact	Condoms do not prevent against this, as the parasite lives in pubic hair or under the skin	Itchiness in the genital area	See a chemist to get some pubic lice treatment Hot wash all bedding and clothes just in case
Bacterial	Chlamydia Gonorrhoea Syphilis	Body fluids	Condoms used correctly, every time	Most people experience no symptoms; sometimes pain while urinating, different colour/ smell to discharge	Antibiotics can get rid of a bacterial infection. A serious case may need more than one dose.
Viral	HIV/AIDS – human immunodeficiency virus HSV – herpes simplex virus (herpes) HPV – human papilloma virus (warts) Hepatitis B and C	Skin to skin contact, Body fluids	Condoms can help but are not 100% effective. The skin outside the condom might be infectious	Some people experience no symptoms. Pre-diagnosis: HIV – flu-like symptoms HSV – herpes sores HPV – warts Hepatitis – fever, nausea, jaundice (yellow skin and eyes)	There are treatments now for viral STIs Immunisations for babies for hepatitis B

Chlamydia is an STI caused by the bacteria *Chlamydia trachomatis*. It can affect both males and females. It can be passed on by unprotected vaginal, oral or anal sex with an infected person. Many people who are infected with the bacteria do not realise they have it, as there are often **no** signs or symptoms.

Symptoms that females with chlamydia *may* experience include:

- burning or pain when passing urine
- unusual vaginal discharge
- pain in the lower abdomen
- bleeding or pain during or after sex
- bleeding or spotting between periods.
- sore or inflamed throat.

Without treatment, females can get chlamydial infection in their cervix, uterus and fallopian tubes. These infections are referred to as pelvic inflammatory disease (PID) and can lead to infertility. Chlamydia can be transferred to a baby during childbirth.

Symptoms that males with chlamydia *may* experience include:

- whitish or yellow discharge from the penis
- burning or pain when passing urine
- irritation or soreness around the urethra.

If left untreated, males can develop inflammation in the prostate gland, swelling in the testes and, possibly, infertility.

Chlamydia can be easily detected with a urine sample or throat swab and treated with antibiotics. Legally, any sexual partners need to be informed – they have a right to know! Make sure you practise safer sex during treatment (using a condom correctly, every time).

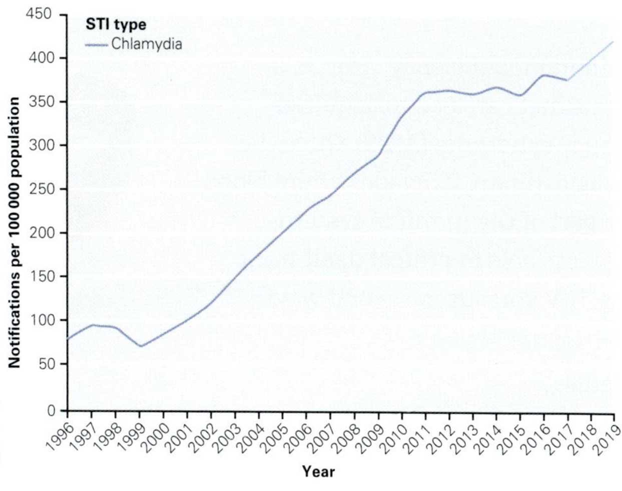

Figure 5.24 Rates of Chlamydia infections 1996–2019

INVESTIGATION

CHLAMYDIA MYTHS AND MISCONCEPTIONS

Purpose

To expose the myths and misconceptions about Chlamydia

Materials

You will need the Chlamydia myths in Figure 5.24.

Chlamydia myths

- Only women can get Chlamydia. True or false?
- You would know if you had Chlamydia. True or false?
- You can only catch Chlamydia once. True or false?
- Chlamydia will go away on its own. True or false?
- Chlamydia can be caught from the toilet seat. True or false?

Weblink Check out these websites if you need to check the myths.

Method

Discuss the following in your group:

1 Whether the statements are true or false.
2 Propose a reason why there are still myths and misconceptions about STIs.
3 Justify whether here is still a stigma associated with STIs. Is it a comfortable topic to talk about in your friendship group?
4 How do the myths and misconceptions impact the rates of Chlamydia in Figure 5.23?

Discussion

1 As a class create a strategy to make talking about STIs easier.
  a Consider overcoming stigma, myths and misconceptions
  b Consider whether age, gender, geographical, cultural and religious factors would have an impact on a strategy.
  c Consider the physical, mental, emotional, financial and sexual factors that might impact a strategy.

HIV

blood-borne carried in the blood

HIV, or human immunodeficiency virus, is a **blood-borne** virus that can lead to acquired immune deficiency syndrome (AIDS). Once the virus is in the bloodstream, it invades white blood cells, which are part of the immune system, leaving the body less able to protect itself from other infections. HIV spreads in several ways:

- unprotected sexual activity
- sharing needles
- from mother to baby during birth or breastfeeding.

Shutterstock.com/leolintang

Figure 5.25 The red ribbon has come to signify the fight against HIV infection.

HIV can replicate itself to continue to invade more white blood cells until there is little chance of the immune system fighting off 'opportunistic' infections such as flu or pneumonia. Once people reach this stage they are said to have AIDS. Most people infected with HIV feel well for long periods of time, and a blood test is often required to detect that someone has HIV or is HIV-positive.

Weblink
HIV in Australia 2020.

FAST FACTS

1 The number of new HIV cases continues to fall in Australia.

2 Australia is attempting to end transmission of HIV by 2025.

Although there is no cure for HIV/AIDS and no vaccination to prevent infection, there are many new forms of antiviral medications that can halt the disease's ability to multiply. Practising safer sex and always using condoms during vaginal or anal sex is the best way to reduce the risk of getting or transmitting HIV through sexual contact.

FACE TO FACE What to say?

If one of your friends told you they thought they may have contracted an STI, what advice could you offer? Who should they speak to and where could they obtain some assistance?

INVESTIGATION ⇨ HIV AROUND THE WORLD

Purpose

To look at the differences between HIV transmission across the world

Materials

Access the World Health Organization HIV/AIDS fact sheet.

Weblink
WHO HIV/AIDS fact sheet.

Method

1. Identify the area of the world that has the highest rates of people living with HIV. Consider why this might be the case.
2. Even though there is HIV/AIDS prevention work done all around the world, there are still approximately 37.7 million people living with HIV in 2020. Why do you think this is?
3. What makes certain groups of people more vulnerable to HIV/AIDS than others? Consider:
   a. young women and girls
   b. pregnant and breastfeeding women
   c. men who have sex with men
   d. indigenous peoples
   e. rural and remote people
   f. people of low socio-economic status (low income or low educational outcomes)
4. See if you can find the rate of transmission in your particular area.

Discussion

As a class, discuss why HIV/AIDS is still in the news. Consider the differences between Australia and overseas.

During COVID-19 travel from country to country was limited. Do you think this will impact the HIV statistics?

CONCEPTION, PREGNANCY AND BIRTH

Along with STIs, sexual activity can also lead to pregnancy.

FAST FACTS

1. The latest census figures state that teenage pregnancies are decreasing.
2. In 2017 there were approximately 6,600 live babies born to teenage mothers.
3. There are more babies born to Indigenous teenage women than non-Indigenous teenage women.
4. There are more babies born to teenagers living in rural or remote areas and areas of low socio-economic status (SES) than in urban and high-SES areas.

Weblink
Australian Institute of Health and Welfare Teenage mothers

In order to be informed and make good decisions, it is important to understand how you can become pregnant.

People with ovaries are born with about 40 000 immature eggs, or ova, already in each ovary. Over the course of their lifetime, usually just one egg is released on average every 28 days. It is released into the fallopian tube and is helped along by small fine hairs, cilia, on its journey to the uterus. If the egg (ovum) is not fertilised during this time, the uterine

menstrual cycle the monthly changes that occur in people with ovaries after puberty in preparation for ovulation and possible pregnancy

conception fertilisation of a human ovum (egg)

wall that has been building up in anticipation of receiving a fertilised egg will break down and be released as blood through the vagina. This is known as the **menstrual cycle**.

If the ovum meets a sperm during its journey, then fertilisation, or **conception**, can occur. A sperm may penetrate the egg and form a new cell – a blastocyst. During male ejaculation, 50 to 500 million sperm can be released. If the ejaculation occurs in or near the vagina, the sperm can enter and start the 'race' to the egg. The sperm have tails that allow them to swim from the vagina, through the uterus into the fallopian tubes. Sperm can survive in the vagina and uterus for up to seven days, but the egg lasts only for one day. Therefore, if unprotected sexual intercourse has occurred one to seven days before ovulation, pregnancy can occur.

Weblink Fertilisation

Only one sperm can fertilise one egg. The fertilised egg may separate into two on its journey to the uterus; when this happens, identical twins are formed with identical genetic material (from one egg and one sperm). If something happens and this process doesn't fully complete separating, you may get conjoined twins, where part or parts of the baby's bodies are fused. The babies are born joined and may require separating after birth. If two eggs are released at the same time and both are fertilised, fraternal (non-identical) twins are formed. The sex of the baby is determined by the sperm. If the sperm that joins the ovum contains the X chromosome, usually a female baby will develop. If the sperm contains a Y chromosome, usually a male baby will develop.

Shutterstock.com/Tinydevil

Figure 5.26 Conception: when an ovum meets a sperm.

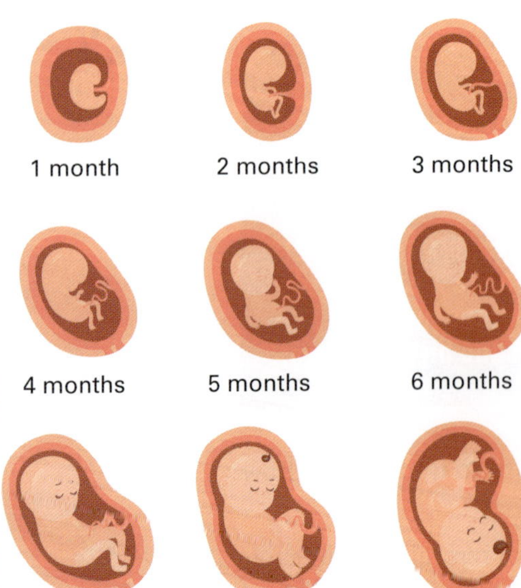

Shutterstock.com/Zhanna Markina

Figure 5.27 Growth of the foetus to 40 weeks

The fertilised egg continues to divide and, on reaching the uterus, implants itself into the uterine wall. It is now known as an embryo. The placenta is an organ that forms during pregnancy and is also attached to the uterine wall. Through the umbilical cord, the placenta provides nourishment to the developing baby and is also able to remove waste products. The placenta releases hormones and offers some protection against infection for the baby to continue to grow and develop. The developing baby is contained within a fluid-filled bag called the amniotic sac.

The baby continues to grow for 40 weeks. When the baby is ready to be born, labour begins with contractions of the uterus. Either the cervix dilates and the baby moves, ideally with its head down, through the cervix into the vagina, or the lower abdomen is surgically opened to perform a caesarean section. After the baby has been born or removed, the placenta (also known as the afterbirth) is expelled.

CONTRACEPTION

Contraceptive methods are used to prevent pregnancy (see Table 5.4). There are different types of contraception, and choosing the right type is a personal decision. There are four different methods of contraception:

- hormonal methods
- barrier methods
- timing methods
- permanent methods.

Contraception is a shared responsibility of both people involved in the sexual relationship.

9780170463102

Table 5.4 Some types of contraception

Contraceptive	Advantages	Disadvantages	Effectiveness	STI and HIV protection
Barrier methods				
Condom: a latex or polyurethane sheath that covers the erect penis and prevents semen entering the vagina or anus	Readily available, cheap, only needs to be used during sexual activity	May break during sexual intercourse	98 per cent when used correctly every time	Helps to protect against bacterial STIs and HIV
Female condom: a lubricated condom for women to insert internally	Only needs to be used during sexual activity, is easy, can be inserted earlier, does not restrict the penis	Are difficult to get and are expensive, may make some noise during sexual intercourse	95 per cent when used correctly every time	Helps to protect against STIs and HIV, provides more protection around the vulva for women and skin around penis for men
Hormonal methods				
Hormonal intra-uterine device (IUD): small plastic T-shaped device that is inserted into the uterus. There are two types of IUDs: those made of copper and those that release hormones (such as the Mirena IUD). The hormonal IUD thickens the mucus around the cervix, preventing sperm from accessing the uterus. It may also affect ovulation.	Stays in place for five years, may make periods lighter or disappear	Expensive, needs to be put in by a professional, hormonal side effects	99 per cent	No protection against STIs
The combined pill: a pill that contains oestrogen and progesterone; it prevents an egg from leaving the ovary	Can regulate menstrual cycle and alleviate period pain and acne	Need a doctor's prescription, has to be taken at the same time every day or the effectiveness is reduced	99 per cent if taken at the same time daily	No protection against STIs

Images, from top: Shutterstock.com/Yannick FEL; Shutterstock.com/nito; iStock.com/Lalocracio; iStock.com/claylib

Contraceptive	Advantages	Disadvantages	Effectiveness	STI and HIV protection
Hormonal methods (continued)				
The mini pill: a progesterone-only pill that makes the fluid around the cervix thicker, stopping the sperm from entering the uterus	Starts working quickly, is a low dose of hormone	Need a doctor's prescription, must be taken at the same time each day, hormonal side effects	99 per cent	No protection against STIs
Vaginal ring: a ring that contains oestrogen and progesterone that stops the ovaries releasing an egg every month, thickens the fluid around the cervix	Stays in the vagina for three weeks, and then out for one week, cannot feel the ring when inserted, usually lighter periods, can help acne	Need a doctor's prescription, is expensive, hormonal side effects, have to remember to change it	99 per cent	No protection against STIs
Implanon: a hormonal implant inserted under the skin on the inside of the upper arm; it can prevent ovulation as well as thicken the mucus of the cervix preventing sperm from entering the uterus	Very effective in preventing pregnancy; can last up to three years	Requires a minor medical procedure for insertion or removal May cause irregularities in periods; may cause headaches, acne or breast tenderness	99.9 per cent	No protection against STIs
Timing methods				
Natural family planning: involves not having sexual intercourse during the time when women are most fertile during their cycle	Doesn't require any chemicals or hormonal treatments	Requires a good understanding and awareness of fertility within the menstrual cycle May be inaccurate in timing of non-fertile days	75 per cent	No protection against STIs

Contraceptive	Advantages	Disadvantages	Effectiveness	STI and HIV protection
Permanent methods				
Vasectomy (males): the tubes that carry the sperm, the vas deferens, are cut under anaesthetic	A permanent method of contraception (though you can get the surgery reversed)	Will need to use an alternative contraceptive method until a sperm count is zero	Greater than 99 per cent	No protection against STIs
Tubal ligation (females): the fallopian tubes are blocked to prevent sperm and egg meeting Cauterised Tied and cut Banded	A permanent method of contraception (though you can get the surgery reversed)	May be a medical risk during the procedure	Greater than 99 per cent	No protection against STIs

Images, from top: Alamy Stock Photo/BSIP SA; Alamy Stock Photo/Science Photo Library

Most contraceptives do not prevent the transfer of STIs. The only way to be 100 per cent sure of not getting pregnant or contracting an STI is to practise abstinence by not being sexually active at all.

Weblink
Follow the weblinks to learn more about contraceptive options.

SUPPORT NETWORKS

It is essential that you don't feel that you are going through adolescence alone. There are support networks that can assist you to come to terms with the many changes that are taking place at this time. Having a strong support base will allow you to develop and accept who you are. Strong bonds with your family, friends and teachers constitute informal support networks that you can access any time to talk about your feelings. There are also more formal support systems that can provide you with factual information and advice on sexual health and relationship issues. The following are some formal organisations that can assist you in times of need.

- **headspace** is Australia's youth mental health foundation, which helps young people aged 12 to 25 who are going through a tough time.
- **Kids Helpline** is a counselling service for Australian children and young people aged between 5 and 25 years.
- **Youthbeyondblue** offers advice and support in a number of youth health priority areas such as depression, bullying and gender preference.
- **The Action Centre** provides medical appointments for young people and information about STIs, contraception, sexual identity and relationships.

Weblink
Follow the weblinks for information and support websites.

- **Melbourne Sexual Health Centre** provides information on STIs, contraception and relationships.
- **ReachOut.com** is an online youth mental health service for young people aged 14 to 25.

1 If you have chlamydia and don't get tested, describe what can happen and what the long-term implications might be.
2 Describe how fraternal twins are made. Describe how identical twins are made.
3 Justify why the condom (male or female) is an effective form of contraception, if used the correct way every time.

1 Propose a reason why not having any symptoms of an STI might be a problem.
2 Discuss why there is still stigma attached to STIs, despite the fact that there are so many people infected.
3 Discuss how you could bring up the conversation about using a contraceptive with a sexual partner. Consider when is the right time.

1 Discuss how a pregnancy would impact your identity. Does the impact change depending on the age of the mother?
2 The United Nations says that young people have a right to comprehensive Sexuality Education.
   a Discuss what this means to you.
   b Discuss what comprehensive sexuality education should look like in schools.
   c What suggestions would you give your school that would improve the sexuality education program that is delivered?

Quiz
How can I have sex safely?

9780170463102

CHAPTER 5 REVIEW

1 Identify the differences between sex, gender and sexuality. Why do so many people get them mixed up?

2 Explain why identity is individual, fluid and never completed. Discuss how others might influence someone's identity.

3 Discuss how stereotypes influence adolescents' identity formation. Give six examples.

4 Explain what a respectful relationship is. Give five examples of how someone might be treated in a respectful relationship.

5 Describe what needs to be considered when contemplating a sexual relationship.

6 Propose ways that you can maintain your safety when developing more intimate relationships.

7 Describe how power can impact a relationship.

8 Define the differences between the terms 'discrimination', 'harassment' and 'abuse'.

9 Consider the impact gender-based violence has had on Australian communities.

10 Explain what consent is. Propose a strategy to ask for consent. Identify what you would do if the person you asked said 'no'.

11 Discuss how the conversations about respectful relationships have changed in the last few years. Consider any examples you have witnessed.

12 What are STIs? Identify one STI and provide a description of the infection, its signs and symptoms, ways of contracting the infection and effective treatments for the infection.

13 Consider how different contraceptives might suit different people. Give some specific examples.

14 Devise a list of three organisations you or a friend could contact if you had any questions about anything in this chapter.

COMMUNITY AND RELATIONSHIP SAFETY

Shutterstock.com/GoodStudio

IN THIS CHAPTER

You will consider factors that contribute to your safety and that of others at home, at school and in community settings. In most cultures, the onset of puberty is the time when individuals move rapidly towards sexual maturity and associated adult responsibilities. Consistent with that major change in social roles and the associated rites of passage, young adults must learn to make safe choices in order to develop lifelong positive behaviours.

By the end of the chapter, you should be able to:

- recall strategies to deal with challenging and unsafe situations
- rehearse contingencies or alternatives that can lead to positive outcomes
- implement strategies and options for managing situations where your own and others' safety may be at risk
- plan for realistic responses to scenarios where peers are encouraging risk-taking behaviours
- plan and practise responses you could use in emergencies where you may be required to administer first aid
- investigate what happens when power in relationships isn't equal and relationships aren't respectful
- propose actions that could be taken when relationships aren't respectful
- develop strategies to increase your assertiveness
- practise 'street safety' and recognise potentially unsafe situations
- think, act and be safe.

HOW CAN I IMPROVE MY KNOWLEDGE OF FIRST AID AND REMAIN SAFE WHILE USING IT? 255

HOW CAN I INCREASE MY COMMUNITY AND STREET SAFETY?

Quiz
Pre-chapter

Before you start, take the pre-chapter quiz to find out how much you already know.

As you move through secondary school you will develop a greater sense of independence and will start going to sport, school, parties and other places on your own, rather than with the help of a parent. You will probably develop improved social skills, better communication skills and enhanced problem-solving skills, which will be refined as you move into adulthood. However, this is also a time when peer pressures, societal expectations and differing rates of maturity can increase the risks you may face.

Fairfax Syndication/Melissa Adams

Dreamstime.com/Hongqi Zhang

iStock.com/Birdland

Getty Images/Weston Colton

Figure 6.1 How can you stay safe in each of these situations?

FACE TO FACE Safety check-up

1 With the person next to you, discuss a couple of situations where you have felt unsafe.

2 For each of the situations you discussed, what have you done to improve your safety?

3 Consider situations where you may be required to perform first aid on any family/community members and how prepared you might be to handle this situation safely. Discuss these with your classmate.

There are two basic facts that underpin all new protective behaviour programs in Australia:

- Everyone has a right to feel safe all the time.
- No matter how terrible you feel, or how difficult a situation may be, you should be able to find a responsible adult or trained health professional to talk to.

These two facts ensure that everyone can experience personal safety. Young people need to recognise when they are not feeling safe and develop skills and strategies so that positive action can be taken in getting help when needed.

Young people should identify personal networks of trusted, helping adults to whom they can turn if they are not feeling safe. These might be parents or guardians, relatives, teachers, counsellors or community members such as the police, doctors or sports coaches. It's important to review these 'go-to' people as you grow and go through school, because your level of trust and types of relationships may change. Who are your go-to people?

FACE TO FACE Who would you go to?

- With the person next to you, discuss who you would approach if you had a problem, were in trouble or felt a situation was becoming dangerous.
- List the qualities go-to people have.
- List three qualities that might make you a go-to person for others.

Worksheet 6.1

How many times have you avoided action or said nothing because you thought 'nothing can be done anyway' or 'no matter what I say or do, nothing will change'? This is an example of internalising key negative messages such as 'failure is inevitable' or 'this problem is unsolvable'. A person chosen as a go-to person or 'trusted other' should take action to make sure that the young people reporting to them, and seeking their help, feel safe.

FAST FACT
In more than 80 per cent of cases of abuse, the adult is known to the person experiencing the abuse. Review your network and remove anyone you no longer trust.

Young people in Australia spend large amounts of their recreational time at shopping centres, movies, attractions (fun parks, live music venues), sporting venues, parties and other public spaces, and at friends' houses. Trouble can often start at one of these places. With forward thinking, teens can enjoy each others' company in a range of settings in ways that ensure everyone feels respected and has a good time. Meeting in a shopping centre is fine, but when individuals start making it uncomfortable for others in the area, problems are bound to occur.

Newspix/Glenn Miller

People feel safe when they are around friends, family and teammates they can trust to look out for them. This tends to ensure a high degree of physical and emotional safety. However, you have probably also been in situations or places where you did not feel safe.

Figure 6.2 Anti-social behaviour is hurtful, unsafe and needs to be stopped via positive strategies, otherwise the police might become involved.

FACE TO FACE What is unsafe?

With the person next to you, discuss what feelings might indicate you are placing yourself at risk and are potentially 'unsafe'.

Worksheet 6.2

FACE TO FACE Trouble at parties

The teenagers in the image below held a party at a friend's house while her parents were away.

Getty Images/Ghislain & Marie David de Lossy

Figure 6.3 These teens look like they're having lots of fun together. Discuss what could go wrong in this situation, and how it could become unsafe for one or all of those involved.

As a class, discuss the following questions.

1 Identify five things that can go wrong in this kind of situation.
2 Propose why you think some teenagers and adults drink… excessive amounts of alcohol.
3 At some parties, there will be people who you don't really know all that well and who may be outside your immediate friendship group. Why should caution be taken when interacting with them?
4 Prepare a checklist you would need to tick off if you were planning to have a safe house party. When completed, swap your list with another student to check for similarities. Discuss any differences and add to your checklist if you think your partner's suggestions are appropriate.

RECOGNISING UNSAFE SITUATIONS

Young people are good at picking up signals that suggest a situation could be unsafe. These signals may include:

- someone acting suspiciously or exhibiting strange behaviour
- someone directly seeking you out or confronting you
- people carrying implements that can be used as weapons
- someone standing or moving in a way that suggests they may become aggressive
- groups of unknown people making a lot of noise
- someone looking dangerous or taking unnecessary risks
- feeling symptoms such as a faster heart rate, increased sweating, tightening in the stomach and/or feeling sick.

FAST FACT

Take greater responsibility in relation to your own health by planning ahead, thinking about what could go wrong and considering safer options or 'opt out' actions.

According to a recent forum conducted by the New South Wales police, students at secondary schools feel 'at risk' in situations with the following characteristics:

- discrimination – sex-based, racial and homophobic
- harassment or intimidation – usually groups (or gangs) of young people who target girls and boys
- physical fighting – most often boys fight boys
- strangers in the street making comments, asking for money or acting unpredictably
- problems at a party – gatecrashers, alcohol, drugs and fights; this often results in parties being cancelled and young people having nowhere to go other than public places such as parks or shopping centres
- a situation getting out of control – at the football, outside shops at night, on railway stations, at parties, drinking or drug taking
- groups in cars – P-platers making comments and following a group of teens as they walk along the street, asking them to get in the car, driving dangerously, or drinking and driving
- gangs – being stood over or threatened, or having things stolen
- being alone – boys fear being bashed; girls fear being dragged into cars, raped or abducted.

iStock.com/PeopleImages

Figure 6.4 Many people are afraid of what can happen to them when they are walking alone, especially at night.

CASE STUDY → AUSTRALIA'S YOUNG DRINKERS REPORT UNWANTED SEXUAL ATTENTION AND VIOLENCE

Identify

More than 70 per cent of young women who drink at risky levels report receiving unwanted sexual attention from their peers, while more than a third of young drinkers have witnessed serious violence.

Understand

In the largest survey of Australian adolescent regular drinkers to date, published in *Public Health Research and Practice* on Wednesday, researchers from seven universities spoke to 3465 people who identified themselves as risky drinkers, aged 14 to 19, about their experiences with other drinkers …

Most young women surveyed reported experiencing unwanted sexual attention from other drinkers (71 per cent of respondents), being harassed in public places (42 per cent), placed in a state of fear (33 per cent) and left alone in unsafe situations (31 per cent).

In contrast, the harms identified by young men were more likely to relate to aggression: they were more likely than their female peers to report being pushed or shoved when drinking (42 per cent), yelled at or verbally abused (38 per cent) or being physically hurt (17 per cent) …

Australians are delaying their first drink for longer. According to AIHW data, the average age of a young person's first drink has risen from 14.8 to 16.1 between 1995 and 2016.

Young people are also reporting drinking less: surveys by national responsible drinking organisation Drinkwise found the number of people aged 18 to 24 who reported drinking to excess declined from 31 per cent to 21 per cent between 2016 and 2018 …

Drinkwise CEO Simon Strahan said the survey data was 'alarming', although he stressed underage drinking and young people drinking to excess is on the decline in Australia.

'Our [young adult drinking] campaign, The Internet Remembers, found that the greatest concern for young drinkers, aged between 18 and 24 years of age, is the actions that they regret after excessive drinking,' he added. 'This includes concerns such as vomiting (26 per cent) or making a fool of themselves (24 per cent).'

Source: 'Australia's young drinkers report unwanted sexual attention, violence' by Mary Ward, *The Age*, 4 December 2019.

Discuss

1. With the person next to you, discuss what you believe to be the top three factors leading to violence associated with underage drinking.
2. Suggest two possible reasons why people aged 18–24 are engaging in less excessive drinking than they were 10 years ago.
3. Advertising that depicts what can go wrong when teens drink excessively, using what are known as 'shock tactics', is far more effective at reducing numbers likely to engage in this behaviour than advertising the physical harms associated with excessive drinking. Do you agree?
4. It is not unusual to read stories in the newspaper or hear on the news about teen parties getting out of control. Quite often these parties are unsupervised, poorly planned and involve the consumption of large amounts of alcohol.
   - a Find three reports or media articles focusing on teen parties that have gone wrong.
   - b After considering all three, list the factors that contributed to the parties getting out of control and becoming unsafe or dangerous.

9780170463102

FACE TO FACE Safe partying

Critique local services that support and provide advice on 'safe partying'.

1 Search advice offered by the police, local councils, teenage support groups and other sources. You should search for state police 'Party Safe Programs', or another great resource is 'Tune In Not Out'.

2 With other students, investigate ways to present and capture tips, advice, contact information and other vital information offered by these services.

CASE STUDY  TUNE IN NOT OUT

Identify

Tune In Not Out is a program of the Australian Lions Drug Awareness Foundation (ALDAF). Check out the 'TINO' website and enter 'safe partying' as a search topic to gain access to multiple videos, fact sheets and firsthand experiences associated with alcohol and parties.

Australian Lions Drug Foundation

Weblink
ALDAF

Understand

Heaps of people love to party. If you're thinking of having a party, attending a party or having a night out, there are a few things you should consider. There are tips that you can keep in mind to party safely, including facts about drugs and emergency contacts.

We like to party

Life would be pretty boring without the occasional party or night out. But partying can mean different things to different people, and you should always try to stay safe when you're partying or planning on having a big night out.

Things that could go wrong

Dangers you could run into when partying include:

- drink-spiking
- violence
- unsafe sex
- drug and alcohol overdoses
- drink driving (or being the passenger of a drunk-driver).

Ways to party safely

You can party safely by:

- having a plan before you go out, including how to get help if needed and transport options. This might involve choosing a designated driver who doesn't consume any alcohol or other drugs.
- ordering your own drinks, and keeping them with you. This is a good tactic to avoid drink-spiking as it will make it much harder for someone, whether it's a stranger or someone you know, to get near enough to your drink to contaminate it.
- avoiding getting too drunk by spacing your drinks out, drinking something without alcohol in it and drinking slowly and sensibly
- avoiding fights, and trying to resolve them without becoming aggressive
- remembering to practise safe sex strategies if you decide to have sex.

Party safely at your place

If you're having a party at your place, try to have some plans to deal with unexpected stuff. Know what you will do if someone drinks too much or if gate crashers turn up. It's a good idea to have a couple of house rules to keep people safe (as well as making sure your place doesn't get trashed). Make sure you know the laws in your state or territory relating to giving alcohol to people under the age of 18. In a number of states, such as Victoria, New South Wales, Queensland and Tasmania, it is an offence for a person to supply alcohol to someone who is aged under 18 years on private property without parental consent.

You should also let your neighbours know you're having a party, and be ready to deal with their concerns and keep them happy. This is particularly important when it comes to keeping the party noise down to a minimum. A quick way to annoy your neighbours is playing loud music at 3 a.m.

You can register a party or event with the police. For information on how to register, search 'Party Safe' for your state.

Drugs and partying

People at parties or on nights out sometimes take drugs. Find out more about drugs to learn about some of the common effects of the different kinds of drugs. If you're having a party and you don't want guests to consume drugs (particularly if it's a big one with people that aren't your close friends), it's worthwhile saying that up front on the invite.

If you choose to take drugs at a party or on a night out, try to let someone know what you're taking. This way a friend can help you if something goes wrong.

Emergency contacts

If something goes wrong, it's important to keep in mind the contacts that you may need. If there is an emergency where someone is injured, someone gets really unwell or if you feel that the party or your night out is getting out of control, don't hesitate to call Emergency Services.

- Emergency Services (for emergencies only) 000
- Poisons information centre – 13 11 26

Remember, if an ambulance is called to assist someone at a party, they are not legally obliged to inform the Police of their call out and in all situations that are unsafe and requiring emergency assistance, time is critical. Rather than waiting to see if 'things get better', it is advisable to make the emergency call and take action to improve the situation and people's health and wellbeing.

Source: Tune In, Not Out. Adapted from an original fact sheet produced by ReachOut Australia, which can be found at au.reachout.com

Discuss

1. Summarise the laws in your state around supplying alcohol to people under the age of 18 at parties.
2. Discuss why some teens avoid calling the ambulance service or police when a party is getting out of control and some people may require medical assistance.
3. Discuss any other plans you would need to consider if you were going to a party at a friend's house next weekend, and you knew their parents weren't home.
4. Can you think of any other strategies to add to this list to ensure a safe party?

The Australian Drug Foundation has researched the effects of teenage alcohol consumption and revealed the following findings:

Worksheet 6.3

- Different parts of the brain develop at different rates as we grow, depending on whether we are male or female. The pre-frontal cortex houses the part of the brain that controls rational thinking. This part of the brain does not begin to mature until age 19 and only fully matures by around age 21 in females and age 28 in males. So damage to the pre-frontal cortex during its development can have lifelong consequences for memory, personality and behaviour.

- Drinking alcohol during teenage years can cause permanent brain damage. (See Chapter 1, pages 18–19 for the short- and long-term effects of alcohol.)
- Alcohol can affect a teenager's social development if they start drinking at an early age. They may turn to alcohol as a form of coping with problems and be more open to experimenting with other substances. The learning difficulties caused by teenage drinking can result in poor school performance and an increased risk of social problems, depression, suicidal thoughts and violence.
- A hangover can be just as damaging to the brain as heavy drinking by reducing a person's ability to learn and recall.

FAST FACT

People who first use alcohol before age 15 are five times more likely to misuse alcohol than those who first use alcohol at age 21 or older.

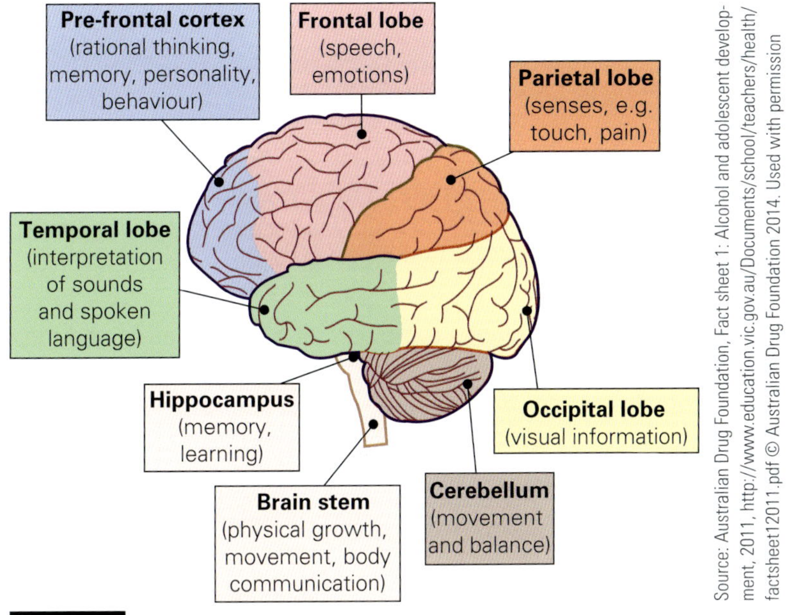

Source: Australian Drug Foundation, Fact sheet 1: Alcohol and adolescent development, 2011, http://www.education.vic.gov.au/Documents/school/teachers/health/factsheet12011.pdf © Australian Drug Foundation 2014. Used with permission

Figure 6.5 Areas of the brain

FACE TO FACE Communicating with your school community

In pairs, discuss the information and facts around the dangers of consuming alcohol that you might consider placing in the school newsletter.

- State the format you would use to make sure it grabs people's attention.
- Discuss if you might consider using any statistics.
- List three images you might consider using, and justify their selection.

Worksheets 6.4 and 6.5

Impaired brain development resulting from alcohol and drug consumption during secondary school increases the risk of impaired decision-making, unsafe choices and potential lifelong negative behaviours, associations and outcomes in the future.

WELLBEING CHECK IN

GRATITUDE JOURNAL

This activity has been developed in collaboration with Dr David Bakker of MoodMission

Identify

While it's important to stay safe, it's also important to keep a level head and not feel in danger all the time. It can help to stay aware of both the bad and good things.

Understand

There are many good things that happen to us every day. But sometimes we don't take notice of them, especially if we're preoccupied with the bad stuff. When people feel down and depressed, they can have a tendency to only pay attention to the negatives. This is called a cognitive bias. Research has shown that this bias can be corrected for by deliberately paying attention to positives.

Practise

1 Take a moment to think back over the past 24 hours. What has happened in that time?
2 Now try to think of three things that you are grateful for. They could be things that happened to you, opportunities you had, people you saw or help you received. They might not be big things, and could be simple day-to-day things that your life would be worse without. Write down your three things.
3 Make a plan to write down three more things tomorrow, and the day after, and the day after, for a week. You could get a special diary just for this, or use a journalling app. Keeping a practice of gratitude journalling every day has been shown to improve mental health and wellbeing.

Reflect

Sometimes we don't stop and just notice the everyday good happening around us. Did you come up with anything that surprised you? Or anything that you didn't realise you were grateful for before you stopped and really thought about it?

WAYS TO AVOID TROUBLE

There are many strategies young people may employ to help them avoid unsafe situations.

1 **Focus: distractions can lead to danger**

- Don't listen to music or talk on your mobile when walking or driving, as this will reduce your awareness of what is going on around you.
- Stay awake on public transport, at parties and in public spaces – you cannot sense danger if you're asleep.
- Messing around with friends can cause distractions, decreased levels of concentration and an inability to process important environmental cues.

CASE STUDY ⇨ CYCLING AND MOBILE PHONE USE

Identify

It's common knowledge that it's against the law to pick up your mobile phone while driving – but did you know it's also illegal to ride a bike and use a phone at the same time?

Understand

The issue was brought to the fore after WA Police posted a photo to Twitter of a $400 fine handed to a cyclist for using her phone while riding a bike in the Perth suburb of Armadale ...

Traffic Enforcement Group Inspector Mike Sparkman said under WA laws in place since 2011, motorists and cyclists were treated the same when it came to mobile phone usage ... However, he said it was not illegal to use a phone with headphones while cycling, 'as long as you're not holding the phone up to your ear or you've got it on loudspeaker and you're riding one-handed.'

Inspector Sparkman said cyclists should park their bike safely in order to make or receive phone calls.

'You should be off the road so you're safe,' he said. The same law applied to mobility scooters, but not to people pushing prams.

'We actually had an issue about five or six years ago where a lady pushing a pram was given an infringement and that was withdrawn because when we reviewed the legislation, it didn't fall within the legislation,' he said. 'But then again, you're a pedestrian ... and you are a vulnerable road user ... and you've got to be fully aware of your surroundings.'

Acting road safety commissioner Iain Cameron said laws around mobile phone usage also applied to people riding horses on the road. Cyclists using footpaths were not exempt either.

'The reason that is illegal is because it's dangerous and risky to the rider themselves ... or to someone that they may ride into,' he said. 'Essentially a pedestrian being hit by a cyclist will be severely injured, particularly if that's an elderly person ... so it is about the risky behaviour and the fact that it's dangerous.' Mr Cameron said a common complaint from those fined was that they did not know the rules.

'But the road rules are published, everyone does a road rules test when they get their driver's licence,' he said. 'It is everyone's responsibility to know the road rules they think will be relevant to them.'

Shutterstock.com/Alessandro Romagnoli

Figure 6.6 Wearing headphones greatly increases the likelihood of motorists and cyclists being involved in serious accidents.

Discuss

1. New wireless headphones make it possible for people to ride bikes and conduct conversations on their mobile phones, which are fixed to their bikes. Many argue that this is like using hands-free phones in motor vehicles and should be allowed. Do you agree?
2. Some people believe that cyclists should be allowed to use footpaths, as this reduces the likelihood of them coming into contact with motor vehicles. Others argue that this creates greater safety issues. Discuss your thoughts on this issue with a classmate.

Scout Kozakiewicz

Figure 6.7 Using a mobile phone while crossing a road places the user at risk of serious injury.

Scout Kozakiewicz

Figure 6.8 A quick, one-second glance at their phone while driving at 60 km/h means this driver will travel 17 metres without looking. What could happen?

FACE TO FACE Headphones

1. Discuss times when you have seen people wearing headphones or earphones that could have resulted in decreased safety. Outline the potential outcomes.
2. List three ways wearing headphones while driving can be dangerous.
3. Discuss a time when you have seen young people 'mucking around' in a way that could have placed them in danger.
4. As a class, debate the following statement: 'It should be illegal to wear headphones while in control of a vehicle or bicycle.'

2. **Stay with friends**
   - Having friends with you creates an immediate support network you can call upon if things go wrong.
   - Friends look out for each other; they also tell each other things they probably wouldn't tell people they are not close to.
   - Sometimes young people get swept into group activities. It is important not to be pressured into doing things you are not comfortable with, and friends can help you to 'stand firm'.
3. **Constantly monitor your environment**
   - If you are at a party, never leave your drink unattended (someone might add something to it).
   - When you are walking home, pay attention to others around you and keep your head up. This allows you to see what is going on and positively identify people at a later stage, if necessary.
   - When you are at a sporting venue, be careful where you place your backpack. If you do not maintain visual and physical contact with it, it is an easy target for someone to take.
   - When you are shopping, pay attention to others around you.

4 **Watch out for strangers**

- Cross the road if you think you are being followed. If the person continues, run somewhere where there are people and tell someone what is happening. Phone home and ask to be picked up or call the police.
- Never get into a vehicle with an unknown person.
- Do not get into a lift with someone you think is following you; wait for another one.
- Use your mobile to contact a go-to person, parent or police.

iStock.com/MachineHeadz

Figure 6.9 Always trust your gut instincts. If you think you are being followed, call someone to let them know where you are and that you feel unsafe.

5 **Always take your (charged) mobile phone and let trusted people know where you are**

- Carry your phone at all times.
- Use your phone to inform go-to people when you are leaving a party, the movies or a training session, or catching public transport, a taxi or rideshare option.
- When getting into a taxi or Uber, note the number plate and driver's ID, which should be on the dashboard or in the app on your phone. If using a rideshare service, check the number plate before getting in the car. Give clear directions where you want to go. If you are not on the agreed route, stop the taxi at a safe place.
- If you and your friends are travelling home late at night, text each other when you get home. If someone doesn't 'check in' when expected, call their home on the landline or call their parents.

6 **Be a mate**

- Stay with your group – especially in large crowds at parties, concerts and festivals. If you need to leave the group (to get food, look for a friend or go to the toilet), do so in a group of two or three.
- Avoid arguments – logic and rational thought are decreased by alcohol or illicit drugs.
- If a friend is seriously affected by alcohol or illicit drugs, call an ambulance immediately. Ambulance officers are not required to inform the police, so don't be scared about this happening.

7 **Think it through**

When considering personal safety, use the following three steps to minimise the risk of harm in difficult situations:

- **Think** about where you are going, what you are likely to be doing and who is going to be there. Go through a list of 'ifs' and 'I wills' to consider how things might eventuate. This will mean that you have considered 'if this happens, I will …' for various situations that may arise. For example, 'If someone asks me to drink, I will …' or 'If a friend gets intoxicated, I will …'

Worksheet 6.6

- **Evaluate** how things are going and listen to what your body is telling you about symptoms such as increased heart rate, sweating or an 'uneasy feeling'. In most cases, if you don't feel right, the situation probably isn't right and requires action. Quickly decide if the situation is risky but under control, or if it is potentially harmful and you should remove yourself from the risk.
- **Act** quickly to remove yourself from any unsafe situation and tell or take friends as well. Having planned ahead will help you to respond in a safe and positive way. In a similar way to a reflex action, planning and going through scenarios prior to participating in an event or activity will allow you to act automatically when you pick up cues from what is going on around you.

FAST FACT

Along with vision, hearing, smell, taste and touch there is what is often described as a 6th sense: intuition. This is sometimes referred to as having 'inner consciousness' or 'inner feelings'. Always act on your intuition if you sense that something is unsafe.

Figure 6.10 Drinking excessive amounts of alcohol can impair decision-making skills, often resulting in fights.

Worksheets 6.7 and 6.8

UP AND MOVING Alternative endings

1 With two or three other classmates, write a 'sliding doors' script that looks at street situations where potential conflict can arise. One 'door', or ending, results in a combination of verbal, physical and emotional abuse or violence; the other 'door' results in the conflict being defused and not escalating into abuse or violence.

2 The script should then be role-played for the rest of the class to view and comment on. In particular, what strategies were put in place that resulted in a positive ending or pathway?

FAST FACT

A driver who has consumed cannabis or amphetamines is at the same risk of being involved in an accident as a driver with a blood alcohol concentration (BAC) of 0.10 or above. This is twice the legal limit. The legal limit for drivers (not probationary/P-platers) is 0.05. If your BAC is between 0.10 and 0.125 your speech will be slurred and your balance, vision, reaction time and hearing will be impaired. Your motor coordination will also be significantly impaired, and you'll experience a loss of judgement.

SOCIAL MEDIA SAFETY

Any social network has both advantages and disadvantages associated with it. Online social platforms such as Facebook, Instagram and the rapidly growing TikTok are popular amongst young people and teenagers. Used responsibly these platforms keep people connected, allow for rapid information sharing and allow school, local and broader communities to engage with each other.

TikTok allows users to edit videos and lip-sync, as well as allowing them to do what they like to create their own funny, unique stories and showcase their talents and personalities. TikTok's editing features make it easy for users to create professional-looking videos, further encouraging creativity. Sounds like a lot of fun and potentially positive.

Because of the large and diverse user base, social network platforms have many potential risks associated with their use. These include:

- inappropriate content such nudity, pornography and explicit material
- 'challenges' and dangerous behaviours that can be life threatening
- on-line bullying

TikTok and the other social platforms have features that allows users to filter, report spam, offensive comments and specific keywords, and block accounts, but these are very easy for people sharing information to get around and disguise.

Under no circumstance should you reveal personal information like name, address, phone number, family details, or age/date of birth as online predators may use it to If someone you don't know tries to contact you online or if you believe something is suspicious about and on-line post or request, tell a trusted friend.

CASE STUDY

TIKTOK REACHES 7.38 MILLION USERS IN AUSTRALIA (FEBRUARY 2022)

Identify

Digital 2022 Australia says TikTok is the most downloaded mobile entertainment app. According to the annual report, TikTok now has 7.38 million Australian users over the age of 18, with a reach of 37%–55.3% were female and 44.7% male.

Understand

TikTok's emergence comes weeks after Viberate's *State Of Music* report stated that the app 'is the new SoundCloud when it comes to launching new acts'.

Facebook remains the biggest social media platform in Australia with 15 million users (or 57.9% of the total population), followed by Facebook Messenger and Instagram. Instagram now has 12.8 million monthly users after 16% (or an extra 1.8 million users) local growth in the past 12 months. Snapchat had 7.25 million users.

But TikTok users spend more time on the app than any other platform: up 40% to an average of 23.4 hours per person, per month. In comparison, Facebook is down 3% to 17.6 hours while Instagram had a 2% jump to 8.3 hours monthly.

Shutterstock.com/Caftor

Figure 6.11 Time spent on social media by 16–64 year olds is 1 hour and 57 minutes per day.

According to *Digital 2022 Australia,* the number of Australians adopting social media is up 4.6%, to a total of 21 million users, or 83% of the population (25.9 million as of January 2022).

That's almost one million additions in the last 12 months.

The greatest leap was for podcasts, up 35% to 54 minutes, with playing games up 24% to 1 hour and 7 minutes, and television watching (including streaming) increasing by 7% to 3 hours and 44 minutes.

With a need for interactivity in isolation, three in four Internet users aged 16–64 (23.6 million) played games, using a gaming console for an average of 1 hour 7 minutes per day, a 24.3% increase.

As a result, Australian digital media spend on video games rose by 25% to $1 billion. The spend on mobile apps and in-app purchases in 2021 reached $2.8 billion. With 24% of Australians aged 16–64 saying that they discover new brands, products, and services via social media ads, Hootsuite head of Australia and New Zealand, Letrecia Tippett, encouraged advertisers to take note of the increased branding opportunities in this space.

'As social media usage in Australia steadily increases, it is more important than ever before for companies to embrace social media to strengthen their connection with their customers', Tippett, said.

'As Australians navigate through yet another challenging year of the pandemic, we encourage businesses to make use of the data from our Digital 2022 report to help them navigate which social media platforms will be most efficient in connecting them with their audiences'.

Source: *TikTok most downloaded mobile entertainment app in Australia, reaches 7.38m users, by Christie Eliezer, The Music Network, 11 February 2022, https://themusicnetwork.com/tiktok-usage-australia-2022/.*

Discuss

1 Investigate social media use in your class and summarise the main type of platform used (TikTok, Facebook, Instagram, Snapchat, etc.) and the type of content that is usually accessed.
2 Contrast the amount of time you spend on social media to the amount of homework you do per day (take an average).
3 'FOMO' stands for 'fear of missing out'. Propose how this emotion contributes to people acting in ways they normally would not. Discuss any similarities between 'FOMO' and peer pressure.
4 As a class debate the following 'if people are influenced by inappropriate content or encouraged to engage in dangerous behaviour on social media it's their own fault – they should know better'.

TEENAGERS ON THE ROAD

Learning how to drive and getting your driver's licence can be exciting, but driving also comes with responsibilities and risks. Young drivers are over-represented in road crashes that result in injuries and death compared with all other age groups. Human factors have been identified as the main causes of road accidents, even when the weather is fine, lighting is good and roads are in sound condition.

Regardless of driver age or experience, there are four factors that contribute to most road accidents:

- excessive speed
- failure to comply with road rules
- non-use of seatbelts
- alcohol or substance abuse.

When inexperience and overconfidence are combined with these risky behaviours, the risks of accidents and road trauma increase markedly. Extensive supervised driving practice in a range of conditions during the learner phase has been shown to significantly reduce the risk of road accidents. Many states are now mandating over 100 hours of documented, supervised and varied driving practice before learner drivers are eligible to go for their licence.

iStock.com/londoneye

Figure 6.12 Getting your licence makes you more independent, but also gives you more responsibility.

FACE TO FACE Driving hours

Discuss the **cognitive skills** a learner driver with 120 hours of practice would have developed compared to a driver who has only had 80 hours of supervised practice.

cognitive skills skills that deal with thinking, problem assessment and problem-solving and are generally related to understanding what a situation requires

Young drivers are often impaired while driving; this may be due to alcohol and drug use, fatigue, distraction or any combination of these. Alcohol impairs a person's ability to make decisions. For example, the ability to judge stopping distances and response times is negatively affected by alcohol and drugs. In fact, even being tired leads to a dangerously similar effect on the roads: if you've been up for 17 hours (e.g. getting up at 6 a.m. and going to bed at 11 p.m.), your driving will be as impaired as it would be if you had a BAC of 0.05. The combined use of different drugs and alcohol leads to an extremely high risk of road trauma.

FAST FACTS

A recent study of risky driving among Australian teens by the Australian Institute of Family Studies indicates that:

- 1 in 6 learners had exceeded the speed limit by between 10 and 25 km/h on a recent trip and 1 in 4 had driven when fatigued – two behaviours commonly implicated in serious road crashes
- 1 in 10 teens had also been the passenger of a driver who was under the influence of either alcohol or drugs.

INVESTIGATION

HOW CAN WE REDUCE THE NUMBER OF ROAD ACCIDENTS YOUNG PEOPLE ARE INVOLVED IN?

Purpose

To collect information around common road trauma events and see if any patterns emerge.

Method

1. In groups of three, investigate the age, gender, road user type (driver, passenger, pedestrian, cyclist or motorcyclist), types of crashes and level of injury caused by accidents that young people are involved in. You can also use the questions below to guide your research.
2. Allocate research tasks to each group member to complete, then collate the findings of your investigation.
3. For this investigation, young people are classified as teenagers up to 25 years old. Your local motoring authority should have much of the information you are seeking, along with government agencies.

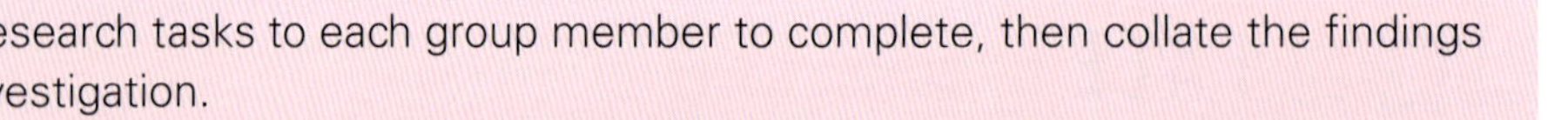

Weblink
VicRoads
Regional Roads Victoria

Discussion

1. In terms of road crashes, which gender is represented the most? Provide possible reasons for this.
2. Are more cyclists or motorcyclists involved in road crashes? Why might this be the case?
3. Are people aged 15–25 more likely to be involved in road accidents than those aged 55–65? How do you account for any differences?
4. Discuss two safety features car manufacturers have recently (within the past 10 years) included to minimise the effects of road accidents.
5. What strategies aimed at young people has your state government put in place in an effort to reduce the road toll for drivers under the age of 21?
6. In a group, brainstorm at least three new strategies that would promote safer road use attitudes and behaviours by young people.
7. List five strategies you would consider using if you are in a car and you think the situation has become unsafe due to 'triggers' you are sensing around you. Outline what these 'triggers' or warning signs might be and provide a strategy to deal with each one.
8. 'Driverless cars should decrease the amount of road accidents because AI and computers can avoid more potential dangers than human drivers.' Discuss or debate this statement as a class.
9. As a group, present your findings to the rest of the class via a multimedia presentation or an 'infomercial'. Provide recommendations as to how young people can reduce their risk of becoming one of the statistics the group has investigated.

9780170463102

FAST FACTS

1 Almost 80 per cent of P-platers have engaged in a form of risky driving during their last 10 drives.

2 The most common forms of risky driving were speeding (up to 10 km/h over the limit) and driving while tired.

3 About 4 per cent of teens have driven under the influence of drugs or alcohol in the past year.

4 An even greater one in 10 teens have been the passenger of someone driving under the influence in the past year.

Source: Growing Up in Australia: The Longitudinal Study of Australian Children (LSAC) Annual Statistical Report 2018, Vol. 9 –, December 2019 © 2020 Australian Institute of Family Studies.

FACE TO FACE Risk scenario

Young people are often pedestrians – more so than older people, many of whom drive. Brainstorm a scenario in your community where a young person might be placed at risk as a pedestrian moving around local streets. What are the potential risks? Suggest some ways these risks can be minimised or removed.

Worksheet 6.9

Hoon drivers

The colloquial term 'hoon' refers to anyone who drives in a manner that is anti-social. Typically, this involves high-performance or highly modified vehicles fitted with aftermarket sound systems that often blare music to attract attention. Quite often hoons engage in dangerous behaviours such as 'burnouts', illegal street racing and reckless driving.

Fairfax Syndication/Angel Wylie

Figure 6.13 Hoon behaviour is often about trying to impress onlookers.

FAST FACTS

1 Drivers are held responsible for any accident they cause while they are fatigued. Commercial and truck drivers must take mandated rest breaks to avoid fatigue and document rest periods in a logbook.

2 Pedestrians with a blood alcohol concentration at or over 0.15 are 15 times more likely to be involved in a road accident than those with a blood alcohol level of zero.

INVESTIGATION

Purpose

To investigate the impact of hoon drivers on the local community.

Method

1 Brainstorm questions you could ask members of your local community to gauge their thoughts on drivers in their area. Some examples could include:
- Have they ever encountered hoon drivers in the community?
- Do they feel safe on the road?
- Do they think there need to be more laws in place to tackle hoon driving?

2 From your brainstorm, choose five questions to create an interview.
3 Find three different people in your local community to interview: try to get three different people of different groups e.g. a grandparent, a teacher, a young person.
4 Collate your results, then analyse them to answer the following questions.

Discussion

1 Identify how prevalent hoon driving is in your local community.
2 Suggest three strategies to tackle hoon driving in your community.
3 Discuss why hoon driving is a problem.
4 Use the internet to find articles about 'hooning', the negative effects of such behaviour and the consequences of being caught hooning.
5 Research the consequences of hooning in your state compared with at least two other states or territories. Which anti-hoon laws do you think are the most effective in eliminating this type of behaviour on the roads?
6 Summarise the findings of at least two articles, and compare these to your interview findings by creating a short report.

Worksheet 6.10

1 Consider why it is important to discuss your concerns with a 'go-to person' rather than trying to deal with them by yourself.
2 Discuss why alcohol can have a much more negative effect on the brain of a teenager than on someone who is in their thirties.
3 Why is it important to pre-plan and consider the 'what-ifs' when considering going to a party, sporting event or other social gathering?

1 Violence and trauma caused by 'one-punch' or the 'coward's punch' has been highlighted by various campaigns. Why is education and media attention so important to all of these campaigns?
2 Try to recall a situation where you have witnessed someone so focused on their mobile phone that they have potentially placed themselves in danger. Write a brief account of what happened.

9780170463102

3 If the pre-frontal cortex is damaged by excessive alcohol drinking between the years of 15 and 21, what long-term/irreversible problems might result?

4 Consider how excessive alcohol consumption as a young person can contribute to long-term depression.

1 Propose why P-plate drivers have special conditions applied to them when it comes to alcohol consumption and driving motor vehicles. Research the proportion of P-plate drivers who are involved in motor vehicle accidents compared to other drivers, and refer to this in your response.

2 Find a news article that highlights anti-social behaviour by teens. Write a summary of the article that clearly outlines the following points:

- where the teens were gathered
- what unsafe or antisocial behaviour the teens were engaged in and who might have been affected
- the social, emotional and financial costs associated with their behaviour
- 'healthy' alternatives the teens could have been involved in.

Quiz
How can I increase my community and street safety?

WHAT CAN I DO TO ENSURE MY SAFETY WHILE IN RELATIONSHIPS?

As you progress through school you will come into contact with many people, and you will develop multiple relationships. Some of these will be superficial, but others will be more meaningful, significant and possibly **intimate relationships**. Meaningful relationships with people at school, at work and in the communities you belong to are important in developing your sense of self, connectedness and emotional wellbeing.

intimate relationship a deep emotional connection with another person

For any relationship to become intimate there needs to be openness, honesty and trust. These are the key ingredients of any respectful relationship. Clear and consistent communication is an essential part of safe intimate relationships. Sexual intimacy and personal intimacy are two very different forms of intimacy that can become blurred when clear boundaries are not agreed upon.

The decision to become sexually active is personal and is based on many factors. You should not feel pressured by your partner, friends, the media or other outside influences. The type and degree of intimacy a couple shares is their collective choice. Some young adults have a high degree of sexual intimacy, while other intimate relationships involve no sexual activity at all.

Remember that if something doesn't feel right, it probably isn't. Think very carefully about what you feel comfortable and safe doing. Communicate openly with your partner about what you want from your relationship. Someone trying to get you to do something you're not comfortable with is unsafe.

FAST FACT

NO means NO. You have the right to set your own sexual boundaries and say yes to some things and no to others. You also have the responsibility to respect another person's decision to say no to some things and not question or try to force them to change their mind.

123RF.com/dragon_fang

Figure 6.14 Public displays of affection can have a varied impact on bystanders and people who observe such behaviour.

Having sex without your partner's consent is a crime. You might consent to kissing, hugging, touching and massaging but not want to engage in sexual intercourse. Just because you consent to some forms of sexual activity doesn't mean you consent to all of them. If you no longer feel comfortable during sexual activity, it's okay to change your mind and say no. (See also Chapter 5, page 211.)

ABUSIVE RELATIONSHIPS AND SEXUAL HARASSMENT

In Years 7 and 8 you learnt about respectful relationships and the need for a balance of power in relationships. If one person abuses their power, they can place others at risk. All types of abuse – physical, emotional and sexual – are harmful. Young people need to develop strategies to protect themselves and keep themselves safe in these potentially dangerous situations.

If you feel unsafe, the first thing to do is to remove yourself from any future harm. Consider talking to one of your go-to people or trusted others. There are a few essential strategies young people should practise and develop to protect themselves from unsafe situations:

- thinking and planning ahead
- personal safety strategies
- techniques for being **assertive**.

assertive
being forthright and up-front about your wants and needs while still considering the rights, needs and wants of others

Sexual harassment is a form of discrimination based on unwanted sexual advances or actions. This can involve sexual actions or advances that make you feel uncomfortable, or offensive sexual jokes, emails, messages or behaviours. Every school and workplace has policies and practices clearly stating that sexual harassment is against the law and unacceptable.

The Equal Opportunity Commission in each state deals with instances of sexual harassment that are brought before it. You should never feel that it is your fault or that

FACE TO FACE Sexual harassment

1. List at least three 'warning signs' in this image that show that this could be an unsafe situation for the individual on the left.
2. What feelings might they have that indicate they might be unsafe?
3. Discuss this statement with a classmate: 'Sometimes what women wear sends the wrong messages to men.'

Thinkstock/diego cervo

Figure 6.15 What in this image tells you that this is an unsafe situation?

you deserve to be sexually harassed; it is important to report it to your go-to people. No one has the right to make you feel uncomfortable about your gender or sexual orientation via sexual harassment or discrimination.

CASE STUDY  HOW CAN WE MAKE PEOPLE FEEL SAFE?

Identify

When Plan International surveyed hundreds of young women in Sydney, they found that 9 in 10 felt unsafe at night ... Many women are harassed in the most public of places, like public transport or a busy city street. Often there are witnesses, and often they do nothing.

Understand

In a 2018 report, Plan noted: 'For the most part, witnesses just stand by: they do little or nothing to help, and girls and young women feel that there is little point in reporting harassment to the authorities because they believe the authorities have neither the will nor the power to do anything about it.'

So why exactly are people so reluctant to intervene when they're seeing women being harassed?

Claire Tatyzo, who runs bystander intervention workshops for YWCA Australia, says there are a few things going on. Firstly, there's something called bystander apathy. It's the sense we have in group situations that someone else will intervene, when often no-one does. In some situations, intervening may feel unsafe ... And, sometimes, people are afraid of reading the room wrong and looking silly, Ms Tatyzo says. Instead of giving yourself reasons to avoid intervening, think about the reasons you should do something, she says ...

Here are a few practical ideas for intervening in problematic or questionable situations.

1 Call out poor behaviour

If you witness poor behaviour, call it out. It might be a sexist joke. It might be someone harassing a woman at a bar. Often, you don't have to say much. It could simply be something like 'That's not funny, mate' or 'That's not OK', two phrases mentioned in a recent Victorian campaign.

'It might shut it down, or it might make that person think twice before doing it again. It might also lead to a discussion about why that [behaviour] is harmful,' Ms Tatyzo says.

2 Create an interruption

By interrupting an interaction, you can provide an opportunity for a woman to leave or get some help. It doesn't have to be confrontational either.

Here are a few approaches you can use.

- The 'old friend' method: One man in one of Ms Tatyzo's workshops saw a woman being harassed at a bus stop. He went up to her, pretending to know her, and said, 'Hey, Sally' which created an opening for her to leave.
- The 'I'm lost' method: Another non-confrontational approach is to create an interruption by asking for directions.
- The 'checking in' method: It's obvious, but if you spot someone in a potentially problematic situation, you could ask them if they're OK.

3 Get help

In many situations, the best thing you can do if you spot something concerning is to get help from the professionals.

'It could be police, security, or if you are on public transport, notifying the bus driver that something's going on. That is something we would certainly encourage,' Ms Tatyzo says.

Discuss

1. Discuss why it is important to have the courage to call out poor behaviour and not ignore it.
2. Consider another two approaches that could be used to create an interruption to stop uncomfortable behaviour.
3. Identify who you would consider seeking out to discuss inappropriate behaviour you witness at school. List at least three different people.

Worksheet 6.11

FACE TO FACE Respect

In groups, consider the below questions:

1. Discuss three reasons why some people discriminate against or bully others.
2. List five negative outcomes associated with discrimination.
3. 'Valuing diversity' is a term often heard. What does this actually mean, and how can respectful relationships contribute to this positively?
4. Excluding others from what you are doing is a form of bullying. Discuss two situations where you have witnessed exclusion occurring. Explain what you think it means to be 'inclusive' in your actions.
5. Suggest what you could do if you became aware one of your friends was being harassed or discriminated against.

PEER PRESSURE

Peer groups are groups of friends who are all about the same age. You might have a peer group in school, in a sporting club and a community group. Having a group of friends is an important part of being a teenager. Your friends have an impact on your decisions about both safe and unsafe behaviours.

Peer groups can be positive if they are supportive and understanding, if they value and respect the rights of others and are engaged in social and responsible behaviours. Positive peer pressure makes you feel better, healthier, happier and good about what you are doing. However, some peers engage in risk-taking and antisocial behaviours that do not take into consideration the rights of others, and so they become a negative influence. Negative peer pressure can make you feel unhappy, uncomfortable, 'not right' and unsafe about what you are doing.

Why do some people remain in a group that makes them feel uncomfortable and engages in risk-taking behaviours? The need to feel connected to a peer group and be 'accepted' is a strong influence, and this sometimes outweighs a person's ability to detach themselves from the group and its behaviours. Teenagers may lack the confidence to be assertive and say no to activities they know could be potentially dangerous and put them at risk. This influence is known as 'peer pressure'. Sometimes, young people lack experience in dealing with 'difficult' situations and may not fully understand how things may eventuate, and so find themselves involved in the behaviour that can easily lead to unsafe outcomes.

iStock.com/apomares

iStock.com/FatCamera

iStock.com/FatCamera

Shutterstock.com/oliveromg

Figure 6.16 Positive peer pressure can influence people to exercise and socialise together with positive outcomes for all involved.

ASSERTIVENESS

You have a right to be assertive. Assertiveness is a communication skill that ensures your needs are met and your feelings are understood in your relationships with others. Assertiveness empowers you to speak up for the rights of others and to air your views in situations where you can have a positive (or negative) influence. Learnt and practised, assertiveness improves self-esteem.

Video
Assertiveness: What is the difference between aggression and assertiveness? How can assertiveness help you stay safe? Watch the video and start the discussion.

Table 6.1 Three types of communication

Passive	Assertive	Aggressive
Violates your own rights	Respects your own needs and rights	Violates the rights of others
The needs of others are given priority	Respects others' needs and rights	Your own needs are given priority

WELLBEING CHECK IN

ASSERTIVE COMMUNICATION

Identify

Sometimes it's hard to speak up and say what we want, especially when we're scared of the other person's reaction. However, there are ways of being assertive that reduce the likelihood of upset and led to win-win scenarios for you and the other person.

This activity has been developed in collaboration with Dr David Bakker of MoodMission

Understand

Assertiveness is a skill that can be practised and developed over time. As shown above, it's very different to being aggressive, because you're trying to reach a win-win situation rather than one where you win and they lose. Once you start communicating assertively, you will gain more confidence and self-respect, which will help you continue being assertive with others.

It's easy to forget that others may have different assumptions or knowledge about a situation, so any assertive communication always starts with establishing the facts. Then it's important to be open about how you are feeling. Once the other person sees things from your perspective, it will be much easier for them to be receptive to your request. Finally, you may like to remind them about the rewards of doing what you're suggesting.

Practise

Think of something you've been wanting to ask or have been scared of being assertive about. Now note down:

1. What are the facts about the situation?
2. How do you feel about it?
3. What do you want to ask for or say no to?
4. What will the other person get out of it?

Reflect

Can you see yourself saying those things to someone else? Maybe you can make a time to try it out! You might be surprised at their reaction, especially if you take time with the first two steps, as they are the most important.

People who are assertive engage in the following behaviours:

- communicate their rights, opinions, feelings and needs in an open, honest and confident manner while also considering the rights and feelings of others
- acknowledge that everyone has the right to their own opinions (e.g. 'What are your thoughts on this?')
- listen to what is being said and use receptive listening
- use a firm and relaxed voice while speaking fluently and without hesitation
- adopt an open body stance and often use open hand gestures
- distinguish between fact and opinion (e.g. 'You might think that … but actually I feel …')

9780170463102

- typically put themselves at the centre of what they say by using 'I' statements such as 'When you behave like this, I feel …' or 'I need you to stop being so possessive because every time I …' or ' I don't think it's fair the way you guys always gang up on …' or 'I will not allow you to force yourself onto me like that …'.

FACE TO FACE Supporting a friend

Discuss a situation where you have stepped in for a friend and provided them with support that they otherwise would not have received. What might have happened if you weren't there or didn't have the courage to be assertive on your friend's behalf?

Worksheet 6.12

WELLBEING CHECK IN

SELF-ESTEEM AND SELF-BELIEF

Identify

Sometimes we're our own worst critics and can be pretty hard on ourselves. There's nothing wrong with celebrating our own achievements and feeling proud of little things.

This activity has been developed in collaboration with Dr David Bakker of MoodMission

Understand

Self-esteem is how we think about ourselves. Having high self-esteem means believing that you're confident, capable and loved, while low self-esteem leads to thoughts like 'I'm not worthy' and 'no one likes me'. Low self-esteem is a major contributor to mental health problems like anxiety and depression, so it's important to improve our self-esteem where possible. Having good self-esteem isn't arrogance. You're not saying you're better than anyone else – you're just proud of who you are. Research has shown that reminding ourselves of our strengths and things we've done well can improve our self-esteem.

Practise

1. Take a moment to think back over the past 24 hours. What have you done with that time?
2. Now try to think of three things that you have done well during that time. They may not be huge, amazing achievements, and they might feel like parts of your simple day-to-day routine. For example, maybe you made yourself or someone else breakfast. Or maybe you got to school on time. Write down your three things.
3. Now take a moment to think about why you were able to do those things. Think about what personal qualities you have that helped you achieve them. For example, you could say that you are helpful if you helped someone, or that you are careful if you made sure that something bad didn't happen. Write down the three personal qualities that correspond to your three things done well.

Reflect

Was it hard to come up with things? If so, why do you think that is? Would it be easier to come up with good things about someone else? Do you think it would be harder if you had higher standards for what you consider to be 'good'?

1. List what you believe to be the top three expectations people should have when it comes to respectful relationships.
2. Propose what schools can do to increase the opportunities for students to feel 'included' and a part of the school community.

1. Recall a situation where one of your friends said something that you felt was inappropriate, but you didn't speak up. Write down a couple of things you could have said to make your friend aware that you didn't appreciate what they were saying, or think it was appropriate.
2. Discuss ways peer groups can have a negative influence on group members.
3. Provide three examples of 'positive peer pressure'.
4. Outline at least four strategies you could use to stop you from behaving negatively or acting antisocially as a result of peer pressure.
5. Search the internet for different types of assertive behaviours.
   a. You could start by looking up 'consequence assertion', 'discrepancy assertion', 'empathic assertion' and 'negative feeling assertion'.
   b. Summarise what some of these different assertiveness styles involve.
6. Justify the style of assertion that suits you best and are most likely to use when required to do so.

1. Design a poster that communicates five ways to become a 'Better Bystander' when it comes to harassment or uneven use of power.
2. a. Use the internet to find anti-discrimination laws that exist in your state. Summarise these in point form.
   b. Find out about the policies your school has in place to protect students and staff from discrimination. Summarise these in point form.
3. Respectful relationships enhance the wellbeing of staff and students in your school. Search the internet for 'health-promoting schools', then design a poster that lists 10 common characteristics of a health-promoting school.

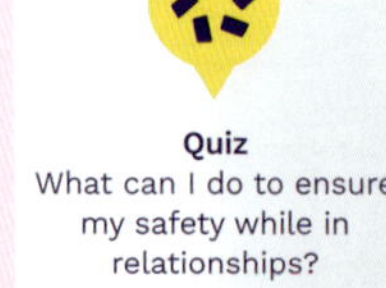

Quiz
What can I do to ensure my safety while in relationships?

9780170463102

HOW CAN I IMPROVE MY KNOWLEDGE OF FIRST AID AND REMAIN SAFE WHILE USING IT?

People are often seen in hospital emergency departments because they sometimes overestimate their abilities, make poor decisions or are pressured by peers into doing things they otherwise wouldn't. Talking about what can go wrong, practising basic first aid and having the courage to do something when accidents and emergencies occur are things that everyone needs to do.

In every first-aid situation, try to minimise the risk of infection being transmitted to yourself, the injured person and any bystanders. There are standard precautions to take, such as putting on gloves and wearing an apron and face mask if available. In first-aid situations such as resuscitation, resuscitation masks are recommended. It may be preferable for the partner or the parent of the patient to do the rescue breathing if you do not have a resuscitation mask available.

Before first aid	During first aid	After first aid
Wash your hands.	Use a resuscitation shield, if available.	Safely dispose of any used dressings, bandages and disposable gloves.
Use plastic or disposable gloves.	If available, wear gloves and ensure that they are not torn.	After removing disposable gloves, always wash your hands thoroughly with soap and water.
If you have cuts or wounds on your hands, ensure that they are covered by a waterproof dressing before applying gloves.	If you come in contact with body fluids, wash the area immediately with running water and seek medical advice.	

Figure 6.17 The precautions you should take when administering first aid.

THE DRSABCD ACTION PLAN

The DRSABCD action plan (see Figure 6.18) allows first-aid care in any life-threatening situation to be prioritised. DRSABCD will also help in checking for the presence or absence of consciousness and breathing, and in determining the type of basic life support measures required to preserve and/or restore life.

> **FAST FACT**
> One of the most common ways people die from alcohol is by choking on their own vomit. If a person vomits when they are unconscious, they can easily inhale the vomit. An unconscious person has lost basic reflexes such as the reflex to cough things up when they go down 'the wrong way'. If the body cannot get the oxygen it needs, brain damage or death may result.

defibrillation an electrical shock to reset the heart so that it beats by itself

automated external defibrillator (AED) a portable automatic device used in cardiac arrest; the AED automatically analyses the casualty's heart rhythm and sends a shock to the heart to restore a normal rhythm

The last 'D' stands for **defibrillation**, which can be performed using an **automated external defibrillator** (AED). The AED is very simple to use. First aiders simply follow a set of instructions in the form of voice prompts and visual guides.

Once you have used DRSABCD and assessed that a person is conscious (heartbeat and breathing both present), other injuries or conditions may require your attention.

SHOCK

Sometimes the circulatory system cannot supply enough oxygen to meet the demands of the body and an individual can go into 'shock'. This is a potentially life-threatening situation and must be dealt with quickly, appropriately and safely. Shock commonly results from head trauma, heart attack, heat exhaustion, severe dehydration, severe internal or external bleeding, alcohol or drug abuse (poisoning), bites or stings, or accidents.

signs factors that someone else can sense by sight, smell, touch or sound

symptoms factors that a person themselves can sense and comment on, such as pain or feeling nauseated

These are some of the **signs** and **symptoms** of shock:

- pale, cold, clammy skin
- shallow, rapid breathing or difficulty breathing
- anxiety
- rapid heartbeat, heartbeat irregularities or palpitations
- thirst or a dry mouth
- nausea
- vomiting
- dizziness and light-headedness
- confusion and disorientation.

Effective first aid and prompt medical attention when shock occurs can save a person's life. First aiders should follow these steps to treat shock:

- Follow the DRSABCD action plan.
- Lie a conscious person down and keep them warm and comfortable. If possible, raise their legs a little to improve blood flow to the brain, heart and lungs. (Do not raise their legs if a spinal injury is suspected or if moving their legs causes pain.)
- Do not give the person anything to eat or drink, as they may need an anaesthetic in hospital.

> **FAST FACT**
> Shock is the body's response to a sudden drop in blood pressure. At first, the body responds to this life-threatening situation by constricting (narrowing) blood vessels in the extremities (hands and feet), causing cold and clammy skin. If symptoms go untreated, shock can lead to death.

DRSABCD action plan

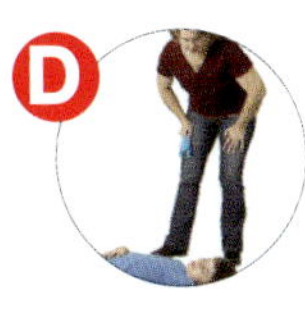

Danger Check for danger and ensure the area is safe for yourself, bystanders and the patient.

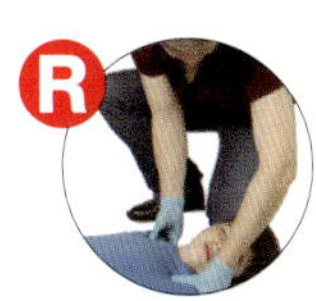

Response Check for a response: ask name and squeeze shoulders. **No response?** Send for help. **Response?** Make comfortable; monitor breathing and response; manage severe bleeding and then other injuries.

Send for help **Call triple zero (000)** for an ambulance or ask a bystander to make the call. Stay on the line. [If you are alone with the patient and you have to leave to call for help, first turn the patient into the recovery position before leaving.]

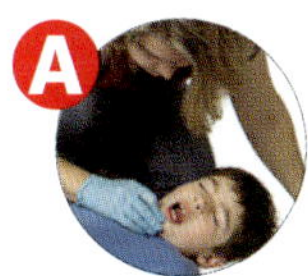

Airway Open the patient's mouth and check for foreign material. **Foreign material?** Roll the patient onto their side and clear the airway. **No foreign material?** Leave the patient in the position found, and open the airway by tilting the head back with a chin lift.

Breathing Check for breathing Look, listen and feel for 10 seconds. **Not normal breathing?** Ensure an ambulance has been called and start CPR. **Normal breathing?** Place in the recovery position and monitor breathing.

CPR Start CPR: 30 chest compressions followed by 2 breaths. Continue CPR until help arrives, the patient starts breathing, or you are physically unable to continue.

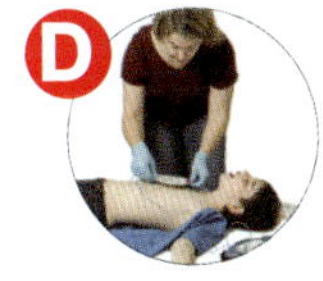

Defibrillate Apply a defibrillator as soon as possible and follow the voice prompts.

In a medical emergency call Triple Zero (000)

DRSABCD Danger ▶ Response ▶ Send for help ▶ Airway ▶ Breathing ▶ CPR ▶ Defibrillation

You could save a life with first aid training • www.stjohn.org.au • 1300 360 455

This information is not a substitute for first aid training. Formal instruction in resuscitation is essential.

St John Ambulance Australia is not liable for any damages or incidents that may occur in the use of this information by other parties or individuals.

Figure 6.18 DRSABCD action plan

Worksheet 6.13

Video
A good way to recall what to do in an emergency and how to effectively apply DRSABCD is to view each step online. Search for 'DRSABCD video' for clips that have been prepared by various first-aid organisations such as the Red Cross, St John Ambulance or Royal Life Saving Association. Watch the video clips and ensure you know what to do – especially when it comes to CPR. Always remember 30:2 compressions:breaths.

- Reassure and encourage them.
- Stay with them until the ambulance arrives.

Fainting occurs because of decreased blood supply to the brain; it can occur when people stand for extended periods of time or get up too quickly. Fainting episodes are usually very brief; the person tends to be semiconscious and progresses to full consciousness within a few minutes. If fainting occurs, treat the person as you would for shock.

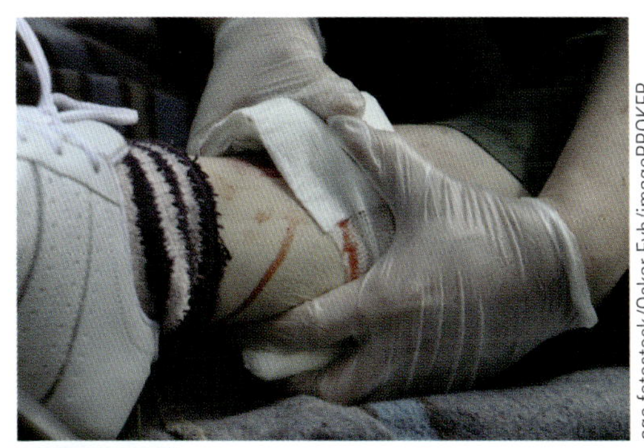

Figure 6.19 Use gloves if available, to avoid direct contact with the person who requires first aid.

CONCUSSION

The brain is free to move a little within the skull, which means it can be shaken by a blow or jolt to the head. This causes concussion, which can temporarily change the way the brain works. Common causes of concussion are traffic incidents, sports injuries, falls and blows received in fights.

AFL Photos/Photo by Darrian Traynor

Figure 6.20 Contact sports such as football and rugby have concussion assessment procedures that protect players.

The following are signs and symptoms of concussion:

- headache or dizziness
- loss of memory, particularly of the event
- confusion
- altered state of consciousness
- blurred vision
- wounds on the head (face and scalp)
- nausea and vomiting.

First aiders should follow these steps to treat concussion:

- Follow the DRSABCD action plan.
- If the person is conscious: reassure them, keep them warm, position them on their back with their head and shoulders slightly raised.
- If the casualty is unconscious: call 000, place in recovery position and monitor their airway, breathing and circulation. If there is blood or fluid from an ear, put the person with the injured side down to drain onto clean dry gauze.
- Keep monitoring the person until help arrives or they are transported to a medical facility.

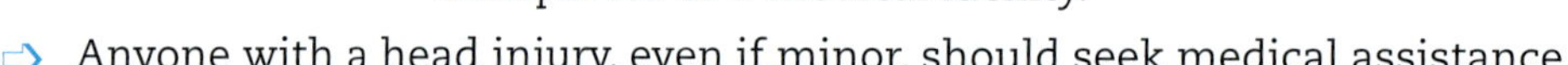

- Anyone with a head injury, even if minor, should seek medical assistance.

Worksheet 6.14

Cerebral compression can develop following head trauma. This occurs when the brain swells after an injury, creating pressure inside the skull. Because this condition is very serious and can require surgery to relieve it, anyone who has had any head injury or blow to the head must be monitored for at least 48 hours after the event. There are numerous stories of young people (usually men) being involved in a fight and receiving a minor blow to the head, then being helped home by mates, lying down on their bed and being found dead in the morning by family or a friend.

INVESTIGATION ⇨ CONCUSSION

Purpose

To investigate concussion and states of consciousness by finding articles about sports concussion and 'street concussion' (that is, concussion that occurs at parties or on the streets walking home).

Method

1 Conduct an internet search for articles that contain information about sport/street concussion.

2 Find at least two sources of information each for sport and street concussion.

Discussion

1. Are there any common themes that appear in the articles?
2. List in order the recommended steps to assist people suffering concussion.
3. Sportspeople who suffer repeated concussions have a higher likelihood of sustaining chronic damage to their brains. What does this mean? Summarise the types of altered brain function that commonly accompany these situations.

Following a concussion, it is best to wait until you are feeling better before you go back to your normal activities. Don't go back to work or school until you have fully recovered. Ask your doctor for advice. Don't return to sport until all the symptoms have gone. Your reaction times and thinking may be slower after a concussion, so you may be at risk of further injury. A second concussion that occurs before the brain recovers completely from the first – usually within a short period of time (hours, days or weeks) – can slow recovery or increase the likelihood of long-term problems.

FAST FACT

It is a myth that people suffering concussion should not be allowed to fall asleep in case they do not wake up. The brain needs rest to recover and repair itself. However, it is important that people suffering any form of concussion are closely monitored for 48 hours after sustaining a concussion, including while they rest or sleep.

DEALING WITH DIABETICS

Someone in your room or year level might have diabetes, but you may not know. There are no obvious signs or symptoms, as most diabetics manage the condition well.

Everyone needs glucose to provide their muscles and cells with the energy necessary to perform and function at optimal levels. Bodies absorb glucose with the assistance of insulin; people who have diabetes either don't produce any insulin, or they produce it only in small quantities. People with diabetes carefully monitor their daily activity levels and intake of foods to ensure there is a balance between insulin and glucose within their bodies. To help them maintain this balance, they either take oral medications or perform daily insulin injections, after checking their blood glucose levels.

FAST FACT

Both hypoglycaemia and hyperglycaemia can lead to people with diabetes feeling light-headed and can result in loss of consciousness.

FACE TO FACE Diabetics

Look carefully at the image.

1. Describe what the first aider has done to assist the unconscious man.
2. She believes the man might have diabetes – what clues might have led the first aider to arrive at this conclusion?
3. Discuss how the first aider could use the next person to arrive at the scene to assist both her and the unconscious man.
4. State what the first aider should do if she notices that the man has stopped breathing.

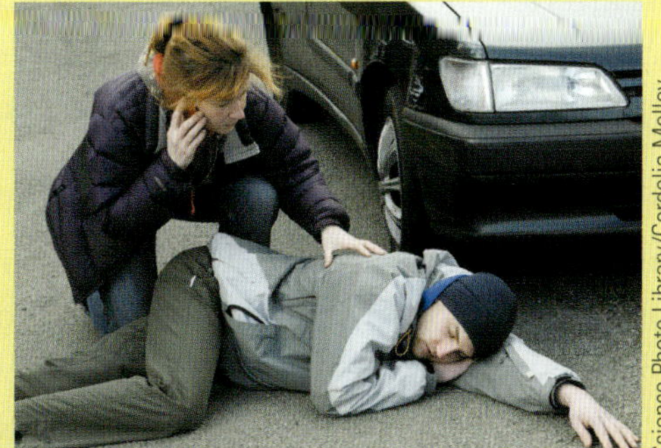

Science Photo Library/Cordelia Molloy

Figure 6.21 Hypoglycaemic or hyperglycaemic – what happens next?

hypoglycaemia abnormally low blood glucose levels

hyperglycaemia abnormally high blood glucose levels

The signs of **hypoglycaemia** or **hyperglycaemia** are very similar, and unless a blood glucose reading has been taken it is difficult to determine if the person's blood glucose is too high or too low. If in doubt, treat them as if they have hypoglycaemia. If this is not the case and they already have high levels of glucose, the extra glucose will not lead to any further harm. Table 6.2 lists the signs and symptoms of hypoglycaemia and hyperglycaemia and also the first-aid steps to be taken in both instances.

Table 6.2 Diabetes action plans

	Hypoglycaemia (low blood glucose)	Hyperglycaemia (high blood glucose)
Signs and symptoms	sweating weakness trembling fast heartbeat confusion irritability hunger headache drowsiness or unconsciousness	more urine output than usual increased thirst dry skin and mouth 'acetone' breath, like nail polish remover decreased appetite, nausea or vomiting fatigue, drowsiness or lack of energy unconsciousness
First-aid steps	Follow the DRSABCD action plan If the person is conscious: reassure them, keep them warm and give them jelly beans, fruit juice or another sugary drink (if you suspect hypoglycaemia) If the person is unconscious but breathing: place them in the recovery position, call 000 or 112 (mobile) and assess their airway Keep monitoring the person until help arrives or they are transported to a medical facility **Do *not* try to give the patient a dose of insulin – this can be dangerous unless the patient's blood sugar level has been tested. If they are conscious and have tested themselves, you can assist them to inject their own insulin.**	

BLEEDING

The cardiovascular system comprises the heart and all the blood vessels, which together make up the circulatory system. Blood flows throughout the body from the heart via arteries and then back to the heart via veins. Blood transports oxygen and nutrients to working muscles and organs, and removes waste and carbon dioxide. It is also responsible for maintaining a constant core temperature.

Severe external bleeding

External bleeding that fails to clot and stop by itself can occur as a result of many situations, including a sporting injury, an outdoor accident, road trauma and physical violence.

First-aid management for severe external bleeding includes PER (pressure, elevation and rest):

- Stop any bleeding by applying a clean cloth firmly to the wound. If blood seeps through the cloth, apply more padding on top of the first cloth. If this added padding gets soaked, it can be replaced, but leave the initial pad in place to minimise contamination. Continue the firm pressure. If gloves are available, use these prior to commencing first aid in an effort to minimise risk of infection between yourself and the casualty.

- If the wound is on the arm or leg, raise the limb above the level of the heart to help slow bleeding.
- If the person is conscious, they may be placed on their back with their head and legs slightly supported and raised.
- Do not apply a tourniquet unless the bleeding is severe and fails to stop with direct pressure or a pressure bandage (if appropriate).

CASE STUDY: A GLASSING, BASHINGS AND OUT OF CONTROL PARTIES IN PERTH

Identify

Police have had a busy weekend in Perth, responding to three out-of-control parties, a nightclub glassing and two bashings.

Understand

Just after midnight on Sunday, a 20-year-old man was smashed in the head with a glass at the Library nightclub in Northbridge during a brawl involving two groups of young men, police said.

The man was taken to hospital by friends for treatment to a 6 cm cut to his forehead.

Police said investigations were continuing and no charges had been laid.

Also, in Northbridge, about 1.30 a.m. a 25-year-old man was knocked unconscious when he was king hit by an unidentified man in a laneway off Aberdeen Street, police said.

The man suffered a broken jaw and was to undergo surgery in Royal Perth Hospital.

Police inquiries into that incident are continuing.

Meanwhile in East Perth about 12.30 a.m. on Sunday a 41-year-old was attacked by two young men who kicked and punched him to the ground and robbed him of his wallet and phone.

He was taken to hospital for treatment of swelling and bruising to his face and head.

About 9.45 p.m. on Saturday the police helicopter and patrol cars were sent to an out-of-control party at Noranda in Perth's northeast, where gatecrashers damaged several cars.

Police later attended another wild party in Cooloongup in Perth's south, where people were brandishing bottles and one man received a head injury requiring him to be taken to hospital.

Just after 1 a.m. on Sunday, officers were called to a party in Gosnells in Perth's east, where they moved people on after gatecrashers caused trouble after being refused entry, police said.

Source: 'A glassing, bashings and out of control parties in Perth', *WA Today*, 19 June 2011.

Discuss

1. Discuss why bleeding resulting from 'glassings' and head trauma needs to be treated with greater care than external bleeding resulting from a graze to a knee.
2. List three reasons why parties get out of control.
3. Hypothesise, if the man who was 'king hit' died in hospital the following day, discuss how the following people would feel:
   - his girlfriend
   - his four-year-old son
   - his attacker's wife
   - his football teammates
   - his attacker
   - his attacker's mates.

Internal bleeding

Internal bleeding is a medical emergency. The signs and symptoms that can suggest concealed internal bleeding depend on where the bleeding is inside the body, but may include:

- pain at the injury site
- bleeding from the mouth or ears
- swelling
- nausea and vomiting
- pale, clammy, sweaty skin
- breathlessness
- unconsciousness.

First aid cannot manage or treat any kind of internal bleeding.

Urgent medical assistance is vital and needs to be sought. Listen carefully to what the person tells you about their injury – for example, where they felt the impact. In the case of a head injury, they may display the signs and symptoms of concussion. If you are one of the first people present when an incident resulting in suspected internal bleeding occurs, you should treat this in the same way as shock.

CASE STUDY

CAR CRASH

Identify

You are out early one morning jogging when you witness a speeding car smash into a parked car outside a friend's house. As you get closer to the two smashed vehicles, you realise that you are the first person on the scene and will need to quickly assess the situation and decide what to do. You are confronted with the following scene.

Fairfax Syndication/Justin McManus

Figure 6.22 Distracted drivers increase their risk of having an accident.

Understand

- The driver of the speeding car is trapped by the airbag, unconscious and bleeding from his legs, which have been badly injured by the front of the car being crushed inwards.
- The woman in the parked car has hit her head on the windscreen and blood is streaming from a gash in her forehead. She seems dazed, confused and is looking very pale. She keeps crying out for help.
- Both vehicles having sustained significant damage and you can smell petrol and see smoke coming from the front of the car that was speeding. The parked car has been hit with such force that it has been moved into the middle of the road.

Discuss

1. List three factors that may have contributed to the driver crashing into the parked car.
2. Consider the injuries that may have occurred in this scenario. List them in order of highest priority to lowest priority.
3. Follow the DRSABCD action plan and outline how you would respond to this emergency.
4. Assume that your friend heard the crash, was woken and has run out to see what has happened. Outline how you could use her during the vital first few minutes to help you.
5. Your friend seems to have gone into shock and starts vomiting. What can you do or say to deal with this added situation?

ALCOHOL AND DRUG ABUSE = POISONING

A poison is any substance that, when introduced into the body, results in illness or injury. Poisons can be introduced into the body through the following means:

- ingested – eaten (e.g. alcohol, foods, medications)
- inhaled – breathed in via respiratory system (e.g. fumes, glues, paints, gases)
- absorbed – via the skin (e.g. contact with plants, fertilisers, pesticides)
- injected – into the skin or bloodstream (e.g. drugs, insect bites, animal stings).

Poisoning may be accompanied by the symptoms shown in Figure 6.23.

First aid for casualties who are showing signs of poisoning needs to occur quickly until medical assistance becomes available. The following steps are vital in assisting people who may be suffering from poisoning:

- Follow the DRSABCD action plan.
- If the person is conscious, reassure them and keep them warm.
- Ask the person or others around them what was taken, how much and when.
- If you think the person has been poisoned by substances other than alcohol, contact the Poisons Information Centre in your state on 13 11 26 or call 000 or 112 (mobile) for advice on treatment.
- If the person is unconscious and breathing, place them in the recovery position and regularly assess their airway.
- Keep monitoring the person until help arrives or they are transported to a medical facility.
- Withdraw if the victim becomes violent.

FAST FACT

Treatment for poisons used to involve inducing vomiting to try to get the poison out of the body. Now, the recommended treatment is to rinse the casualty's mouth and provide them with small sips of milk or water. Why do you believe first-aid recommendations have changed?

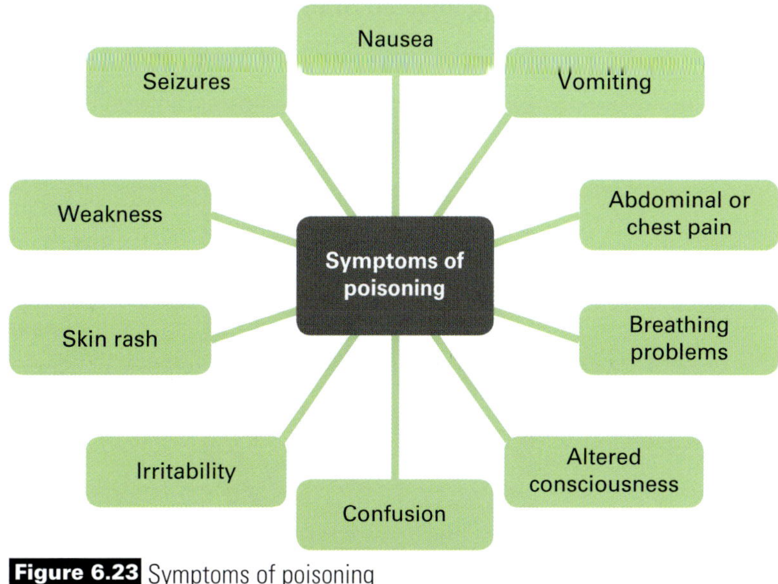

Figure 6.23 Symptoms of poisoning

FACE TO FACE Losing consciousness after a party

Getty Images/Matt Cardy

Figure 6.24 A good time, or is it about to go terribly wrong?

After considering the image, discuss the following questions:

1 List four potential dangers the person has placed themselves in by drinking excessively and losing consciousness on a bench in a public space.
2 The person in the image wakes up the next day in a hospital bed and can't recall much about the previous night. One of their friends visits them and informs them that they slept with at least two people who were at the party and that they were 'off their face'. Discuss the physical, social and emotional consequences of this information.
3 Consider where they would go to find out more about potential risks such as unwanted pregnancy, sexually transmissible infections and how to practise safe sex.

ANAPHYLAXIS

anaphylactic shock a rapidly developing and serious allergic reaction that affects a number of different body systems simultaneously

Allergic reactions to foods, bites, stings and drugs can result in **anaphylactic shock**, which is a life-threatening situation. Anaphylactic reactions result in swelling of the respiratory system, breathing difficulties, nausea, vomiting and respiratory distress. Anaphylaxis can be caused by substances that gain access into the bloodstream by being injected or ingested. An extensive reaction involving the skin, lungs, nose, throat and gastrointestinal tract can then result. Severe anaphylactic reactions can be fatal.

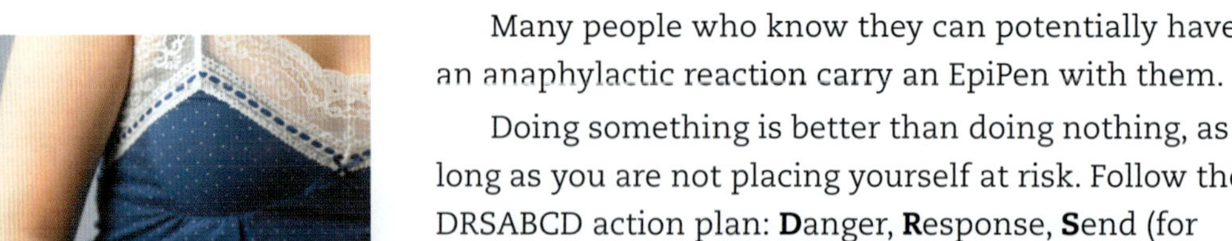

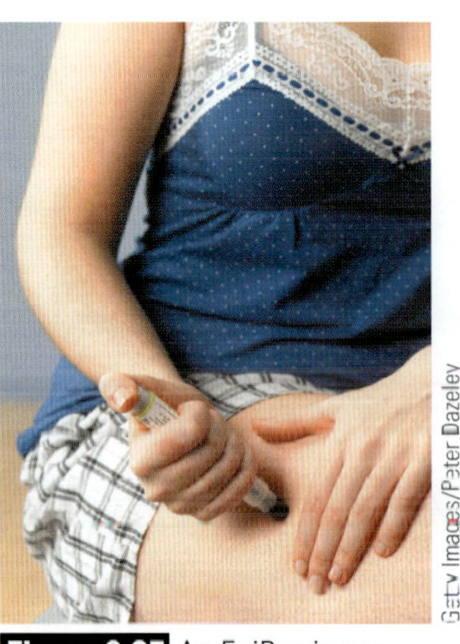

Getty Images/Peter Dazeley

Figure 6.25 An EpiPen in use

Many people who know they can potentially have an anaphylactic reaction carry an EpiPen with them.

Doing something is better than doing nothing, as long as you are not placing yourself at risk. Follow the DRSABCD action plan: **D**anger, **R**esponse, **S**end (for help), **A**irway, **B**reathing, **C**PR, **D**efibrillation.

- Reassure the person.
- Stay with the person and ensure total rest.
- Assist the person to administer their medication (if they have an EpiPen or Anapen) or, if they are unconscious, administer into front of thigh (quadriceps muscle group) and call emergency services.

9780170463102

- Call 000 or 112 (if calling from a mobile).
- Keep monitoring the person until help arrives or they are transported to a medical facility.
- Be prepared to perform CPR.

INVESTIGATION: HOW DO ALLERGIES AFFECT US?

Purpose

To find out more about anaphylaxis, its causes or triggers and suggested emergency response steps.

Method

1 Conduct an internet search to find out how anaphylaxis is different to other allergic reactions:
  a What are the triggers?
  b How do people react?
  c What do people feel and how do they look when having an anaphylactic reaction?

Discussion

1 Summarise the signs and symptoms that show that someone is having an anaphylactic response.
2 How does an EpiPen work to reverse the anaphylactic response?
3 How would you use someone's EpiPen on them to help with the situation if they have lost consciousness?
4 Can you use another person's EpiPen on someone who does not have their EpiPen with them?

1 One of your classmates badly rolls their ankle during a PE class and you believe they are going into shock. List three signs you could use to make this judgement.
2 Why should an unconscious person never be placed on their back if they have a pulse?
3 What is a 'defibrillator' and how can it be used to save lives?

1 Recall if you or a friend have ever had a blood nose. Research and list the recommended steps involved in managing a nose bleed effectively.
2 Alcohol poisoning can occur when individuals consume large amounts of alcohol in short periods of time – this sometimes happens with binge drinking. What should you do if you believe one of your friends is suffering from alcohol poisoning?

3 Use the internet to find a video that demonstrates how to use an AED. Watch the video, then discuss the following questions:
   a How easy do you think it would be to use an AED?
   b How does an AED device help an unconscious person whose heart may have stopped beating?
   c Have you seen an AED? If so, where?
   d Do you think you would be able to help someone who needs external defibrillation?

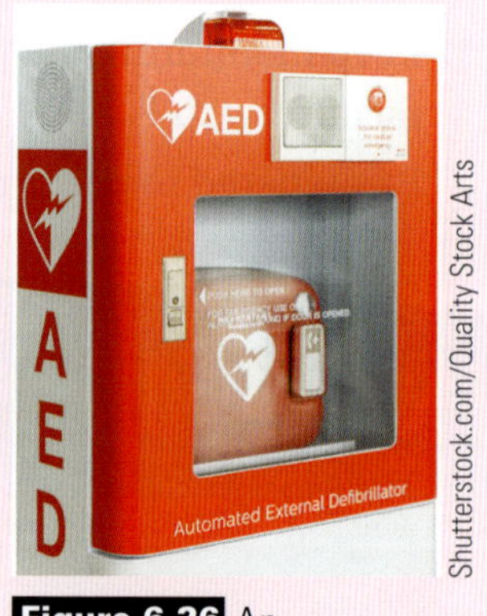

Shutterstock.com/Quality Stock Arts

Figure 6.26 An automated external defibrillator (AED)

1 You are at a party and someone falls and hits their head. You believe they might have concussion. Your friends say the best thing to do is to keep talking to the injured person and not let them fall asleep because their condition will worsen. Research this claim and either confirm or refute your friends' suggestion.
2 Discuss if it is okay to use someone else's EpiPen if a person is having an anaphylactic reaction and they do not have their EpiPen with them. Does the same apply to asthma inhalers?

Alamy Stock Photo/Helen Sessions

Figure 6.27 An EpiPen. Clear instructions for use are given on the side of the dispenser.

CHAPTER 6 REVIEW

1 Young people must identify their own personal network of trusted, helping adults to whom they can turn if they are not feeling safe. These are known as go-to people. Who are your 'go–to' people at school, home and community settings?

2 Consider two situations where your safety might be challenged at school. What strategies would you include as a part of your forward–thinking safety plan?

3 Think about any potential for a non-school setting or situation to become unsafe or risky. List at least two physical outcomes you might experience that contribute to an 'uneasy feeling'.

4 Discuss why it is important to act quickly to remove yourself from situations where you have determined that your safety could be placed at risk and things could go wrong for you or your friends.

5 Alcohol and drug abuse can have long-term effects on the developing brain of young people. Based on your opinion, list five of these in order from worst to least severe.

6 Positive relationships are characterised by both people having equal rights. What does this mean?

7 List two negative physical, emotional and social consequences that might result if violence is used to sort out problems.

8 Emotional, physical and sexual abuse involving young people needs to be stopped via the intervention of a trusted other that you have identified in your support network. What can bystanders do to stop inappropriate and unsafe behaviour?

9 Bullying behaviour is unsafe and needs to be stopped before people are hurt. What are some reasons why bullying occurs?

10 Assertiveness leads to respect for your own needs and rights, and also to those of others. Write down three different assertive statements you could use when being pressured into doing something you do not really want to.

11 When considering young drivers, what factors markedly increase the risk of them being involved in accidents and road trauma?

12 Shock is a potentially life-threatening situation and must be dealt with quickly, appropriately and safely. What are some of the signs and symptoms that someone is experiencing shock?

13 Why do you believe it is important for every student in Australian schools to undertake basic first-aid training as part of their education?

BEING ACTIVE AND ADVENTUROUS OUTDOORS

7

IN THIS CHAPTER

You will discover the role that outdoor recreation and physical activity have played and continue to play in Australian society. You will also explore the qualities needed for leadership and the key factors necessary for safe bushwalking.

Shutterstock.com/GoodStudio

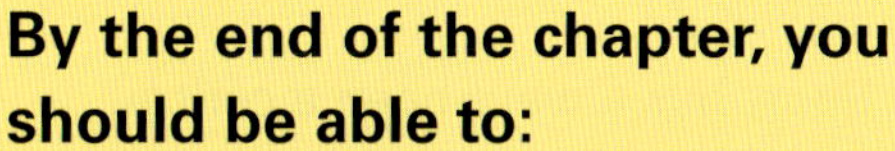

By the end of the chapter, you should be able to:

- recognise the role that outdoor recreation, sport and physical activity has played in the lives of Australians
- explore the impact of media messages associated with outdoor recreation, physical activity and sport
- appreciate the significance of working as a team or group, and of developing leadership and collaborative skills
- refine your skills in increasingly complex situations and know how you can transfer skills from one challenge to another
- evaluate and review your ability to work in a team and how effective you are as a team member
- recall the importance of planning and preparation for bushwalking.

Video
Case study: What is the importance of outdoor recreation, sport and physical activity to these students and why?

WHAT ROLE DOES SPORT, OUTDOOR RECREATION AND PHYSICAL ACTIVITY PLAY IN AUSTRALIAN LIVES?

Quiz
Pre-chapter

Before you start, take the pre-chapter quiz to find out how much you already know.

SPORT

Australia is a unique and diverse country in every way – in culture, population, climate, geography and history. Australians have always been passionate about sport. Observers have commented about the 'sporting obsession' of the Australian people, and many people say that life in Australia would not be the same without sport. Sport has always been a central feature of Australian culture, and enthusiasm for sport is generally thought to be an Australian characteristic.

Alamy Stock Photo/Arcaid Images

Figure 7.1 Sport is a big part of Australian culture.

Sport has always been looked upon as a national pastime that has largely unified Australia and helped bridge race and gender inequalities. Australia's sporting and recreational pursuits date back to the country's original landowners – the First Nations Peoples.

First Nations Peoples way of life dates back some 80 000–100 000 years. Children were encouraged to participate in activities such as climbing, jumping and running, which were enjoyable, but also important survival skills. Gathering and hunting for foods meant that an adept climbing ability and throwing accuracy was important. Mock fights using mini weapons were a popular pastime.

iStock.com/benoitb

Figure 7.2 Traditional customs and traditions such as hunting for food were common in many First Nations tribes.

European arrival in Australia in the late 1700s led to significant changes to the traditional First Nations Peoples way of life, and their customs and traditions were dramatically altered. While First Nations Peoples worked hard to maintain a connection to the land, the European immigrants were more focused on expansion and change. Many First Nations groups were wiped out as a result of this expansion and colonisation, and those who survived the European invasion were forced to take on European forms of sport and leisure.

Early arrivals consisted mainly of convicts from England. After approximately 10 years the first 'free settlers' arrived, bringing new activities to Australia. Hunting remained constant, but new sports such as cricket, rowing and boxing were introduced.

It wasn't until the 1820s and 1830s that water sports were officially organised. Bathing was even a popular pastime in some areas. Clubs were established for boxing in 1814, horseracing in 1825, cricket in 1826, rowing in 1835, billiards and sailing in 1836, shooting in 1842, lawn bowls in 1846 and golf in 1847. During this period in Australian history, from pre-1788 to 1850, sport changed from something that served as a survival means to an activity designed for leisure and culture.

Competitions were held between the 'new Australians' and First Nations Peoples. Any success in overseas sporting competitions boosted national pride. There was also a rise in intertown and intercolonial competition. A new game, Australian Rules, was created. On 7 August 1858, the first game of Australian Rules Football was played between Melbourne Grammar School and Scotch College, near the current site of the Melbourne Cricket Ground (MCG). The game was influenced by Marngrook (a Gunditjmara word for 'game ball').

Australian football was the first sport to be unique to Australia. An increase in competition prompted the standardisation of rules. As a result, several organisations were formed, including the New South Wales Cricket Association in 1858, the Northern (Queensland) Rugby Union in 1874 and the Victorian Football Association (Australian Rules) in 1879.

Australia's first significant international sporting success took place in 1876, when Edward Trickett won the sculling (rowing) championship of the world. Australia was the best in the world at something! The same year a combined New South Wales and Victorian cricket team beat an all-England team in Melbourne. These triumphs stimulated a nationalism and a fervour for sport that dramatically increased over the following years.

Getty Images/DEA/BIBLIOTECA AMBROSIANA

Figure 7.3 Edward Trickett first won in the rowing championship of the world in 1876 on the River Thames in England. This illustration shows him in 1877 when he again won the championship, this time on the Parramatta River in NSW.

Australia's sporting passion can also be linked to the physical environment. The sunny climate and warm seasons enable sports to be played year-round, and there is plenty of suitable land for playing fields. Additionally, the population has always been largely concentrated near the coasts, making water sports accessible. Large concentrations of people in major cities facilitated competition both within and between cities.

WB Worksheet 7.1

Sports historians believe that contemporary Australian sport has developed as a mixture of different games, traditions and ideologies imported here following European settlement. Despite what has been inherited, Australia has placed its own stamp on sport.

INVESTIGATION

AUSTRALIAN RULES FOOTBALL – HOW WAS THIS UNIQUE SPORT INVENTED?

Purpose

Conduct your own research into the origins of Australian Rules Football.

Method

Undertake an internet search to discover the origins of Australian Rules Football and any significant changes that have occurred in the past 150 years.

Discussion

1. Summarise how the game was created in the 1800s, including:
   a. who was involved
   b. how many teams made up the competition
   c. how often games were played
   d. which states and territories played the game.
2. Create a timeline to show how the game has changed and developed since its inception in the 1800s. Ensure you identify changes to the playing field, number of players per team, rule changes, state involvement and any other interesting changes you come across.
3. First Nations Peoples are among some of the most talented AFL players making up teams in today's national competition.
   a. Discuss the role of First Nations Peoples play in creating Australian Rules Football.
   b. Propose why you believe 'Aussie Rules' is so popular among First Nations populations.

FACE TO FACE Aussie sport

As a class, discuss the question: 'In what ways has Australia created its own unique sports culture?'

Australia has a mixed culture and population. Many Australians were born overseas and brought many of their traditions with them when migrating to Australia. Soccer is one example of a sport that grew out of Australia's migrant population.

It was once believed that soccer was only taken seriously in Australia by those of ethnic background (mainly from Europe, where soccer is the main national sport in many countries). This has changed a lot over the past 20 years. Today, soccer has one of the highest participation rates of all the football codes in Australia, and has many supporters. In Australia, soccer now has a premier league; games are televised and famous teams from around the world are invited to come and play.

FACE TO FACE Football codes

Investigate participation in the football codes through the Guardian Australia interactive football statistics.

1 Out of all the football codes, name the code that has the highest participation rate.
2 Consider if the male and female participation rate is similar, and propose a reason why this might change in the next five years.
3 Contrast particpation and spectator rates in AFL.
4 Consider how country of birth affects participation rates in football codes.

Weblink
Battle of the codes: Australia's four football codes compared

Alamy Stock Photo/Aflo Co. Ltd.; Newspix/Gregg Porteous; Fairfax Syndication/Sebastian Costanzo

Figure 7.4 Football is now one of the most popular sports in Australia.

OUTDOOR RECREATION

Outdoor recreation, or outdoor activity, refers to leisure pursuits engaged outdoors, often in natural or semi-natural settings outside towns. Whatever the activity, the common element is enjoyment.

Outdoor recreation activities have these typical features:

- do not involve organised competition or formal rules
- can be undertaken without any buildings or infrastructure
- may require large areas
- may require mostly unmodified natural landscapes.

Outdoor recreation usually occurs in places with these characteristics:

- tracks that are long enough for walking, running, horseriding, cycling, trail-bike riding or four-wheel driving
- reliable winds for sailing, hang-gliding or kite flying
- reliable waves for surfing, windsurfing, paddleboarding or kitesurfing
- sufficient snowfall for skiing, snowboarding and cross-country skiing
- suitable rock formations for climbing
- calm sea conditions to snorkel, scuba dive or fish from a boat or from the rocks.

iStock.com/PeopleImages

Figure 7.5 What other activities would you consider to be outdoor recreation?

Worksheet 7.2

Weblink
Go to the Australian Alps National Parks website and click on the link to the fact sheet, 'Recreation and tourism in the Australian Alps'. Investigate how this area has changed over time.

Examples of outdoor recreation could include a restful day in a beautiful place, bushwalking, having a picnic, fishing, meditating or cycling.

FACE TO FACE Think-pair-share

1 Propose your own definition of outdoor recreation.
2 Summarise any changes to participation in outdoor recreation pursuits in Australia over the past 50 years. Briefly discuss your answer.

beneficial use doing an activity with the goal of physical and social rewards

aesthetic contemplation appreciating the beauty surrounding you

The two primary purposes for outdoor recreation are **beneficial use** and pleasurable appreciation of the outdoor environment.

Outdoor activities include backpacking, canoeing, canyoning, caving, climbing, hiking, hill walking, hunting, kayaking, rafting, water sports, snow sports and horseriding. Goal-directed outdoor activities are mostly physical, but they may also be mentally, emotionally and spiritually rewarding. Outdoor recreation can meet the needs of physical health, self-sufficiency, risk-taking, building social ties (including team building) and the needs of achievement (such as practising and improving challenging skills, testing stamina and endurance, and seeking adventure or excitement).

Being out in nature can be very pleasurable. Your wellbeing and/or spiritual life may be improved through outdoor activities and outdoor-related activities such as nature study, **aesthetic contemplation**, meditation, painting, photography, archaeological or historical research, and First Nations culture. These activities may also be physically rewarding.

iStock.com/SolStock

Figure 7.6 Kayaking is a challenging outdoor activity.

Many people in developed countries believe that nature is only valuable for utilitarian reasons (the way we can use it). They discount the inner wellbeing and/or spiritual benefits of being out in nature.

Outdoor activities might help you find peace in nature, enjoy life and relax. They can be an alternative to more expensive forms of tourism. Outdoor activities are also frequently used for education and team building.

WELLBEING CHECK IN

MINDFUL NATURE WALK

This activity has been developed in collaboration with Dr David Bakker of MoodMission

Identify

Outdoor adventures don't have to be action packed. Even a simple walk can be beneficial.

Understand

As discussed in Chapter 4, mindfulness is a way of focusing the mind and letting go of distractions or unhelpful thoughts. Doing something mindfully increases our appreciation of it because we pay closer attention. If you get distracted from what you're mindfully focusing on, just put the distraction to one side and gently refocus. Research has shown that spending time in nature has positive mental health effects. Studies have found that going for a walk in a natural environment is more beneficial for wellbeing, happiness and stress-reduction than going for a walk in an urban environment. But it can be hard to get these benefits if we're being distracted from our surroundings.

Practise

1. Plan a short walk in a natural setting. It could be in a garden, at a park, in the bush or somewhere else surrounded by trees and plants.
2. Go on your walk without other distractions. That means going alone and leaving your headphones out.
3. While walking, try to pay special attention to your surroundings. Look at the trees and count the branches. Notice their shapes. Notice how many colours you can see in each plant, and compare these with neighbouring plants.
4. If you get distracted by something, first notice that you've become distracted, put the distraction to one side and then refocus on your surroundings. You might have lots of distracting thoughts about yourself, other people, the past, the future, etc. That's okay – in fact, it's totally normal. Refocusing on your surroundings will help you let go of any anxieties or negativity.

Reflect

Describe something about your surrounds that stood out – briefly discuss why. Elaborate on any new 'ponderings' or thoughts you may have had whilst walking. People often learn a lot about themselves when they try to focus on something other than themselves.

PHYSICAL ACTIVITY

Physical activity is important to maintain good overall health and wellbeing. Being physically active reduces the risk of some chronic conditions, helps to control weight and improves mental wellbeing. Some forms of physical activity may also help manage long-term physical problems, such as arthritis and type 2 diabetes, by reducing the effects of the condition.

Worksheet 7.3

hypokinetic diseases conditions that occur as a result of lack of exercise and movement

Recently there has been a decline in physical activity. This is due to changes in transport, the increasingly sedentary nature of many forms of work, and the growing popularity of activities such as watching television or using a computer. The COVID-19 pandemic also impacted on participation in physical activity, particularly a decline in organised sports. Sedentary behaviour has been linked to a rise in people being overweight and obese and an increase in **hypokinetic diseases**. Obesity has been shown to increase an individual's risk of cardiovascular disease, colon and breast cancers, type 2 diabetes and osteoporosis. Chapters 3 and 9 also talk about the benefits of physical activity.

CASE STUDY  NATIONAL HEALTH SURVEY: TYPES OF ACTIVITY AT WORK

Identify

The National Health Survey was conducted by the Australian Bureau of Statistics (ABS) in 2020–2021. The survey focused on long-term health conditions, heath risk factors such as physical activity levels and demographics of Australians.

Understand

People who were employed and worked in the week prior to interview were asked to describe their usual workday. Almost half (49.4 per cent) of people aged 18–64 years who were employed and worked in the last week described their day as mostly sitting, with standing (18.8 per cent) and walking (17.5 per cent) the next most common activities.

People aged 18–24 years were least likely to report mostly sitting (23.8 per cent) on a typical workday.

Men aged 18–64 years were four times more likely than women to report mostly heavy labour or physically demanding work (19.4 per cent compared to 4.4 per cent).

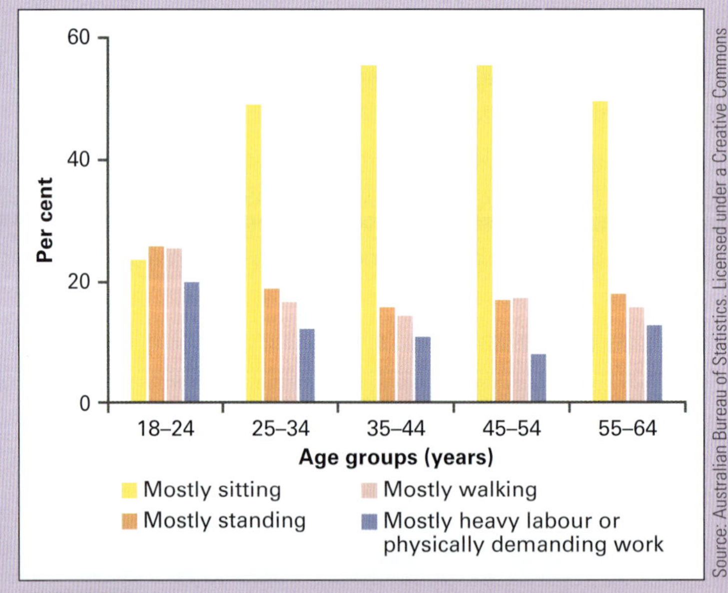

Figure 7.7 Type of physical activity at work on a typical work day

Discuss

1. Summarise what these findings reveal about sedentary behaviour in adults. Find three interesting facts from reading these findings.
2. Search the ABS website to compare the recent survey to ones from the past 20 years. Describe the changes in physical activity and sedentary behaviours that have occurred.
3. Propose what might happen to levels of sedentary behaviour in the next 10 years and why.

Weblink
ABS

Figure 7.8 Decreased physical activity caused by technological advancements have contributed to an increase in hypokinetic diseases.

Shutterstock.com/MicroOne

MEDIA INFLUENCES

The media in Australia has evolved from print-only newspapers to film, television, multimedia digital broadcasting and social media.

In the early days after European arrival in Australia, press reports of sporting endeavours played a role in shaping the way Australia viewed and understood itself. The media continues to influence the way Australians see themselves, and sport in general.

As Australia developed economically, more Australians had time to play and watch sport and more time to read about it. The first newspaper devoted entirely to sport was *The Referee*, published in Sydney in 1886.

The media now has a very powerful influence in society, particularly because of the internet and television. This can have both negative and positive influences on individuals and organisations.

CASE STUDY ⇨ OUR WOMEN ARE THE PRIDE OF AUSTRALIAN SPORT, BUT MEDIA COVERAGE IS YET TO CATCH UP

Identify

For many years the Women's Electoral Lobby (WEL) published statistics purportedly representing the coverage of female sport in what is now referred to as 'mainstream media'. Typically, coverage devoted to women's sport would be in the low single figures with the possible exception of those weeks when a Karrie Webb major, Australian team netball triumph or some other stupendous international achievement caused a blip on the graph where viewing numbers increased.

The release of these figures, accompanied by a harshly worded press release, was intended to name and shame media outlets which were supposedly deliberately ignoring women's sport and failing to provide the coverage it deserved. But, rather than study those single digit percentages and immediately order a centre page spread on the Opals or lead the next day's back page with the Southern Stars, editors would glibly – and, in some cases, justifiably – defend their coverage.

'Demand and supply.' 'We're running a business not a charity.' 'We gave the tennis and swimming a good run.' 'Our readers just don't want to know about women's cricket or football.' …

Beyond #Sandpapergate, the biggest story in Australian sport last year was the continued rise in both the performances and prominence of female athletes.

Australia's best footballer is, unquestionably, Samantha Kerr.

Cricket lovers have spent the summer lamenting Australia's inability to produce elite players – except those watching Ellyse Perry, Alyssa Healy and the other heroes of the recent T20 World Cup triumph. And so it goes through the deeds [of] the various national teams, Stephanie Gilmore's latest world surfing championship and Ash Barty's defeat of [now former] world number one Simona Halep this week, to name but a few.

As former road cyclist and now head of the Victorian Office for Women and Sport in Recreation Bridie O'Donnell said on *The Project* recently: 'At the moment it's the women's time to shine … they've become role models for all young kids.'

So given their achievements and increased recognition, you might expect a dramatic spike in the day-to-day coverage of women's sport. And in competition time there has been a noticeable increase.

The AFLW particularly attracted such prominent headlines in its first season that the cave-dwelling element of AFL fandom, and even a few administrators from competing female sports, grumbled that the new league's profile vastly exceeded its quality. The fact a women's league was accused of getting 'too much coverage' in itself seems something to celebrate. Yet across the board the general media exposure does not yet reflect the rise in prominence of female athletes.

Women Sport Australia's new 'sport photo action award' is a welcome initiative because it recognises that an image of a woman performing on the field is far

Getty Images/Cameron Spencer

Figure 7.9 Ellyse Perry was awarded the ICC Cricketer of the Year in 2019.

Getty Images/Tony Feder

Figure 7.10 Many people believe Sam Kerr is among the best three female footballers in the world.

Alamy Stock Photo/ZUMA Press

Figure 7.11 Stephanie Gilmore is a seven-time world champion surfer.

Getty Images/Chaz Niell

Figure 7.12 Ash Barty was the world tennis number 1 for most of 2019 and 2020.

9780170463102

7

more powerful, and empowering, than one of a female athlete in a skimpy outfit or cuddling a koala (these cheesy shots counted as 'female sport coverage' under the WEL statistics).

Yet there remains a discernible difference in how major male and female sports are prioritised.

Male sport coverage is predictive with an assumption readers/viewers in Sydney and Brisbane want NRL 365 days a year and those in the southern states want a 24/7 all-you-can-eat AFL buffet. So even in the off-season the most trivial 'stories' are covered.

The coverage of female sport is reactive, with immediate events or achievements given greater prominence than in the past, but women's sport is disappearing altogether when there is no 'news hook' or obvious angle.

No doubt the old 'demand-supply' argument would be used to justify the decision to run a story on a fringe Collingwood player with a minor off-season hamstring strain instead of a WBBL player's 50-ball century.

But even if this commercial logic stacks up, you can no longer argue there are far fewer prominent and successful female athletes from which to generate legitimate stories.

Perhaps this disparity in coverage will not change until the generation of young female participants currently flocking to sports grounds become consumers of, and advocates for, the more prominent telling of female sports tales.

Meanwhile, expect to hear about an NRL star with a sore toe two months before the first serious game, and merely hope you'll also get some insight about how the Diamonds perform during their tour of England.

Source: 'Our women are the pride of Australian sport, but media coverage is yet to catch up' by Richard Hinds, ABC News, 11 January 2019

Discuss

1 Discuss what this means, and what the article mentions about sportspeople becoming role models for younger Australians. What qualities are being modelled by our sports professionals?

2 a 'Sandpapergate' is a term that arose during a cricket test series between Australia and South Africa. Conduct your own research to find out what this referred to.

  b How did 'Sandpapergate' tarnish the image of Australian sports performers?

3 Consider how women receiving equal pay in sports like tennis, and most recently soccer, influence greater media attention, and thus promotion of their sport(s).

1 Consider how Australian success in sporting contests help increase the quality and availability of outdoor spaces.

2 Participation in outdoor recreation has multiple purposes and benefits. List at least two of these.

3 Create a table to summarise both the positive and negative influence the media has on people using outdoor spaces.

1 Other than money, list some of the factors that influence people to use the outdoors as part of their professional work – this might include sportspeople, conservationists, rangers, etc.

2 The natural environment provides many intrinsic rewards. What does this mean? Recount one of the times you have found yourself experiencing the same kinds of rewards in your answer.

1 Recall, or research, a First Nations person who is active in the outdoors and promotes their sport, work, organisation or recreational pursuits. Discuss how media coverage of what they do contributes to more people using the outdoor environment.

HOW CAN TEAMWORK BE PROMOTED WHILE INTERACTING WITH OUTDOOR ENVIRONMENTS?

Groups or teams are formed for various reasons, such as for sport or to complete a class project. Whatever the reason, all members of a group or team will have a common goal. It could be as challenging as climbing Mt Everest or as basic as getting from A to B. Often you will need to work in a group or **team** to complete a common task, so it is important to learn to work well as a team member.

team a collection of people who interact with one another, accept rights and obligations as members and share a common identity

People sometimes underestimate the importance of team or group memberships on their lives. Some people go on solo journeys, but most life experiences involve being engaged with others. Groups can be quite varied – for example, a family going for a walk, the crowd at a football game, an Internet discussion group or a group of classmates.

iStock.com/AscentXmedia

Figure 7.13 Safe and successful abseiling relies on a well-developed team.

DEVELOPING A TEAM OR GROUP

In 1965, Bruce W Tuckman, an educational psychologist, described four stages of group development. He looked at the behaviour of small groups in different situations and identified distinct phases that they go through. According to Tuckman, all groups or teams need to experience all four stages before they achieve maximum effectiveness.

The process may not be consciously recognised by a group, but if the stages are understood, it can help the group become effective more quickly and less painfully.

Stage 1: Formation

In the first stage, members' behaviour is driven by the wish to be accepted by others and to avoid conflict. Difficult issues and feelings are avoided as individuals focus on things like team organisation, who does what and when to meet.

At this time, each member is also gathering information and impressions about each other and about the scope of the task and how to approach it. This is a comfortable stage; however, avoiding all conflict means that not much actually gets done.

9780170463102

Stage 2: Difficulties

As important issues start to be looked at, some members' patience will break and small conflicts will appear that are quickly solved or else ignored. Conflicts may be about the work of the group itself or about roles within the group. Some group members will be happy to deal with the issues; others will be uncomfortable about the conflict.

Stage 3: Working together

As the second stage evolves, the 'rules' for the group are set up and the scope of the group's tasks or responsibilities are clear and agreed. Having had some conflict, the members now understand each other better and start to value each other's skills and experience.

Most team members listen to, appreciate and support each other and are prepared to go with the group's decisions; however, some team members may resist a suggested change as they fear that the group might break up or go back to arguing.

Stage 4: Performing

Not all groups get to this stage. At the fourth stage everyone knows each other well enough to be able to work well together. They trust each other to work for the good of the group. Members' roles can change according to what is needed at that moment, without needing discussion. Group identity is strong. Each team member is equally focused on the job and on the team. This means that all the energy of the group can be directed towards the task(s). Do you recognise this process from a group you have been in?

Source: Adapted from http://www.chimaeraconsulting.com/tuckman.htm

Forming	Storming	Norming	Performing
In this stage, most team members are positive and polite. Some are anxious, as they haven't fully understood what work the team will do. Others are simply excited about the task ahead.	Next, the team moves into the storming phase, where people start to push against the boundaries established in the forming stage. This is the stage where many teams fail.	Gradually, the team moves into the norming stage. This is when team members start to resolve their differences, appreciate colleagues' strengths and respect your authority as a leader.	The team reaches the performing stage when hard work leads, without friction, to the achievement of the team's goal.

Adapted from 'Developmental Sequence in Small Groups' by Bruce W. Tuckman, *Psychological Bulletin* 1965, Vol. 63, No. 6, 384–99

Figure 7.14 Tuckman's model has four stages, which are sometimes referred to as 'forming', 'storming', 'norming' and 'performing'.

DEVELOPING COLLABORATION SKILLS

Worksheet 7.6

Collaboration is about working together effectively on a task. It is also about taking actions that respect the needs and contributions of others, and contributing to and accepting the decisions of the group. Collaboration also involves negotiating a 'win–win' solution to achieve the team's objectives.

To function as a team, each person must work to build and maintain the relationships that exist within the team.

iStock.com/DMEPhotography

Newspix/Justin Lloyd

Newspix/Sarah Reed

Figure 7.15 Think of a group you were part of. If it had problems forming, what stage did those problems appear in?

- Collaborative behaviours
  - Giving and receiving feedback from peers or other team members
  - Sharing credit for good ideas with others
  - Acknowledging others' skill, experience, creativity and contributions
  - Listening to and acknowledging the feelings, concerns, opinions and ideas of others
  - Expanding on the ideas of a peer or team member
  - Tactfully stating personal opinions and areas of disagreement
  - Listening patiently to others in conflict situations
  - Defining problems in a non-threatening way
  - Supporting group decisions even if not in total agreement

Figure 7.16 Behaviours showing collaborative teamwork

In addition, the team also needs to achieve its goal or task.

Figure 7.17 These behaviours show effective group work.

DEVELOPING LEADERSHIP SKILLS

Often when you work in a group or as part of a team, decisions need to be made. This becomes more important when the team is placed under pressure or challenged. Frequently, people show leadership qualities when the group is facing a particularly difficult situation. While some people have inherent leadership qualities, leadership is not something a person is born with, nor can it be learnt from books but it does come more easily for some. Leadership is about you reaching within yourself to be the very best you can. It is also about inspiring others with loyalty, trust, commitment and ensuring that everyone works together as a team. Great leaders are confident in their own ability; they know what they want to achieve, are able to motivate others and put the team or group before themselves.

What might be some key aspects of leadership that will help individuals to become leaders of the future?

FACE TO FACE Leadership essentials

1. List some essential points about leadership and discuss them with a partner.
2. As a class, discuss some of these leadership essentials.

Elements of leadership

Figure 7.18 shows some of the elements of leadership that are important to group success.

Elements of leadership

Recognition and reward
Make sure you recognise when people are deserving of praise for their efforts and achievement. People like to know that they have done a great job or performed well.

Personality and character
Be yourself, and be a person who demonstrates trust, honesty, integrity, fairness, empathy and care. Remember to treat people the way you would expect to be treated.

Courage
Be strong when things do not go as planned, and even though sometimes you might be fearful, stick with your gut instincts and remain true to yourself.

Communication
Make sure that everyone clearly understands you and what your intentions or plans are. Sometimes people will interpret things differently, so check their understanding to make sure they are on the same page as you.

Planning
Make sure your planning covers everything from the smallest detail to the most obvious element. In order to have a plan you must also have a goal or vision and know how you are going to achieve that.

Enjoyment
You need to be buoyant and optimistic about what you are doing and ensure that this can be seen by everyone you work with.

Humility
Be humble and modest about yourself. Do not expect people to work with you unless you are also prepared to do the hard work or the more unpleasant tasks. This shows understanding of what is required and what others may need to do. It also helps to unite people in completing a task, because you are leading by example.

Responsibility
You need to accept that you are responsible for your actions. It is inevitable that you will make mistakes; however, if you learn from these then you can improve your performance and be a more effective leader.

Figure 7.18 Elements of leadership: which do you think you have?

Leadership styles

WB Worksheet 7.7

Normally in a group situation, someone will need to demonstrate effective leadership skills so the group can achieve their task. The style that they use and when they use it can have an important influence on how the group progresses. Leadership style is the manner and approach of providing direction, implementing plans and motivating people.

Thinkstock/Comstock Images

Figure 7.19 An autocratic leader makes all the decisions.

There are three main styles of leadership:

- The autocratic leadership style is where leaders tell people in the team what they want done and how they want it done, without getting the advice of the rest of the team. It is a good style to use when you have all the information to solve the problem, you are short on time and your team is well motivated. Often this style is thought of as aggressive: yelling, using demeaning language, leading by threats and abusing the leader's power. It's important to remember that this is abusive and unprofessional. The autocratic style should be used rarely. To gain more commitment and motivation from your team, the democratic style is more effective.

- The democratic leadership style involves the leader including one or more members of the group in making decisions. However, the leader has the final decision-making authority. This style shows strength and a team will respect this.
- The laissez-faire leadership style allows the members of the team to make the decisions, although the leader is responsible for the decisions made. Team members are able to analyse the situation and determine what needs to be done and how to do it. This style can be used when the people in the group are fully trusted to be competent.

Good leaders will use a number of different styles, with one of them usually dominant.

An effective team can often achieve far more than its individual members could on their own. Teams or groups provide opportunities to develop essential leadership skills. Effective leadership can accelerate the group-forming process and reduce difficulties that may arise as a team moves through the different stages.

Dreamstime/Konstantin Chagin

Figure 7.20 A democratic leader involves team members in decisions.

Getty Images/Kevin Dodge

Figure 7.21 Laissez-faire leadership allows all the team members to make decisions.

CASE STUDY → SIR EDMUND HILLARY

Identify

Sir Edmund Hillary and Sherpa Tenzing Norgay were the first people to conquer Mt Everest. The following is part of an interview with Sir Edmund Hillary later in life when he reflects on his own ability as a leader and working as part of a team.

Understand

The interviewer: The qualities you mentioned, soundness and mature judgement, they're the qualities of leadership. So what you're really talking about is how you developed as a leader.

Sir Edmund Hillary: There are some people who are natural leaders, who have the ability to think quickly or choose the right decisions at the right moment. But I think there are an awful lot of us who have to learn how to be a leader, and in actual fact, I believe that most people, if they really want to, can become competent leaders. I think I was the prime example of someone with relatively modest abilities, but I think I learnt to become a reasonably competent leader. Even practice is quite a useful attribute in this respect. As you do more expeditions and more adventures, you get more experience and you know more clearly what to do in moments of emergency. But I certainly never regarded myself as a natural leader.

The interviewer: How do you think you developed this healthy balance between being part of a team and being an individual striver?

Sir Edmund Hillary: I did my best. I certainly had strong individualistic attitudes, and I think probably I was at my best when I was given the job as leader of a project. In other words, I was forced to think ahead and make decisions and make sure that everything was carried out successfully. I don't think I was a very good follower because I think I did have my own personal ideas and I didn't particularly like being ordered around, to tell the truth. On the other hand, as a leader, I was not the type who ordered other people around. I did expect my groups to have good, strong ideas of their own, but for us all to work happily together and I think, on the whole, our expeditions were very happy ones and that we had quite strong team spirit.

Alamy Stock Photo/Everett Collection Historical

Figure 7.22 Edmund Hillary reads his mail in camp after descending from the summit of Everest.

Getty Images/Royal Geographical Society

Figure 7.23 Edmund Hillary and Tenzing Norgay climbing Mt Everest in 1953.

The interviewer: What's the proportion of skill, planning, leadership and luck?

Sir Edmund Hillary: You need all those things, of course. You certainly need planning and you certainly need a degree of skill and fitness, and there's no question at all that you need a little bit of luck. People often say you make your own luck, and I think probably 90 per cent of the luck is self-created, but there is that 10 per cent. You've got to have things right at the right time. If you're heading for the summit, you've got to have a reasonable day for it. And if the weather doesn't treat you right, nobody's going to get there. I guess you'd call that luck. But if you plan things, maybe you're organised so you can wait for another day and put in your push to the summit. But I do believe that a little bit of luck is a good thing to have.

Source: Sir Edmund Hillary, Conqueror of Mount Everest, The Academy of Achievement Interview, 16 November, 1991, https://achievement.org/achiever/sir-edmund-hillary/#interview

Discuss

1 Summarise three key leadership qualities you believe Edmund Hillary displayed. What makes these important?

2 Consider how government funding and public donations can be used to conserve the natural environment to ensure users 'leave no trace'.

3 Explain how trekking through the snow or mountains provided Hillary with a sense of mindfulness.

CASE STUDY  STEVE IRWIN – THE CROCODILE HUNTER

Identify

Australian Steve Irwin became world-famous for his TV series *The Crocodile Hunter*, among other nature programs.

Getty Images/Justin Sullivan/Stringer

Figure 7.24 Steve Irwin was a wildlife enthusiast who showed excellent leadership.

Understand

Steve did not have a science degree but rather learnt 'on the job' while growing up on his parents' wildlife park, which later became Australia Zoo.

Irwin found a fellow animal enthusiast in Terri Raines, who he met in 1991. While honeymooning, the couple filmed crocodiles and it was this footage that would be shown in *The Crocodile Hunter* TV show in 1992. American cable network Animal Planet picked up the series in 1996 for five seasons and numerous specials, which aired in more than 200 countries.

Audiences loved Irwin's enthusiasm and fearless attitude to the dangerous creatures he would come face-to-face with. Irwin's breadth of knowledge about animals made him an educator as well as an entertainer.

In 2006, while snorkelling at the Great Barrier Reef, Irwin was struck directly in the heart by the barb on a stingray's tail and died shortly after.

Irwin is remembered for his many contributions to wildlife education and conservation. Australia Zoo pays tribute to him each year on Steve Irwin Day (15 November).

Discuss

1. Effective communication is an important leadership trait. Discuss the ways Irwin used communication that in your opinion made him a leader.
2. Irwin purchased large tracts of land in Australia, Fiji, Vanuatu and the United States and labelled these as 'conservation parks'. What is meant by this?
3. Steve Irwin and Edmund Hillary both had significant people supporting their quests. Outline how contributed to the leadership shown by Irwin, Hillary and Norgay, and Raines.

UP AND MOVING Carousel debate

Set up the class in a carousel formation by making two circles (an inner and outer circle) of students facing each other. One circle assumes the negative viewpoint and the other circle assumes the positive viewpoint. Begin a debate on each of the topics provided. After one to two minutes, the circles rotate three places in the opposite direction from each other, the viewpoints change and the debate continues.

1. Are leaders born or bred?
2. Autocratic leadership is the best style of leadership.
3. Courage is the most important leadership quality.

1. 'Good leaders need good followers.' Discuss why collaboration is vital in team settings.
2. When groups of people are brought together and expected to work as a team, what causes them to 'norm' and how can this stage be 'fast-tracked' to the performing stage?
3. List at least two different ways you can check that people have understood what you are expecting from them, or communicating to them.
4. Suggest two strategies you can recommend to make the transition through the 'storming' phase of teamwork relatively quick.
5. Propose why some groups do not reach stage 4.

1. What style of leadership do you respond best to on the sporting field and in the classroom? Discuss situations when you might respond better to different types of leadership styles.
2. Sir Edmund Hillary said, 'I was forced to think ahead and make decisions and make sure that everything was carried out successfully.' Summarise a time when you have had to plan ahead in a school setting, and this has contributed to you experiencing a successful outcome.
3. Consider if you have good collaboration skills. In the future you will build on the skills that you develop when completing a group challenge or task. Many employers regard these skills as important and desirable attributes.
   a. Go through the list of collaborative skills in Figure 7.16 and try to identify which ones you are aware of or have demonstrated. Rate yourself on each of the behaviours:

      1 = I do this very well. I am consistent and successful in it.

      2 = I am good at this. With some practice I can make it perfect!

      3 = I am getting better, but still need to work on this a bit more.

      4 = I am not particularly good at this – yet!

1. Choose a leader who has impressed you and summarise the qualities they displayed that made them stand out for you. Can these qualities be learnt and thus improved on? Briefly discuss.

9780170463102

HOW CAN I REFLECT ON AND EVALUATE MY PARTICIPATION IN OUTDOOR ACTIVITIES?

WHAT INITIATIVE AND PROBLEM-SOLVING SKILLS CAN BE DEVELOPED BY PARTICIPATING IN OUTDOOR ACTIVITIES?

Having explored how groups work, it is time to apply those skills in practical situations. You will observe your peers and assess your own group and leadership skills. This is also an opportunity to demonstrate ethical behaviour, which is characterised by honesty, fairness and treating others well.

UP AND MOVING

Through the wringer

This is a very simple but effective team-building activity.

Aim: The aim is to pass everyone through the hoop as quickly as possible.

Equipment: One hoop for each group.

Rules: The clock starts as soon as the hoop is picked up off the floor and stops when the last person has placed it on the floor.

You can have three or more official attempts and you will be given some time in between attempts to plan. The class might be split into two or three groups with a competition to see who is the quickest. To add more complexity, complete the activity with a smaller hoop, or try having the whole team hold hands and not let go while they manouevre through the hoop.

Source: Reprinted with permission from playmeo. To access more interactive group activities, go to www.playmeo.com.

Reflect

1 Why are some teams faster than others?

2 What can be learnt by observing others performing a task?

The maze

This is a more complex team challenge.

Aim: The aim of the challenge is to get all members of your group across the maze one at a time. One person will start at one end of a row of flat spot markers or circles of chalk drawn on the ground. The key is to work out the correct pathway through the maze.

Rules: The pathway will not be in a straight line; it may change direction at any time and will not cross itself or use a marker or spot twice. Once the pathway has been identified it will not be changed. Only one person is permitted to enter the maze or be inside the maze of spots at any point. Every time a spot or circle is stepped on with both feet you will be given the thumbs up (indicating that it is part of the path) or thumbs down (indicating it is not part of the path). Try to complete the maze with the minimum number of errors.

If this is completed easily, try one of these variations to increase the complexity:

- If the wrong spot is stepped on, the person must retrace their steps correctly back to the start and another team member starts.
- Decide on a set number of errors each group is allowed to complete the task (e.g. 30).
- Don't allow any verbal communication once the game has started. This does not include preparation time before the event.

Source: Reprinted with permission from playmeo. To access more interactive group activities, go to www.playmeo.com.

Reflect

1 Is it better to proceed through the maze 'slow and steady' or try to complete it as fast as possible?

2 A common expression is 'sometimes you've got to go backwards to go forwards'. This might sound strange; what do you think is meant by this phrase?

MOVEMENT CHALLENGES

Identify

The way we travel has changed over the course of human history. Sometimes it's good to get back to the way we used to do it.

This activity has been developed in collaboration with Dr David Bakker of MoodMission

9780170463102

Understand

There's a lot of evidence that humans evolved to be good runners. Early hunters ran long distances to tire out their prey and bring it back to feed the group. However, we now travel a lot less by foot and spend a lot less time out in nature, and this can have some unfortunate effects on our mental health. Going for a run outside can provide the double benefit of exercise and being in nature; there's plenty of evidence that both of these things can be beneficial for mental health.

Practise

1. Choose a time when you will be able to go for a run. Maybe after school or on the weekend.
2. Decide how long you want your run to be. This will depend on your fitness level – if you're just starting out, try just a quick 5–10 minute run. If you get exhausted, it's fine to walk the rest of it. Next time you can try running further.
3. Plot your route for your run. You may like to run around your neighbourhood, go to a nature reserve or maybe even find a nearby parkrun event. If you're plotting your own route, you can try using Google Maps to find a good route.
4. Decide if you want to listen to music or something else on your run. You might even want to track your workout using an app, or you could check out the app Zombies, Run!
5. If you feel uncomfortable running in public, try running in a group with friends. And remember, it's common to worry that people are looking at you, but that's usually your anxiety talking. Most people are too preoccupied with themselves to be paying you much attention. They may even be inspired by you!
6. Go for your run! If you're not a regular runner, take it easy and at whatever pace feels okay.

Reflect

Recall how you feel after your run. Do you feel any different to the way you felt before the run? Did you feel anxious about other people looking at you or judging you while you ran? Did you find a way to challenge these thoughts and do the run anyway?

UP AND MOVING

Blindfold, three-legged frisbee golf

This will involve you and a partner working together to complete a set course.

Aim: The aim is to complete the course in the quickest time with the minimum number of throws. This is exactly like a game of golf, except that you are with a blindfolded partner and you are using a frisbee. A score card could be kept to record the number of throws for the whole course.

The course will need to be shorter than a frisbee golf course and, to save time, the class may need to be split and start at different starting points. Alternatively, two courses could be set up with similar difficulty levels and then pairs could swap over.

Equipment: Five or six hoops and markers or poles to represent the holes, frisbees – one for each pair depending on the number of the groups, and bandanas to tie legs together and to blindfold each other.

Rules: Each pair must stay together for the whole task. The blindfolded partner must always throw the frisbee. Each hole (hoop) is completed when the frisbee lands in the hoop. The partner who is not blindfolded needs to guide their partner on where to go and in what direction the frisbee must be thrown.

Reflect

How can you work more effectively with someone who is relying on you heavily because they have some form of injury or impairment?

UP AND MOVING Gladiator

In groups of three to five students, create your own gladiator to go into battle with another group's gladiator. Your team must create a suit of armour and a weapon that will survive a fight and ultimately destroy the opposition.

Aim: The aim is to destroy the other person's armour.

Equipment: Cardboard boxes, newspaper, magazines, glue, tape, cardboard, egg cartons.

Rules: Each team must have approximately the same amount of resources. Depending on the time available for the whole task, a time limit can be set for planning and building and then for the actual battle.

UP AND MOVING Design your own team challenge

Aim: The aim is to design and create a new group activity or challenge.

Equipment: Equipment will vary depending on the challenge being created.

Rules: A group can create or design anything; whether it is physical, non-verbal, mental or involves singing or dancing, it does not matter. Each group will be given a set amount of time to create their challenge.

After the allocated time each team will present their challenge to the whole class. The activity should be one that the group believes it can do better than any other group! Groups earn points for meeting the following criteria:

- no other group can beat them at their activity (+2)
- they can do another group's activity (+1).

HOW TO EVALUATE

It is important to evaluate and reflect on your ability to work in a team. For any of the previous movement challenges or initiatives, reflect on your involvement in the group challenge.

FACE TO FACE Make an assessment

This activity discusses how to evaluate and reflect on your ability to work in a team and to be an effective leader.

One way to evaluate how well you have contributed to a team is to create a self-assessment or peer-assessment tool. If it is too difficult to identify what qualities to assess, this could be determined in pairs, and then you could determine as a class what key components should be included.

Refine the model so that there is agreement among the class. Table 7.1 is just one example of a self-assessment tool.

Table 7.1 A self-assessment tool

Demonstrating personal qualities			
Look at the statements below: ➪ On the scale next to each statement, choose a rating that reflects how frequently it applies to you. ➪ Total your scores after each domain and reflect on how you have scored yourself.	A lot of the time	Some of the time	Very little or none of the time
Developing self-awareness			
I reflect on how my own values and principles influence my behaviour and have an impact on others.			
I seek feedback from others on my strengths and limitations and modify my behaviour accordingly.			
Managing yourself			
I remain calm and focused under pressure.			
I plan my workload and deliver on my commitments to consistently high standards, demonstrating flexibility to service requirements.			
Continuing personal development			
I actively seek opportunities to learn and develop.			
I apply my learning to practical work.			
Acting with integrity			
I act in an open, honest and inclusive manner – respecting other people's culture, beliefs and abilities.			
I speak out when I see that ethics or values are being compromised.			
TOTAL			

Source: Adapted from Leadership Framework: Self assessment tool © 2012 NHS Leadership Academy, http://www.leadershipacademy.nhs.uk/wp-content/uploads/2012/11/NHSLeadership-Framework-LeadershipFrameworkSelfAssessmentTool.pdf

Look at each statement in the table and choose a rating that reflects how frequently the statement applies to you. Total your scores and reflect on what score you have given yourself. If you answered mainly 'Some' or 'Very little or none' in any particular domain, this domain may be an area you wish to develop further. If you often answered 'A lot', check that these are not overplayed strengths. An overplayed strength could be a behaviour you over-rely on and one that might have a negative impact on your performance.

1 Suggest two ways your team's effectiveness could be improved in the challenges.

1 Summarise your contribution to achieving the task/challenge.
2 Explain how effective you were as a group in completing the challenge.
3 Create your own self-assessment tool to evaluate how effective you are as a group member or as a leader.

1 Provide examples of leadership that were demonstrated by the group.

Quiz
Teamwork, interacting outdoors and reflecting on outdoor activities.

CAN BUSHWALKING PROMOTE PHYSICAL, SOCIAL AND EMOTIONAL BENEFITS?

Bushwalking is the Australian word for hiking. It refers to walking through undeveloped land or wilderness on tracks or cross-country through the bush. Australia is fortunate to have many large national parks that preserve scenic and rugged areas in their natural state. Even Australia's largest city, Sydney, is (quite literally) surrounded by huge areas of national park.

123RF.com/Antonio Ribeiro

Figure 7.25 The Australian bush is beautiful and varied, but it needs to be treated with respect and understanding.

Dr Melissa Parker from the University of Queensland has written a book about bushwalking in which she explains how bushwalking has evolved and how a passion for the bush led to the development of national parks in Australia.

'Bushwalking is such a fundamentally Australian pastime, and is enjoyed by so many people, but the stories had never been told', Dr Harper said. '*The Ways of the Bushwalker: On foot in Australia* reveals how bushwalkers have shaped Australian ideas about the land.

'The earliest European arrivals were often quite hostile towards the bush – it was alien to them.

'Bushwalkers really promoted the idea that the bush was beautiful – a place to relax and rediscover yourself. They encouraged people not to be scared of it, but to treat it with respect and understanding.

'Bushwalkers showed that the Australian environment was varied, not monotonous,' Dr Harper said.

Source: 'Australian history shaped by bushwalkers', *UQ News*, The University of Queensland, www.uq.edu.au/news/article/2007/10/australian-history-shaped-bushwalkers

PREPARING FOR A BUSHWALK AND OVERNIGHT CAMP

Before going anywhere in the bush, it is essential to prepare thoroughly to ensure the activity is enjoyable and safe. The amount of preparation required will depend on where you are walking (e.g. in rugged terrain or isolated areas) and for how long.

Most Australian states now have Activity Adventure Standards that provide minimum voluntary guidelines to follow for adventure activities. The Duke of Edinburgh's Award also provides guidelines for completing an adventurous journey.

Table 7.2 lists the skills that each person and group should have acquired before they go bushwalking and stay overnight. Group skills should be learnt as a group so that when you are actually out in the bush, you are prepared for all possible hazards and have procedures in place to deal with these dangers.

Table 7.2 Skills needed for staying overnight on a bushwalk

Personal skills	Group skills
Personal clothing, equipment, footwear and packing of equipment, equipment load	Cooking skills, food preparation, hygiene and cleaning up
Personal safety – sun protection, basic first aid procedures	Tent pitching and campsite selection, camp practices
Food selection	Route planning
Navigation skills	Navigation skills
Physical fitness	Safe walking skills and procedures

Personal equipment, clothing and footwear

If you are going bushwalking, there is nothing worse or potentially dangerous than being unprepared. Personal clothing and equipment does not need to be 'top of the range' but it does need to be effective and do the job it is expected to do.

123RF.com/klotz

Figure 7.26 The ability to find a suitable campsite and pitch a tent are important personal and group skills.

FACE TO FACE Packing for a night

This activity should be done in pairs.

1 List all the essential personal items you think would be needed for an overnight bushwalk.

2 Compare your list with the one provided below and see how many items you remembered and which ones you missed.

This is a generic bushwalking list, which should be added to for specific activities (e.g. depending on conditions – winter, summer):

- Backpack
- Sleeping bag – with hood, packs up small
- Sleeping mat
- **Outer-layer clothing:**
  - waterproof jacket, strong hood, below waist length
  - over-pants
  - sun hat, 30+sunscreen, sunglasses
  - worn-in boots with laces and soles in good shape (no elastic-sided boots)
- **Mid-layer clothing:**
  - 2 wool or fleece jumpers, not cotton
  - 2 long pants (e.g. track pants), not denim
  - long-sleeved top and T-shirt with collar
- **Inner-layer clothing:**
  - thick socks – at least three pairs, not cotton sports socks
  - wool or fleece hat/beanie and gloves/mitts
  - thermal long-sleeved shirt and long johns – polypropylene – for alpine hikes and winter trips
- Water bottles: two 1-litre bottles
- Eating utensils – bowl, knife, spoon and mug – all lightweight (fork and plate optional)
- Scouring pads, tea-towel
- Toothbrush, etc. – no shampoo needed, small deodorant if any
- Toilet paper in snaplock plastic bag
- Notebook, pen, pencil
- Head torch and spare batteries
- Matches in sealable plastic bag
- Personal first aid needs (e.g. blister tape, ankle bandages, asthma inhaler)
- Repair kit: spare string/bootlaces, strong tape, needle and thread
- 5 plastic shopping bags and rubber bands
- 3 large tough garbage bags to line pack
- Hot drinks – powdered chocolate, cup-a-soup and powdered energy drinks
- Food for all meals plus high-energy snack food (e.g. scroggin: peanuts, dried fruit)
- Optional: camera, cards, chess, bathers and tiny towel, chocolate, jellybeans

Source: Adapted from The Duke of Edinburgh's Award in Victoria. Adventurous Journeys, 2008, p. 9

Scaffold

3 Why do some of the items in the clothes list (waterproof jacket, jumpers, socks and hat) give specific information about the type of clothes to be worn, length of material or quality of material?

It is important that you layer your clothing with breathable and moisture wicking fabrics. You may have a T-shirt underneath and then a long-sleeved shirt, then a jacket with a zip. This means that as your body cools or heats up you can easily adjust the temperature by unzipping, rolling up sleeves or adding another layer to maintain a comfortable temperature. This is also useful when you may have to walk uphill, when your body works harder and you become hotter.

Alamy Stock Photo/Image Professionals GmbH

Figure 7.27 Why is it important to layer your clothing?

You should be comfortable with wearing all the clothing you have packed, and you should have used it before, especially footwear. People often buy new boots but don't wear them in properly; this can cause discomfort and blisters, and will make your trip miserable rather than fun. Ideally, boots should provide ankle support, especially if the terrain is demanding and 'off track'.

Packing your backpack

Learning to organise your gear properly before loading your backpack will eliminate forgotten items and help you remove unnecessary luxuries. In addition, efficiently packing your backpack will give you more comfort, convenience and stability.

FACE TO FACE Packing

As a class, discuss what items should be where in your backpack, and why. Then compare your results with the list provided.

Packing	Fitting the pack to your back
Open all buckles and straps Loosen compression straps, fold top down Fold a garbage bag liner down and around the pack top Think of the logical order for your gear: ➩ Bottom: tent, sleeping bag and spare set of clothes jammed into corners ➩ Middle: stove, toilet bag, torch, food and utensils, spare **trail mix** ➩ Top: waterproofs, jumper, hats, gloves, extra water ➩ Top pocket: first aid and personal medication, map, pen, booklet, camera, trail mix for a day ➩ Water under flap – easy access behind mat or in side pocket	Packs are designed to adjust to different back lengths, sizes and shapes, so you will need to fiddle with the straps a bit to get it right for your body. ➩ The hip belt should be firmly anchored on the hip bone, not below it. ➩ Adjust shoulder straps at top and bottom. These straps pull the pack upright onto the hip belt. Weight should not be on shoulders. ➩ Side straps secure the internal load onto the frame. ➩ There should be a gap between the pack and your back to allow for air circulation.

Source: Adapted from The Duke of Edinburgh's Award in Victoria. Adventurous Journeys, 2008, p. 10

trail mix a mixture of nuts, dried fruit and chocolate (also called scroggin)

Packing tips

- No pack is really waterproof, so line it with a large, thick garbage bag (take three spares). You can also use a waterproof pack cover.
- What you don't need during the day goes at the bottom.
- Compress the sleeping bag at the bottom and jam spare clothing (in small plastic bags) around it, not on top of it; fill the corners.
- Heavier items should be close to your spine in the middle third of the pack.
- Load heavy items such as stove fuel bottle or food (in small plastic bags) close to your back on top of jammed-in sleeping bag and spare clothes.
- Equipment needed during the day (such as waterproofs, snacks, water, map, spare jumper, camera, lunch) should be accessible at the top.
- Waterproof all your clothes and sleeping bag in separate bags.
- Place your insulation roll or sleeping mat inside your backpack by flattening it against one side or rolling it up inside your pack and then pack all your items inside it. This way, it won't get wet or torn.

Source: Adapted from The Duke of Edinburgh's Award in Victoria. Adventurous Journeys, 2008, p. 10

Weight

Weblink
How to pack a backpack

Your pack should weigh no more than 25 to 30 per cent of your body weight, so pack carefully and only take essential items. Items that will be shared, such as food, stove/fuel, first-aid kit and tent/tent poles should be spread evenly among the group. This can also include map and compass, trowel and toilet paper, first aid extras and emergency food.

Figure 7.28 Distributing pack weight

You may need to go back through your items and remove anything unnecessary if your pack weighs more than the guideline.

Practise packing

An effective method of practising packing is to have all the personal equipment ready to go and then take turns to pack the equipment in a backpack.

Also practise fitting the pack, taking it off and putting it on again. Ask your teacher to demonstrate the best method. For example, instead of having to bend and place your pack on your back, you can ask a group member to help, or you could place your pack on top of a fallen tree or a rock to help.

FAST FACT

Use your body weight to calculate how heavy your pack should be.

This sample calculation is for a person who weighs 80 kilograms.

25% of 80 kg = 20 kg

30% of 80 kg = 24 kg

When fully packed, the backpack should weigh 20–24 kilograms.

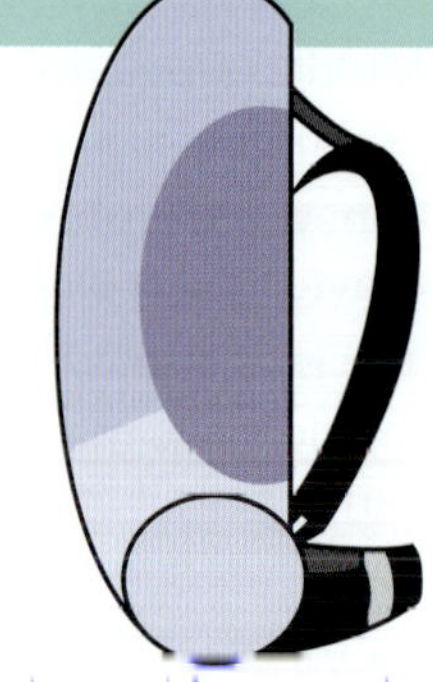

Figure 7.29 How to pack your backpack

Personal safety

Protect yourself from the sun with a wide-brimmed hat, a shirt with a collar and long sleeves, and by applying sunscreen to areas exposed to the sun. Wearing long pants and long-sleeved shirts also helps to reduce scratches and stings. Knowledge of basic first-aid procedures is important. These are explored more thoroughly in Chapter 6.

FACE TO FACE What could go wrong?

As a class, brainstorm some common first-aid problems that could potentially occur while bushwalking and what the best treatment for each would be.

Bushfire emergencies – staying safe

Always inform family and friends of your movements in bushland.

Radiant heat

Being outdoors during a bushfire means you risk exposure to radiant heat, which can kill a human without flames ever touching them. Extreme temperatures from radiant heat cause death from heatstroke where the body's cooling system fails, leading to heat exhaustion and heart failure.

Ensure you continue to stay cool and keep drinking water to stay hydrated. If someone is affected by heatstroke, move them to a shaded area and try to cool them down. Call 000 and seek help immediately.

Seeking shelter

If you need to seek shelter due to a sudden change in your bushfire survival plans, fire agencies say the main priority is sheltering from radiant heat.

- It is recommended you seek shelter in a well-prepared home that can be actively defended, a private bunker built to regulation or at a community shelter or refuge.
- If sheltering in a building, make sure you have a point of exit in every room of the shelter.
- Maintain visibility so you know what's happening with the fire outside.
- Take your survival kit and wear protective clothing to help guard against radiant heat.
- Leaving when a bushfire has arrived is extremely dangerous, but if you have no other option you can go to your local place of last resort – a ploughed paddock or the beach, dam or river – but don't shelter in water tanks.
- Radiant heat can be blocked by a solid object such as a concrete wall or building which creates a barrier between you and the fire.

Driving

This is a last resort. Cars are a very dangerous place to be during a bushfire as they offer very little protection from radiant heat.

Driving somewhere will take longer than you expect and road conditions can become dangerous. There may be road closures, smoke, fallen trees and embers.

- Park behind a solid structure to block as much heat as you can. If this is not possible, pull over to the side of the road into a clear area, well away from debris that may ignite.

- Wind up your windows, close the vents, put on your hazard lights and headlights, leave the engine running and air conditioning on recirculate.
- Get down as low as possible below window level and cover up with a woollen blanket until the fire passes – if you have water, drink it.
- Get out of the car only once the fire has passed.

Food and meals

Depending on how long and demanding the bushwalk is, an approximate guide is to pack 700 grams to 1 kilogram of food per person per day:

- Food needs to be lightweight and compact, easy to cook, nutritious, not perishable, have little packaging and be high in energy for a bushwalking trip.
- The purpose of food is to fuel the body. Allow generous amounts and do not skip meals or snacks. Your body will be working hard and will require more food than usual.
- Tasty food is much more satisfying.
- No tins or glass containers!
- Remove as much packaging as you can – re-pack in snaplock bags, which are light and reusable.
- Avoid buying heavily packaged items. Shop to leave no trace.
- Label the snaplock bags (e.g. Lunch day 1).
- Include foods from the five food groups with a high percentage of complex carbohydrates. (See Chapter 2, pages 69–71.)
- Use variety; avoid junk foods.
- Pack food that is tasty, nutritious and cookable on a two-pot stove.
- Include drinks – both hot and cold, in high-energy, powder form – no UHT drinks (too heavy).
- Avoid instant foods such as 'two-minute noodles' on their own as a meal; they have little nutritional value. Add vegies or tuna.
- Do not pack baked beans and other tinned foods; they are too heavy and create too much rubbish.

Shutterstock.com/oliveromg

Figure 7.30 What can you cook on a camping trip?

- Select food that will survive being crushed and unrefrigerated.
- Remember to pack high-energy snacks of dried fruit, nuts, jubes, chocolate or muesli bars.
- Avoid high-sugar hits; this results in a sudden low blood-sugar level after the instant high.

Source: Adapted from The Duke of Edinburgh's Award in Victoria. Adventurous Journeys, 2008, p. 11

9780170463102

INVESTIGATION

MENU PLAN

Purpose

To design a two-day menu for a camping expedition.

Method

1. Plan a two-day menu that includes two breakfasts, two lunches, one dinner and some healthy snacks.
2. Consider any lightweight/dehydrated options to make your backpack lighter.

Discussion

1. Discuss why it is important not to overpack and bring foods that are not consumed and need to be brought home again.
2. State how you would apply the 'leave no trace' principle while participating in the two-day trek/expedition.
3. Outline the benefit of including variety in your meals while camping.
4. As part of your planning considerations, why should people attending the camp (groups) clearly communicate with each other regarding what they intend to bring?

NAVIGATION

Using a compass

Once you are able to use a map to navigate confidently (whether completing an orienteering course or out in the bush), it is helpful to learn to use a compass as well. That way if you do get lost you have the compass to fall back on. This is an individual requirement that should be shared between group members during the walk. Everyone must be competent before going on a walk.

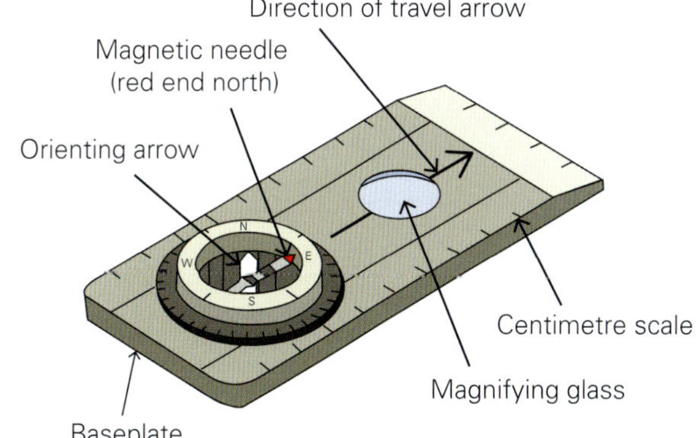

Figure 7.31 Components of a compass

Finding a grid bearing

Step 1: Place the compass on the map with the side edge of the compass connecting your location with your destination. Make sure the direction of travel arrow is pointing to where you want to go.

Step 2: Turn the compass until the orienteering lines on the base are parallel with the north–south lines on your map and the 'N' is pointing to the top of the map. The grid bearing can now be read off the scale opposite the direction of travel arrow (see diagram).

Step 3: Hold the compass horizontally in front of you so that the needle is swinging freely. Then turn

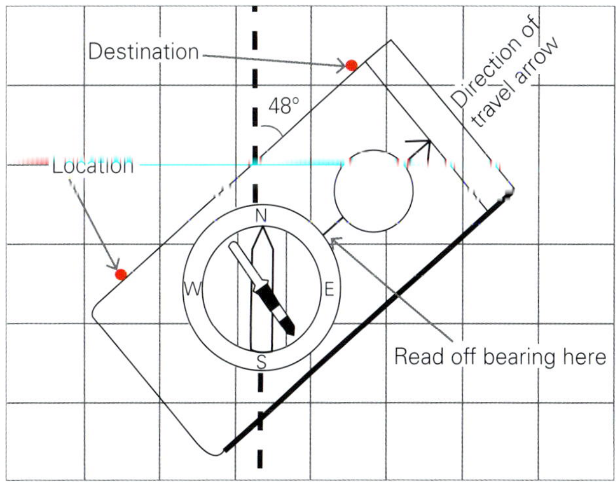

Figure 7.32 Using a map and compass to find a bearing

Adapted from *Land Navigation: Routefinding with Map and Compasses* by Wally Keay, Nicholas Gair, Duke of Edinburgh's Award, Coventry, UK: Clifford Press Ltd, 1995.

Figure 7.33 Taking a bearing

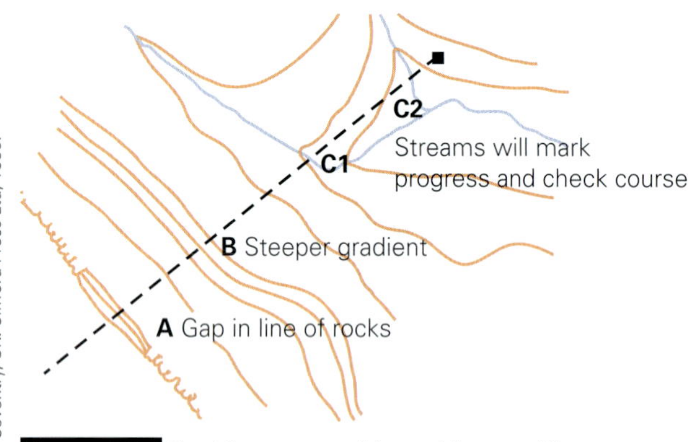

Adapted from *Land Navigation: Routefinding with Map and Compasses* by Wally Keay, Nicholas Gair, Duke of Edinburgh's Award, Coventry, UK: Clifford Press Ltd, 1995.

Figure 7.34 Tracking your position while travelling on a bearing

yourself around until the red end of the compass needle is pointing to the N on the outside of the compass and is parallel to the lines in the bottom.

Step 4: Sight along the direction of travel arrow at some feature in the landscape – it could be a tree or building – then travel towards it. You are now travelling towards your objective. When you reach the point you sighted on, stop and repeat the process until you reach your destination.

These skills will allow you to successfully navigate using a compass with a map. However, if you wish to walk in more difficult terrain or for longer distances, you will need to calculate the difference between true north and magnetic north.

The magnetic variation in Australia is east of true north. The variation is always subtracted from the grid bearing. Detailed maps will have information about the magnetic variation in that area.

Pacing yourself

Another way to ensure that you are following a bearing, especially when in poor visibility, is to know the number of paces you take over a set distance (e.g. 100 metres). If you have practised this skill, you can apply it when it is required to ensure that you are even more accurate with your navigation.

Tracking your position

You should constantly check your position and progress, even when travelling on a compass bearing. This is very important if there are no clear features, such as trees or hills, or if visibility is poor. For example, in Figure 7.34, if the line of rocks at A was in front of you, it would indicate that you were off course. Point B, the steeper slope, would establish your position. The streams crossing your path at C1 and C2 would mark your progress along a bearing and their direction should help you correct your course if necessary.

SAFE WALKING

So that members of your group do not get lost or left behind, the group should have experienced walkers at the front and rear. These people should keep everyone together, ensuring that no one passes the front person and no one falls behind the rear person. Always walk at the pace of the slowest walker in the group and place this person near the front to ensure that they are not left behind or that pressure is not placed on them to 'catch up'. This will ensure a safe, steady pace. Plan regular drink and snack stops, which can be marked on the route card, and appoint a timekeeper who ensures the group keeps to the times allocated for breaks. It is better to have regular drink breaks

9780170463102

and short stops than long breaks that could upset the rhythm of the group and result in you getting cold.

Physical fitness

A reasonable degree of fitness is required to undertake a bushwalk because of the additional weight being carried in your pack. The route will also have an impact, especially if there is a lot of climbing. Therefore, it is important that you improve the endurance in your legs and develop your cardiovascular fitness in the weeks before the walk.

Alamy Stock Photo/Bjorn Svensson

Figure 7.35 Experienced walkers should be at the front and rear of the group.

CAMPSITE SELECTION

FACE TO FACE Choose a campsite

123RF.com/Ina Van Hateren

Figure 7.36 Is this spot a good site for a camp?

1 Look at the photo and discuss in pairs whether it shows a good place to camp.
2 Individually or in pairs, brainstorm factors that should be considered when selecting a campsite. Draw a table with three columns:
   » In column 1, describe the factor.
   » In column 2, explain why this factor is important.
   » In column 3, rank your factors from most important to least important.
3 Use the web to search for relevant factors to complete the task.

Scaffold
Selecting a campsite

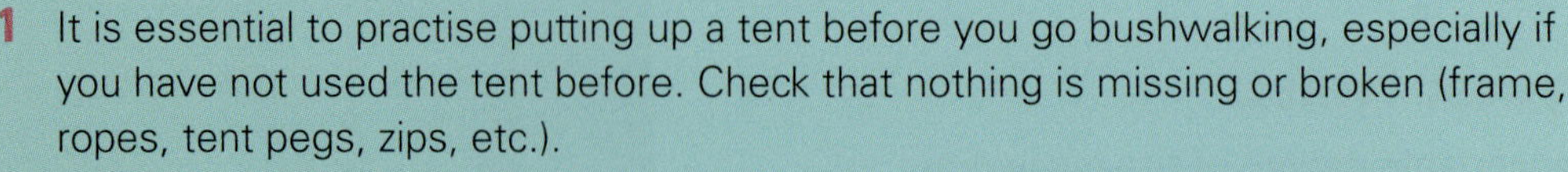

UP AND MOVING Pitching a tent

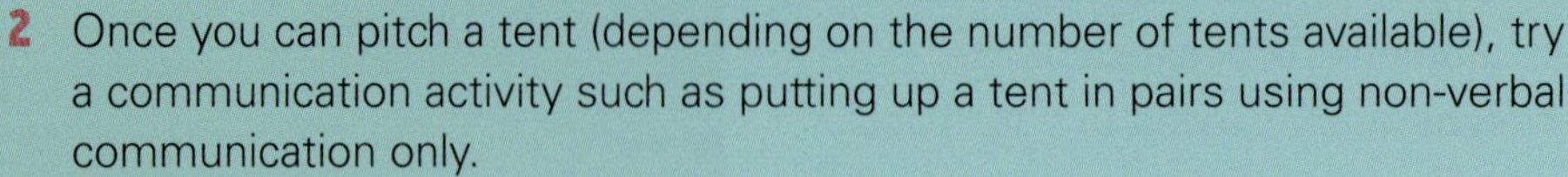

1 It is essential to practise putting up a tent before you go bushwalking, especially if you have not used the tent before. Check that nothing is missing or broken (frame, ropes, tent pegs, zips, etc.).

2 Once you can pitch a tent (depending on the number of tents available), try a communication activity such as putting up a tent in pairs using non-verbal communication only.

Worksheet 7.9

Figure 7.37 A camp stove in use. This style is lightweight and durable and excellent for bushwalking. It uses methylated spirit as fuel.

COOKING AND HYGIENE

Before preparing food, you must wash your hands with soap or a disinfectant gel. Ensure your stove is at least 4 metres away from any tents before lighting it. Table 7.3 shows the basic set-up to prepare for cooking and to ensure hygiene and safety.

Table 7.3 Cooking at a campsite

Preparation
Set up on stable surface, ground or table, not a narrow bench. Set up a minimum of 3 metres clear around stove site (e.g. nowhere near tents). Before lighting stove, have all cutlery, plates, food and water ready at cook site. Take burner to fuel bottles, not bottles to stove. Tighten fuel bottle lid immediately after use. Once the stove is alight, do not move it or leave it unattended. Dig a hole at edge of campsite beneath trees if possible for discarding washing water (no food scraps).
Safety and hygiene
Wash your hands before cooking. Never blow on flame. Stay with your stove once alight and never move stove when alight. Never remove wind shield when stove is alight.
Washing up
After the meal, scrape all food out – no food scraps in wash water. Soak: half fill a dirty pot with water to soak before you start eating. First wash: use bark or leaves for initial scrub of pot, bowl, cutlery, black under pot; tip water into wash-water hole. Second wash: heat clean water in scrubbed pot, pour into plastic wash bowl; scrub whole stove, bowls and cutlery with soap steel wool pad; wash outside of pots last. Rinse soap with water bottle into wash-water hole or spread into bushes away from campsite. Dry everything completely to avoid bacteria and mould growth. Check all equipment and utensils are packed again.

Source: Adapted from The Duke of Edinburgh's Award in Victoria. Adventurous Journeys. 2008, p. 14

Stove set-up

Practise putting together a stove without lighting it. Trangia camp stoves are excellent for efficiency and safety – practise putting one together and taking it apart while blindfolded!

ROUTE PLANNING AND EMERGENCY PROCEDURES

Route planning is an essential skill, and like all other skills it improves with practice. The first step is to plan a route that will satisfy the purpose of your journey. This will depend on many factors (such as experience or age), so it is important to select a route based on the capabilities and requirements of the group.

Alternative routes and escape routes

Planning an alternative route is important in case of poor weather; it allows you to change your route so that you can avoid the worst of the weather and still reach your destination. An escape route will not help you reach your destination, but it will help you escape any immediate predicaments.

FACE TO FACE Escape route

1 As a class, provide some examples of when a group may need to use an escape route.
2 Discuss circumstances where it would be better to stay where you are than to use an escape route.

General considerations for route planning

It is important to decide on the length of the walk at the start, and then divide the route into days. Start your day early to reduce your chances of finishing in darkness. Plan to do major climbs early in the day.

Online activity
Bibbulmun Track

FACE TO FACE Class discussion

List three reasons it would be important to make any major ascents early in the day.

Estimating journey time

Estimating journey time is central to route planning and will be determined by a number of factors:

- physical fitness of the group
- distance to be covered
- height to be climbed
- proportion of journey on tracks or off track
- conditions underfoot
- weather conditions.

Allow for the weight of a pack and for height climbed. Some basic estimates for a fit walker with a full pack and camping gear:

- 15 minutes for every flat kilometre
- an additional four minutes for every 30 metres of ascent. (If the contours on the map are spaced 10 metres apart, this is four minutes for every three contours.)

Planning the route

Depending on the length of the bushwalk, your resting places to camp at night will have an important role in selecting your route. If you are walking in a national park, campsites will be set, which may determine your route options. If you are planning to climb a major peak, then this must also be considered in relation to the horizontal distance that you have to travel.

Scaffold
Route card

Completing a route card

It is important to fill in a route card before you go. A route card is an important statement of your intentions and it provides detailed information regarding your route that can be passed on to relevant authorities before you go. Table 7.4 shows a sample route card.

Divide the journey into sections or legs at natural divisions in the route and where possible by the main checkpoints, or way marks, on the route. Sometimes legs can also be where there is a major change of direction, by ascent or descent, or when there is a change from one method of route finding to another, such as following a path to climbing a gully. The checkpoints or way marks must be physical features on the ground that cannot be confused.

Table 7.4 A sample route card

Aim of trip: Four-day coastal hike, climb Mt Arapiles, search for caves									Group name: Class 9C, Bunbury HS Address: Railway Pde, Bunbury Phone: 07 8341 XXXX	
Day Tues	Date 7/7/2020	Day no. 1		Names of team members:						
Leg	Grid reference START Seaside 783 755	General direction/ bearing	Distance (km)	Height climbed (m)	Walking time (min)	Exploring, rest, meal time (min)	Total time for leg	Estimated arrival time	Setting out time 9am Details of leg	Escape/notes
(a)	(b)	(c)	(d)	(e)	(f)	(g)	(h)	(i)	(j)	(k)
1	To North shoulder 777 735	SW then S	2.8	80	50	10	60 min	10.00	Walked path towards North Point, then path S to Sadler	
2	To Shooting hut 766 734	W	1.3	50	25	10	35 min	10.35	Followed path by Sadler and North Point, then by wall	
3	To Ingleborough 741 735	W	2.9	300	1 h 25 min	30	1 h 55 min	12.30	Path to allotment wall then climbed to summit	
4	To Little Ingleborough 743 735	E then S	>1.4	–	20	10	30 min	13.00	Retraced path to Swiss Trail then path towards cairn	
5	To Swallow Hole 737 723	S and SW	1.4	–	20	10	30 min	13.30	Take SW fork at junction with Running Girl path	
6	To Macoby Cote 732 705	S and SW	1.9	–	30	–	30 min	14.00	Follow path or site bend	
7	To									BAD WEATHER ROUTE: follow proposed route to Solbar then Long Cove to Old Road
8	To									
	Totals:		11.7 km	430 m	3 h 50 min			16.10	Supervisor's name, phone number Ian Ferguson 07 8340 XXXX	

Adapted from The Duke of Edinburgh's Award

CASE STUDY

STRANDED AT SEA: THE SCHOOL CAMP FROM HELL

Understand

A school camp turned into a nightmare for a group of students stranded at sea in waves of up to five metres.

Six students and two teachers from a Secondary College in Warrnambool at the camp in Anglesea took to the water in kayaks.

Swept from their small boats soon after they set out, they then spent almost an hour in the heaving waters of Bass Strait, a kilometre off shore.

Transport Safety Victoria investigated the incident, which took place on a day with a severe weather warning.

Fairfax Syndication/Joe Armao

Figure 7.38 What would you do if you were caught in winds that carried you out to open and dangerous waters?

The group was rescued by off-duty lifeguards, who noticed that the kayaks had been swept out to sea.

TSV's director of maritime safety, Peter Corcoran, said he had 'serious concerns' about the actions of those involved.

WB Worksheet 7.5

'Despite the weather warnings for this week, we're dismayed to see people continuing to put themselves and others at risk', he said.

'It is difficult to understand how any boater would consider the severe weather warnings in place to represent safe boating conditions'.

The school's principal said the lifesavers had only rescued two students.

'We chose to hold this activity at Anglesea front beach because lifesavers were present', she said.

'This incident occurred early in the morning, well before [the bad] weather set in.

'Our staff monitored weather reports throughout the morning to ensure students were nowhere near the water by the time the weather changed', she said.

Weather data from the Bureau of Meteorology shows that at nearby Aireys Inlet, the wind was gusting at 89 km/h at 6.30 a.m. on Tuesday and peaked at 109 km/h at 9.30 a.m.

An Education Department spokesman said schools have to conduct a risk assessment before every excursion.

Mr Corcoran said it was fortunate that the group was wearing lifejackets. Their kayaks were not recovered.

Discuss

1 Outline why the students needed to be rescued. List these in order of priority.
2 Discuss how the potential tragedies could have been avoided. Describe and list the key elements that were lacking in this school expedition.
3 This advice is often given by camp organisers: 'Prepare for the worst, but expect the best'. Discuss what you believe is meant by this and how it may contribute to improved safety.

CHAPTER 7 REVIEW

1 Discuss how Australia has created its own unique sporting culture.

2 Summarise how First Nations Peoples contributed to the sporting landscape in Australia.

3 Identify what outdoor recreation means for you.

4 How does the media influence outdoor recreation in both a positive and negative way?

5 Consider what you enjoy about being part of a team.

6 Summarise the four stages of group or team development that groups experience. Suggest ways the norming phase/stage can be facilitated by a team leader/coach.

7 List the two best behavioural assets you have when working in a team, and why.

8 Name four elements of leadership that are crucial to the success of a team.

9 Discuss why an effective leader or coach should be able to take on any of the three leadership styles depending on the circumstances at the time.

10 Is it necessary for everyone in a camping group to have their own first-aid kit? Briefly discuss. What about specific considerations such as EpiPens and other medications – who should be responsible for these?

11 State why you believe it is important to undertake a few practice walks before going on a hiking camp. Some people suggest simply filling a backpack with items of a similar weight to those likely to be carried during the trek/hike and do laps of the oval at lunchtime. What are your thoughts?

PARTICIPATING AND PERFORMING IN GAMES AND SPORTS

8

IN THIS CHAPTER

You will explore games and sports, the importance of rules in games and the different types of games and sports you can play. You will then investigate the movement and tactical skills you can use to play games and sports, and how to analyse and improve your movement and tactical skills as well as those of other participants.

Shutterstock.com/GoodStudio

By the end of the chapter, you should be able to:

- identify the essential elements of games and sports
- compare different types of games and sports
- examine differences between fundamental and specialised movement skills, gross and fine movement skills, and open and closed movement skills
- investigate the use of tactical skills in games and sports, including tactics, strategies and game plans
- analyse, adapt and refine movement skills and tactical skills in games and sports
- analyse, adapt and refine practice and feedback to enhance your performance and the performance of others in games and sports
- transfer and adapt skills and strategies from previous experience to new movement situations
- modify games to vary complexity, including modifying rules, number of players, space and equipment
- reflect on how ethical behaviour, fair play, codes of behaviour, sportspersonship, teamwork, motivation, confidence, leadership and collaboration skills can affect participation in games and sports.

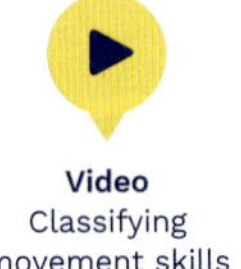

Video
Classifying movement skills

WHY PARTICIPATE AND PERFORM IN GAMES AND SPORTS?

Quiz
Pre-chapter

Before you start, take the pre-chapter quiz to find out how much you already know.

games recreational activities that involve achieving a goal (solving a problem) and have a set of rules that determine what can and can't be done in achieving that goal

Being physically active is important for physical development and health promotion. One way to be physically active is to play **games** and **sports**. Participating in games and sports allows you to adapt and refine a range of movement skills, movement concepts and movement strategies in challenging movement situations. While you are playing with others, you are also helping improve their skills. Games and sports provide opportunities for enjoyment and social interaction as well as for teamwork, leadership and collaboration.

INVESTIGATION ⇨ PARTICIPATION IN ORGANISED SPORT

Purpose

To research changes in children's participation in organised sport.

Method

Organised sport provides opportunities for physical activity, enjoyment and social interaction for children. Report on how often children are participating, which sports are most popular, changes in participation patterns (e.g. whether children today are participating in the same sports as children were five, 10 or 15 years ago) and possible future trends in participation (e.g. whether some sports may become more popular). A good place to start researching this is the Australian Bureau of Statistics. https://www.abs.gov.au

Weblink
Australian Bureau of Statistics

Discussion

1 Identify the five organised sports participated in most by children now.
2 Summarise any significant sports participation trends you have discovered.

sport a special type of game that is competitive, requires you to use movement skills and has a wide following

9780170463102

HOW ARE GAMES AND SPORTS DIFFERENT?

It can be difficult to distinguish between games and sports. Games are generally less structured and competitive than sports, with participation being seen as more important than performance. Rules are often changed in games to suit participants and allow for success. Sports are generally more serious, with organised training, coaches and a governing body. Players adhere to official rules and these are not often changed. Below are some essential elements that all games have, and some extra elements that sports have.

GAMES

Games are recreational activities that involve achieving a goal (solving a problem). They have a set of rules that determine what can and can't be done in achieving that goal.

Games have four essential elements

A goal
There is something you or your team are trying to achieve. For example, scoring more goals than your opponent in netball, or hitting the ball in a way that means your opponent can't hit it back in tennis.

Attitude
Players accept the rules of the game and play by those rules. For example, when you play a game of golf you don't pick up the ball and throw it down the fairway. You abide by the rules and use a club to hit the ball – otherwise you are not playing the game.

A means to achieve the goal
There are only certain ways that you are permitted to achieve the goal. For example, in netball you can pass the ball but you can't run with it; in tennis you can hit the ball back with your racquet but can't catch it and throw it back.

Rules
Rules restrict what you can and can't do in the game. They often stop you from using easier movements, aiming to challenge your skills in the game. For example, in Australian Rules Football (AFL) you are not allowed to throw the ball, so you have to handpass or kick; in rugby you must pass the ball backwards, which is much more difficult than passing the ball forwards.

Figure 8.1 The four essential elements of games

SPORTS

Sports are a special type of game that are competitive, require you to use movement skills and have a wide following. They often have organised structures and rules, so that you have to use a range of movement and tactical skills to be successful. They have all the elements of games (goal, means, rules and attitude) but also have three more elements: movement skills, a wide following and stability.

Figure 8.2 Sports have all the elements of games plus three additional elements.

Additional elements of sports

Wide following
Other people must follow and participate in the game for it to be a sport. A game you make up at school, even if it is a great game, is not yet a sport, because not enough people know about it. You may have played 'speedball' or 'dodgeball' in physical education classes, or 'handball' or 'down ball' at recess or lunchtime, but these are probably games, rather than sports such as netball or cricket, which have many people who play and watch them.

Stability
People have followed and played the sport for a long time. For example, people have participated in and followed gymnastics and hockey for a long time, whereas playing with a yo-yo is an activity that has been popular at certain times but has not maintained its popularity consistently.

Movement skills
Sports involve movement skills, whereas a game may not. For example, chess and Monopoly are games, not sports, because movement is not important in them. In addition, sports are not games of luck or chance. For example, bingo and roulette might be games, but they are not sports because they do not require movement skills.

Fairfax Syndication/Danielle Smith

iStock.com/funky-data

Figure 8.3 A sports team will have uniforms and proper equipment to play (left). Games are less structured than sports (right).

Shutterstock.com/imtmphoto

Shutterstock.com/Alex Bogatyrev

Figure 8.4 Playing one-on-one basketball with your friends is a game (left), but netball is a sport as well as a game (right).

1 What are the four essential elements of games?

1 Identify whether the following are games or sports. Reorder the table in your workbook or in a Word document.

Tennis	Cricket	Bat tennis
Soccer	Checkers	Table tennis
One-on-one basketball	Poker	Chess
Netball	Hockey	Football

1 Clearly state how games are different from sports.

Quiz
How are games and sports different?

HOW DO RULES INFLUENCE GAMES AND SPORTS?

Both games and sports have rules. Rules limit what can and can't be done in the game to present challenges to be overcome. There are two types of rules in games and sports: primary rules and secondary rules.

PRIMARY RULES

Primary rules determine how the game or sport is played and won. If the primary rules are changed, the game becomes a different game or activity. For example, if you allow players to tackle each other in basketball, the game is no longer basketball. If you allowed players to throw the ball forward in rugby, the activity wouldn't really be rugby anymore.

SECONDARY RULES

Secondary rules are rules that can be modified without changing the nature of the game. For example, if you played soccer with five players on each team rather than 11, the game would still be soccer. If you changed the number of overs in a game of cricket from 50 to 20, the game is still cricket.

Newspix/Colleen Petch

Figure 8.5 Officials (umpires and referees) enforce the rules in many sports.

Rules in games and sports • Provide a structure for the game • Present problems for you to solve using your skills	
Primary rules • Identify how the game is played and how the game is won • Are what make hockey 'hockey' and soccer 'soccer'	**Secondary rules** • Can be modified without changing the nature of the game • Are open to interpretation
Examples • No handling of the ball in soccer • No throwing the ball in Australian Rules Football	**Examples** • Tie-break in tennis • Size of the ball in soccer

Figure 8.6 How rules function in games and sports

Primary and secondary rules in a game

Participate in a game during class.

1 When your teacher explains the rules of the game to you, try to identify which are primary rules and which are secondary rules.
2 As a group, decide on a secondary rule that you might change and discuss what effect this might have on the game.
3 Try playing the game with the change to the secondary rule. Did it change the game as you expected?
4 Decide on a primary rule and try changing that when you play the game.
5 What effect did changing a primary rule have on the game?

1 Define secondary rules.
2 Provide an example of a primary and secondary rule from a game or sport you are familiar with.

1 Discuss why a coach would not want to change a primary rule of a game during practice.

2 Research the rules of three sports. For each sport, identify two primary and two secondary rules and explain why you have classified them as primary or secondary rules.

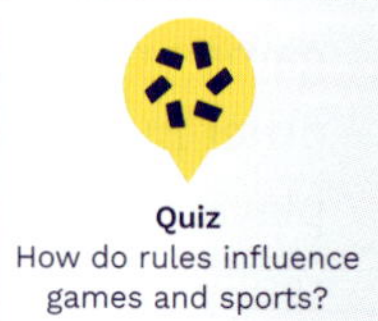

Quiz
How do rules influence games and sports?

WHAT TYPES OF GAMES AND SPORTS EXIST?

Games and sports can be classified according to the interactions between the players or the tactical similarities between them.

INTERACTION BETWEEN PLAYERS

Most games and sports involve other people; however, some sports rely much more on cooperation and teamwork than others. This means that games and sports can be classified by the amount of interaction between the players.

Individual games and sports

Individual games and sports involve practising and performing by yourself in competition with others (e.g. tennis, martial arts, triathlon, skateboarding, gymnastics, figure skating).

Shutterstock.com/Stefan Holm

Dreamstime.com/James Boardman

Figure 8.7 Triathlon is often an individual sport, while lacrosse is a team sport.

Team games and sports

Team games and sports involve individuals working together in competition with other teams (e.g. football, netball, soccer, cricket, lacrosse, hockey).

TACTICAL SIMILARITIES

Games and sports often involve similar tactical skills. This means that what is learnt in one game or sport might transfer to another. Understanding the similarities between different games and sports helps you to participate in a greater range of games and sports. There are four categories of games.

Invasion (or territorial) games

In invasion games, opposing teams attempt to invade their opponent's territory. The aim is to score points either by shooting into a target or goal (soccer, hockey, netball), or by moving the ball into a designated scoring area (rugby or Ultimate Disc). Gaining and keeping possession of the ball/puck/frisbee is crucial to attack the opponent's territory.

In attack, a primary strategy is creating space to allow players to move the ball and execute movement skills more effectively. In defence, occupying or blocking up space and defending the goal area are critical tactical skills. Examples of invasion games include football, basketball, lacrosse, rugby union, rugby league, water polo and netball.

Net/wall games

In net/wall games, players or teams aim to send an object into an opponent's area so that the opponent cannot return it. Players are separated by a net (tennis, badminton, volleyball) or use a wall with alternating hits by players (squash, down ball, racquetball). In some games players use implements to hit the object (tennis) and in other games they use their hand (volleyball). The primary strategy is to place the ball away from your opponent. Examples of net/wall games include tennis, badminton, squash, table tennis, volleyball, newcomb (a variation of volleyball) and down ball.

Striking/fielding games

In striking/fielding games there is a batting team and a fielding team. The batting team tries to score as many points or runs as possible. Players score runs by running between bases or wickets or by striking the ball to a boundary before the fielding team return the ball. The fielding team occupies positions throughout the field into (or through) which the batting team must strike the ball to score runs. Most striking/fielding games use implements to strike the ball. The primary strategy of the batting team is to hit the ball away from the fielders, to maximise time to score runs. Examples of striking/fielding games include rounders, baseball, softball, T-ball, cricket and kickball.

Target games

In target games, the aim is to get an object as close as possible to a target (e.g. darts, archery, lawn bowls) or get the object in the target in as few attempts as possible (golf, croquet). In unopposed target games, your shot does not affect your opponent's (golf, archery, ten-pin bowling). In opposed target games, there is interaction with the opponents' object or the target (lawn bowls, bocce, croquet).

Table 8.1 Games and sports with tactical similarities

Game category	Primary strategy	Game examples
Invasion (territorial)	Gain and maintain possession of the ball	Soccer, netball, hockey, football, lacrosse, korfball, rugby union, rugby league, gridiron, Ultimate Disc, hurling, Gaelic football, team handball
Net/wall	Place the ball away from your opponent	Tennis, bat tennis, racquetball, badminton, table tennis, volleyball, squash, down ball
Striking/fielding	Hit the ball away from the fielders	Rounders, baseball, softball, T-ball, cricket, kickball
Target	Be accurate in relation to the target	Golf, archery, ten-pin bowling, darts, curling, lawn bowls, bocce, croquet

Source: Adapted from M Spittlo, *Motor Learning and Skill Acquisition: Applications for Physical Education and Sport*, Palgrave Macmillan, 2013, p. 326

Worksheet 8.2

Fairfax Syndication/Graham Tidy

Fairfax Syndication/Vince Caligiuri

Newspix/Brett Costello

Alamy Stock Photo/David Burton

Figure 8.8 Rugby is an invasion game, squash is a net/wall game, softball is a striking/fielding game and archery is a target game.

1 Provide three examples of individual sports and three examples of team sports.

2 Provide one example of a game or sport in each of the following categories: invasion (territorial), net/wall, striking/fielding and target.

1 Describe the main strategy of each of the four categories of games: invasion (territorial), net/wall, striking/fielding and target.

Quiz
What types of games and sports exist?

1 There are a range of cultural games developed and played by Australian First Nations Peoples across different traditional country, custodian, language, and nation groups across Australia. Examine one of these games in detail by writing a 300–500 word summary of the background of the game and description of how the game is played.

Weblink
Yulunga Traditional Indigenous Games

WHAT TYPES OF SKILLS ARE THERE IN GAMES AND SPORTS?

Participating and performing in games and sports requires a range of skills. In games and sports, the term 'skill' describes two things:

- A task you perform

  A skill is a task that is learnt through practice or experience in order to achieve a goal. For example, kicking a ball and performing a forward roll are skills.

- Quality of performance

  Players with skill can achieve the goal of the task more efficiently and effectively. For example, a skilled cricket fast bowler can bowl the ball consistently with accuracy, speed and fluency.

BEING SKILLED

Skilled players in games and sports can perform effective movements when required by the situation in the game. Being skilled in games and sports requires movement skills to successfully move, as well as effective tactical skills to decide when and how to use those movement skills:

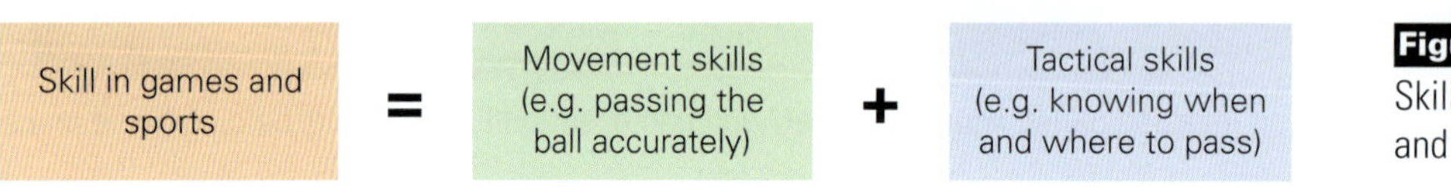

Figure 8.9 Skill in games and sports

Movement skills

Movement skills are learnt behaviours that require you to move your body to achieve a goal. For example, the goal could be to move quickly, control a ball or use a racquet to hit a ball. Some sports and games have more movement skills than others; some only require a few movement skills, or repetition of the same movement skill.

Tactical skills

Worksheet 8.3

Tactical skills are the decisions and actions you make in a game to gain an advantage over your opponent. They involve knowing when and where to use your movement skills. Examples of simple tactics in games and sports include hitting the ball away from your opponent in a striking/fielding game such as cricket, or moving into space to receive the ball in an invasion game such as netball.

MOVEMENT SKILLS

Understanding different types of movement skills in games and sports helps you to analyse, adapt and refine your own skills, as well as help others improve their movement skills.

Classifying movement skills helps us determine how best to learn and instruct particular movement skills. Movement skills can be classified in four different ways:

- whether the skill is fundamental or specialised
- precision of movement (gross to fine)

9780170463102

- organisation of the skill (discrete, serial or continuous)
- stability of the environment (open to closed).

Figure 8.10 Being skilled in games and sports requires players to move effectively and also to make effective decisions.

Shutterstock.com/LittlePanda29

Fundamental and specialised movement skills

fundamental movement skills movements that provide the foundation for other, more advanced movement skills

Fundamental movement skills are basic movements that are the foundation for more advanced and specific movement skills. You learn these skills when you are young and they allow you to develop new skills. For example, a two-handed striking action to hit an object is the foundation skill for more specific movement skills such as batting in cricket and baseball, passing in hockey and hitting a golf ball.

There are three broad types of fundamental movement skills, as shown in Figure 8.11.

Locomotor skills	Manipulative skills	Stability
Moving your body through space, such as moving from one side of the room to the other.	Controlling an object, such as a bat, racquet or ball.	Controlling the body in terms of balance.
Locomotor skills include walking, running, jumping, hopping and skipping.	Manipulative skills include throwing, catching, kicking and striking.	Stability skills include balancing, bending, twisting and agility.

Figure 8.11 The three broad types of fundamental movement skills

As you continue playing a sport, the skills can become more specific. For example, you might use a basic kicking skill when you play soccer for the first time, and then refine this skill and add a variety of different kicks, such as passes, crosses, centring, corners, one-touch passing and shooting for goal.

Specialised movement skills are movements that are specific to a particular game or sport. They are more advanced versions of fundamental motor skills or combinations of fundamental motor skills that you apply to a specific sport. For example, pitching in softball is a throw that is specific to softball, and a cover drive in cricket is a striking movement that is specific to cricket.

Mastering fundamental movement skills and developing a range of specialised movement skills helps you to participate and perform in a wide range of games and sports.

Figure 8.12 Running is a fundamental movement skill.

Newspix/Kit de Guymer

FAST FACT

In manipulative skills, you control an object, such as a bat, racquet or ball. To do this you apply force to cause the object to move. Force is energy that is applied to an object to make it change speed. For example, in hitting a ball, if you apply a large amount of force, the ball will travel faster. If you apply force to the object anywhere but through the centre of the object, it causes rotation; that is, it makes the object spin. This is how you can create different types of spin, such as backspin, topspin and sidespin on a ball in games and sports like tennis, cricket and football.

FACE TO FACE Throwing

With a partner, come up with five specialised movement skills that develop from the fundamental movement skill of throwing.

Gross and fine movement skills

Sometimes in games and sports you have to move with precision and sometimes you have to move with force or power.

Gross movement skills use your large muscle groups and are less precise (e.g. running, jumping and throwing). Fine motor skills use your smaller muscle groups (e.g. hands and fingers) and require you to use precision, accuracy and control (e.g. gripping a racquet or bat, or throwing a dart).

Many skills require you to use both fine and gross movement. For example, the jump shot in basketball involves gross movements of the large arm and leg muscles but also precise control of the fingers to release the ball with the intended spin and correct angle.

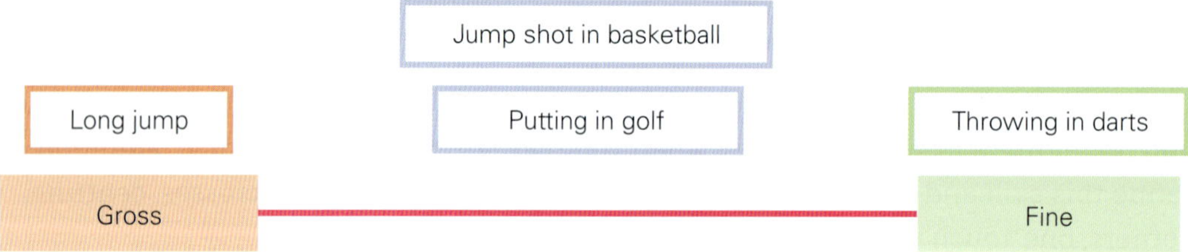

Figure 8.13 Movement skills classified by movement precision

Discrete, serial and continuous movement skills

Discrete skills have one clear movement with a beginning and end. They are movements you perform quickly, often in less than a second (e.g. kicking, throwing and hitting a ball).

Continuous skills have no clear beginning or end and involve repetitive and rhythmic movements performed over and over again. They can take some time for you to complete (e.g. running, cycling and swimming).

Serial skills involve a series of discrete skills linked together in a specific order, so combine some aspects of discrete and continuous skills (e.g. triple jump or a dance routine).

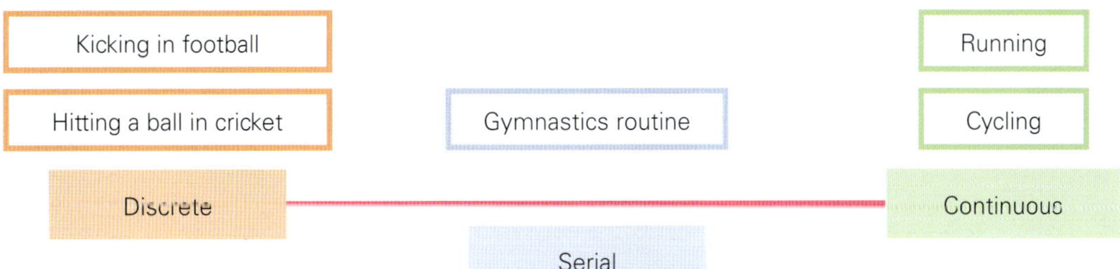

Figure 8.14 Movement skills classified by skill organisation

Figure 8.15 A gymnastics floor routine is a serial movement skill.

Open and closed movement skills

Open skills are skills you perform in an unpredictable and changing environment. This means you have to react and adapt the skill to things going on around you, such as the movements of your teammates and opponents or the movements of the ball (e.g. passing in rugby, batting in cricket and dribbling the ball in a game of soccer).

You perform closed skills in a stable and predictable environment and attempt to produce a consistent and efficient movement that you can repeat in a similar way (e.g. dart throwing, basketball free-throws and high jump).

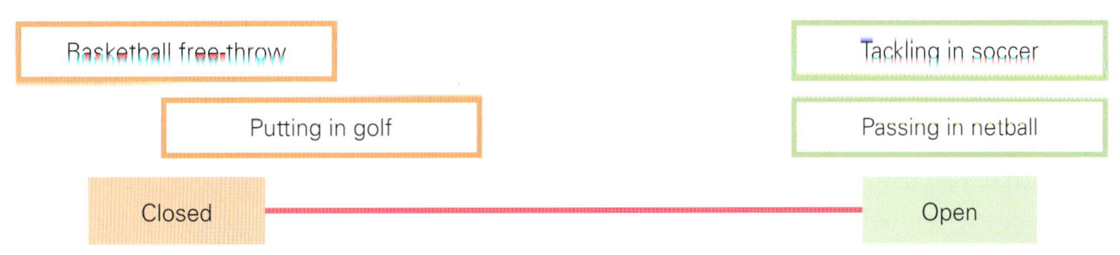

Figure 8.16 Movement skills classified by the stability of the environment

Newspix/Glenn Daniels

Fairfax Syndication/Dean Osland

Figure 8.17 The basketball free throw is a closed movement skill, whereas netball passing is an open movement skill.

UP AND MOVING Comparing open and closed movement skills

1 Sometimes we can modify an open movement skill to be more like a closed movement skill. For example, in striking/fielding games we can hit the ball off a tee (T-ball) or we can react to a pitch from a pitcher that could vary in speed and spin (softball). Why would we do this?

2 Participate in a game of T-ball and then softball, or practise hitting in each of these games. In which game is it easier to hit the ball?

3 Consider why a coach or teacher would modify the hitting skill from open to closed.

Worksheet 8.4

TACTICAL SKILLS

You use tactical skills to gain an advantage over your opponent. In team games and sports, tactics and strategies can help you work together as a team to improve your performance. This helps you become less focused on your own individual performance outcomes and more aware of the roles of other players and on working together for the team.

Tactical skills include tactics, strategies and game plans.

Tactics

A tactic is a plan within a game or sport to gain an advantage. Tactics can be used by an individual or team, and different types of games have different tactics. Tactics change regularly depending on the game and game situation. For example, the weather conditions, opponent, ground size, score and time left in the game can all influence the tactics that you use. Examples of tactics include a soccer player deciding to pass rather than shoot when being actively defended, an AFL footballer deciding to kick long when under pressure and a long-distance runner deciding to make a break with one lap to go.

Strategies

A strategy is a plan of action for a season or several competitions. You use it to take advantage of the strengths of your team and make up for the weaknesses. For example, a netball team with a tall goal shooter might decide to use long passes into the circle, or a soccer team with good midfield players might work on passing the ball rather than kicking long.

Defensive strategies often rely upon one of two defensive structures – one-on-one defence or zone defence. In one-on-one defence, each player is responsible for defending an opposition player. In a zone defence, each player is responsible for defending a specific space on the field. Both of these approaches aim to prevent the attacking team from having space to move the ball freely.

Game plans

A game plan is a basic structure and style of play for a specific game, and involves applying your strategy and tactics. This plan is usually based on a team's strengths and weaknesses, as well as the strengths and weaknesses of the opponent. An example is having players on your team tag specific opponents.

Game plans often incorporate team rules or set plays:

- Team rules are instructions for players to follow (e.g. not passing the ball across goal in defence).
- Set plays are developed for specific situations in the game (e.g. players on a soccer corner have specific places to run to when the ball is kicked, or basketball players who run in a predetermined pattern on an attacking play).

Worksheet 8.5

CHECK YOUR THOUGHTS

This activity has been developed in collaboration with Dr David Bakker of MoodMission

Identify

It's easy to believe that events and situations make us feel certain emotions. But lots of research shows it's actually the way we think about the situation.

Understand

Cognitive behavioural therapy (CBT) is one of the most common and effective types of psychological therapy used to help people overcome a huge range of issues, including depression and anxiety. One of the core principles of CBT is that thoughts come before feelings. If we change our thoughts, we can change our feelings, and can deal with situations better. For example, if before a game you're thinking to yourself 'I'm going to mess this up', you'll start feeling anxious and probably not play as well. If you think, 'This might be tough but I'll have fun and try my best', you'll feel more confident and be more focused. CBT isn't just about 'thinking positive', though. It's important to understand the positive and negative points of any situation. However, sometimes we get bogged down in the negatives, and need some CBT strategies to remind us of the positives.

Practise

1 Recall last time you were feeling sad, anxious or another strong negative emotion. Note the situation briefly; for example, what you were doing, where you were, who was there.
2 Now write down the emotions you were experiencing, for example, sadness, anger.
3 Next, write down the thought that was going through your head, such as 'They're all going to laugh at me' or 'I can't do anything right'.
4 Take a moment to consider that thought. Is there anything you could have thought instead? Write down this alternative thought below; for example, 'It's hard, but I'll be okay'.
5 State any emotion(s) that you would feel if you thought this alternative thought.

Reflect

The next time you feel bad, ask yourself, 'What's the thought that got me here?' We can sometimes notice that these negative thoughts follow similar patterns. Once we know the pattern, it can be easier to see the negative thoughts coming and anticipate dealing with them.

1 Provide an example of a locomotor skill, a manipulative skill and a stability movement skill.
2 State the difference between gross and fine movement skills.
3 Provide an example of a discrete, serial and continuous movement skill.

1 Explain what it means to be skilled in games and sports.
2 State why it is important to develop fundamental movement skills.
3 Discuss why we classify movement skills into different types.
4 Describe the differences between tactics, strategies and game plans.

1 Compare tactics, strategies and game plans in a game. This activity involves you playing some different games with your classmates and comparing the tactics, strategies and game plans in those games.
   a As a class, choose an invasion game. Play this game and discuss the main tactics, strategies and game plans that are being used.
   b Next, choose another invasion game to play. As you are playing, distinguish the main tactics, strategies and game plans that are being used and compare these with the tactics, strategies and game plans that you used in the first invasion game.
      - What are the similarities? Are there any differences?
      - Why might these similarities and differences exist?
   c Finally, choose a net/wall game to play. As you are playing, distinguish the main tactics, strategies and game plans that are being used and compare these with the tactics, strategies and game plans you used in the two invasion games.
      - What are the similarities? Are there any differences?

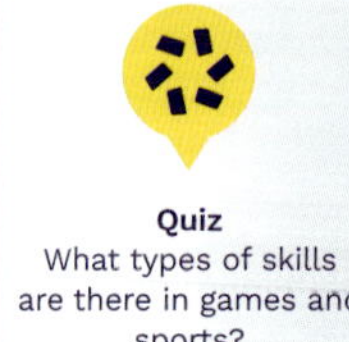
Quiz
What types of skills are there in games and sports?

HOW CAN I ANALYSE SKILLS IN GAMES AND SPORTS?

To improve your performance in games and sports, you can analyse the skills that you use and how you perform them. Performance analysis involves observing and collecting data on performance to provide feedback to coaches and players. This can then be used to make decisions about how to improve movement or tactical skills. Assessment of performance can be subjective or objective.

SUBJECTIVE PERFORMANCE ASSESSMENT

Subjective performance assessment relies on personal judgement and observation of performance. This means that judgements on the performance could vary from one observer to another. Examples of subjective analysis in sports include judges' scoring in diving, gymnastics and surfing.

OBJECTIVE PERFORMANCE ASSESSMENT

Objective performance assessment relies upon measuring movement outcomes such as number of successful shots, distance thrown or speed of the pitch. It could also include the use of statistics (e.g. in tennis the number of forced and unforced errors). Objective performance assessment is more reliable and less influenced by personal judgement.

Performance analysis can be used to assess both movement skills (motion analysis) and tactical skills (games analysis). Both forms of analysis can use simple observation of performance or make use of video-based equipment and technology. There are a range of performance analysis programs available that can perform both motion analysis and games analysis for sports, such as Dartfish, Swinger, SportsCode and Siliconcoach.

MOTION ANALYSIS

Motion analysis (or biomechanical analysis or technical analysis) involves observing the movement skill to analyse technique. This can be subjective (simply watching a player move) or objective (measuring angle of release, height of release and force). Analysis of movement skills is important to diagnose strengths and weaknesses, assess current performance levels, identify areas for improvement, develop training approaches, design practices and identify feedback to provide to players.

You can analyse movement skills by directly observing or watching skill performance, or by using video, which is becoming even more accessible through the use of smartphones, tablets and digital cameras. Video can be analysed using video analysis software; many analysis programs are available as apps for smartphones and tablets, making video analysis available for use at all levels of sport. Video analysis software allows you to edit, manipulate

Newspix/Gregg Porteous

Figure 8.18 Analysing movement skills is important for improving performance.

and make notes on the movements. For example, you can split screen the movement to compare different observation angles, overlay a video of one skill over another, track the movement of projectiles such as the ball, analyse the skill frame by frame and calculate distances, speeds, angles and velocities.

CASE STUDY ⇨ MOTION ANALYSIS AND TECHNIQUE

Identify

Motion analysis involves analysing a player's movement skill to analyse technique. This can help to identify how to improve a player's skill. Steve Smith is a successful batter for Australia who has adapted his technique to be successful in international cricket.

In analysing a skill you should consider the specific requirements of the skill. In some games, movement outcome (scoring runs) is more important than movement form (aesthetics of movement). In other words, technique is more important than style. In other sports, such as gymnastics and dance, movement form (aesthetics) is more important. This means that there might be some variability or differences in technique in some games and sports, even at elite levels. For example, comparing two elite footballers, tennis players or golfers, you will notice subtle differences in their cover drive, backhand or putts.

Getty Images/Stu Forster

Figure 8.19 Steve Smith

Understand

When Steve Smith took to the field in the Ashes Test in Perth at the end of 2013, he was a middle-order batsman worried about his future in the side. With a batting average of just 33 and a solitary Test century to his name from 14 matches, he seemed far from cementing his spot in the side. But part way through his innings, after being peppered by a series of short deliveries from the English fast bowlers, Smith made a change to his game that would alter the course of his career. Before each delivery, he began taking a step back and across his stumps. It was a radical change to his action that he had never practised before.

Smith went on to make his first hundred in Australia that game and since adopting this new technique, which has become far more pronounced over time, he has gone on to become the world's top-ranked batsman and has even drawn comparisons to cricketing legend Sir Donald Bradman. While Smith's new technique is often derided as unconventional and even ugly, for sports biomechanics expert Rene Ferdinands, it's anything but. 'When you look at it from a biomechanical point of view, it makes perfect sense,' he said.

Discuss

1 Discuss how analysing your technique through motion analysis can help you improve your skill in sport.

2 Consider if there is one 'perfect' technique or 'correct' movement pattern for a skill that works for all players in a sport, or whether players need to adapt their own capabilities by developing a technique that works for them.

The analysis of movement skills needs to be systematic. It should follow several steps, including preparation, observation, evaluation, diagnosis and correction.

Preparation and planning

Prepare and plan for the skill analysis by finding out about the skill and its critical elements. For example, watching the skill performed by others, reading books and journals, and searching the internet.

Observation

Plan a systematic observation of the skill. Many movement skills happen very quickly, leaving you little time to pick up the critical elements of the skill. Watching the skill several times and from different angles and different distances can help. Video can also be useful to replay the skill several times.

Detection and evaluation

Look for and identify the potential errors in the movement skill and evaluate the quality of the movement. Is the skill being performed correctly?

Diagnosis

Diagnose what is causing any errors you identify. Some possible causes of errors include movement technique, not paying attention to the right cues, selecting the wrong movement and not understanding the skill or the instructions.

Correcting errors

Once you have detected and diagnosed the errors, the next step is to help the learner to correct the error. This is often through providing feedback to the learner or providing a practice activity for the learner to work on their skill.

Source: Adapted from Knudson, D., & Morrison, C. (1997), *Qualitative Analysis of Human Movement*, Champaign, IL: Human Kinetics, p. 27

Analyse your performance using motion analysis

1 Choose a movement skill from a game or sport and analyse your performance and that of a partner. When performing the analysis make sure that you follow all the steps necessary for motion analysis. You might video the movement skill performance and use video analysis software or an analysis app.

2 Once you have analysed the movement skill, discuss how to improve your performance and provide peer feedback to your partner about their movement skill.

3 Develop a practice activity to improve your movement skill and practice the skill.

4 After practising the skill, perform the skill analysis again and see if your practice has been effective at improving your performance and that of your partner.

Worksheet 8.6

GAMES ANALYSIS

Tactical analysis (or games analysis or notational analysis) can be used to assess the use and performance of movement skills and tactical skills within the game. Tactical games analysis helps identify the tactics of each team. It also helps identify the effectiveness of different tactics in the game so that they can be modified.

In games analysis, you record the skill performance activities of players, such as each time a player passes the ball, the effectiveness of each pass, the number of shots at goal, the effectiveness of those shots and the number of forced and unforced errors. Thus, performance analysts can provide statistical information about how often specific skills were performed and the effectiveness of those skills, as well as an analysis of tactics or broad patterns or styles of play. Games analysis was previously done with a pen and paper, but video analysis and coding games using video analysis software is becoming much more common.

In general, it is better to start with a simple games analysis system and gradually add other actions, rather than begin with an overly complicated system. In basic games analysis systems you should code information about who the player was, where the action happened, the action the player performed and the time in the game at which it happened.

Figure 8.20 A simple process for games analysis

As an example, if in hockey or soccer you were interested in analysing attacking patterns, it would be important to code who the player performing the action was, the player's location on the pitch, the skill (e.g. a pass) and when it happened in the game. The information from tactical analysis helps players and coaches to plan, modify, adjust and review their tactics before, during and after games.

Identify

Ash Barty triumphs in a hard-fought final against Karolina Pliskova at Wimbledon to become the first Australian ladies' champion in 41 years.

Understand

Barty today became the 2021 Wimbledon ladies' champion after overcoming Karolina Pliskova in three sets. In a stunning display of artistry, guile and composure, Barty seized her 6-3 6-7(4) 6-3 victory in five minutes under two hours. Barty, who claimed a first Grand Slam title at Roland Garros two years ago, is the first Australian to lift the Venus Rosewater Dish since her idol Evonne Goolagong Cawley won the second of two Wimbledon titles in 1980. She is only the third Australian woman to triumph in the Open era, joining Goolagong Cawley (also the 1971 winner) and 1970 champion Margaret Court. It was history that could easily have overwhelmed many other players. But from the outset, Barty was a picture of calm.

Getty Images/Xinhua News Agency

Figure 8.21 Ash Barty, 2021 Wimbledon ladies' champion

8

The world No.1 couldn't have wished for a better start as she tallied early winners off both her forehand and backhand, while damaging with her dazzling serve. Holding her own serve to love as she opened, Barty then broke the Pliskova serve in the second game. Something special appeared to be unfolding as the Australian calmly consolidated – Barty broke again as she claimed the first 14 points of the match. While her nervy opponent would eventually find some rhythm, it was the top seed who held all early authority; despite dropping two of her own service games, Barty maintained pressure on Pliskova's biggest weapon, breaking serve again in the sixth game and soon closing out the set. But with four of their seven previous matches extending the full distance, Pliskova would inevitably challenge the world No.1.

After service breaks were exchanged in the second set, Barty broke again in the 11th game to serve for The Championships. A hint of nerves finally appeared to set in as Pliskova pressured; she pushed it to a tiebreak, the Czech then levelling the match. "(To) just keep fighting", Barty explained of the mindset she summoned at that critical juncture. "Karolina is an incredible competitor. She brought out the very best in me today and it was an exceptional match right from the start. I knew that I had to bring my very best level". That Barty's eventual 30 winners were just three more than Pliskova's – and her 29 unforced errors just three less than 32 from the Czech – highlighted an incredibly tight contest. But the third set belonged to Barty, who applied scoreboard pressure as she broke in the second game then consolidated. From there, the world No.1 maintained her lead. While there was a double-fault as Barty served again for the title at 5-3 in the decider, a superbly placed ace set up a championship point. The Australian seized it after a Pliskova error and was soon celebrating with tears of joy.

'Dream come true: Ash Barty seizes Wimbledon crown', by Vivienne Christie for Tennis Australia London, UK, 11 July 2021, https://www.tennis.com.au/news/2021/07/11/dream-come-true-ash-barty-seizes-wimbledon-crown.

Discuss

Take on the role of a performance analyst or coach – analyse a game or sport of your choice and provide a feedback report suitable for coaching staff and/or players about the performance in the game.

Pick a movement skill or tactical skill to emphasise in your analysis. For example, depending on the game or sport you choose, you could analyse the team's set-play performance (corners, frees, penalties), a team's passing performance (effective and ineffective passes), a player's forced and unforced errors or the shots played by an individual.

To do this, you could watch a game live or use video footage and use either a pen and paper or a software program or app.

1 Write a feedback report for the coaching staff about performance in the game. Try to make your report as informative and engaging as you can. For example, present the data in a visual format such as graphs and tables and provide recommendations for the coaching staff. When writing the report, consider the following questions:
   a What did the analysis tell you about the team's performance?
   b What recommendations would you provide to the coach?

1 List, in correct order, the steps in movement analysis.
2 Distinguish between motion analysis and games analysis.

1 Classify the following as subjective or objective performance assessment:
- Runs scored in cricket
- Score on a gymnastics routine
- Number of shots on a golf hole
- Distance thrown in javelin
- Time taken to run the marathon
- Score on a dive

1 Search for apps and resources for performance analysis in sports and then explore how to use them. Research options could include Dartfish, Swinger, Sports Coach, Siliconcoach and SportsCode. Report on one of these apps and what it is used for in sport.

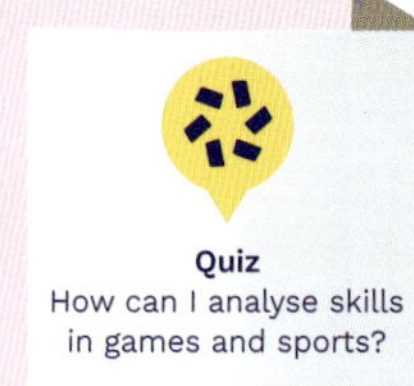

HOW CAN I ADAPT AND REFINE SKILLS IN GAMES AND SPORTS?

Once you have analysed your performance, you need to practise to adapt and refine your movement and tactical skills. Improving your skills and those of others will help you and your team when participating and performing in games and sports. Important principles of skill acquisition include improving specific skills, effective application of practice, use of feedback, adapting skills through transfer of learning and modification of games to match the needs of the learner.

IMPROVING MOVEMENT SKILLS

Skill acquisition is the study of how you can learn and develop movement skills. This includes exploring the changes that you go through when you learn a movement skill and how learning activities such as practice and feedback can influence your learning.

PRACTICE

Practice is the most important factor in acquiring movement skills. If you don't practise skills, you will not develop and refine them. In general, the more you practise a skill, the better you should become. Therefore, the **amount** of practice you do is critical, but the **type** of practice that you do is also important. Two factors that influence the effectiveness of practice are the distribution and the variability of practice.

Distribution of practice

Distribution of practice is how you schedule practice sessions (e.g. the number and length of sessions per week) and how you schedule practice within a session (e.g. how much rest between each time you practise the skill).

Massed practice	Distributed practice
Longer sessions less frequently	Shorter sessions more often
Less rest in practice sessions	More rest in practice sessions

Figure 8.22 Massed vs distributed practice

Distributed practice is practice that is spread out. This involves shorter sessions more often, with more rest in each session. For example, if you had four hours a week to practise golf you could practise in four one-hour sessions a week with 30 seconds' rest between each hit of the golf ball.

Massed practice is practice that is grouped together. This involves longer sessions that occur less often, with less rest in each session. For example, when practising golf you could have one four-hour session a week with five seconds' rest between each hit of the golf ball.

It is better for skill acquisition to have distributed practice sessions rather than massed practice sessions. For example, it is better to have three one-hour sessions a week than one three-hour session a week. How much rest you need within a practice session is probably related to the type of skill you are practising.

- For shorter discrete skills (kicking, passing, shooting), massed practice with less rest is effective because you get more practice attempts in the same amount of time and are less likely to become fatigued.
- For longer continuous skills (swimming, running, cycling), distributed practice with more rest is effective in order to reduce fatigue.

Alamy Stock Photo/Jack Sullivan

Fairfax Syndication

Figure 8.23 A golf shot benefits from massed practice; running benefits from distributed practice.

INVESTIGATION ⇨ WHICH METHOD OF PRACTICE IS BEST SUITED TO LEARNING A MOVEMENT SKILL?

Purpose

To explore the effects of massed and distributed practice on learning a movement skill.

Materials

- Example movement skill: Paddle ball hit to target (this activity could be completed with a different movement skill instead). Participant is to hit (strike) a tennis ball with their non-preferred hand using a paddle so that it will hit a target as close to the bullseye as possible. The ball is to be bounced off the floor by the participant prior to striking it with the paddle. The participant is to complete 30 trials. The distance from the target should be around 5 metres. The centre of the target should be approximately 1 metre above the floor.
- Equipment: Paddles (bat tennis bats), tennis balls, wall targets (e.g. concentric circle targets such as archery targets stuck to wall), stopwatches, cones (to mark hitting distance), tape measures (to measure hitting distance)

Method

Your class will practise a movement skill using massed and distributed practice and compare them to determine which practice schedule led to better performance. Half the class should practise using a massed schedule and half should practise with a distributed schedule.

- Massed practice: 5-second rest between each hit (so one member of the group has 30 tries, then a second member has all of their tries, etc.).
- Distributed practice: 30-second rest between each hit (so members of the group each have one try, then each have a second etc.).

Complete the activity in small groups to collect the data (e.g. in pairs or groups of four). Every student in the group can complete the activity in turn while other students record their scores. Collate the class scores to get a class average for each practice schedule.

Record your score on each hit so you can get an overall score (e.g. if using the archery target, it would be a score of 10 for hitting the bullseye through to 1 for the outermost circle and 0 for missing the target completely). Add your score to the class data and compare the average scores for massed and distributed practice.

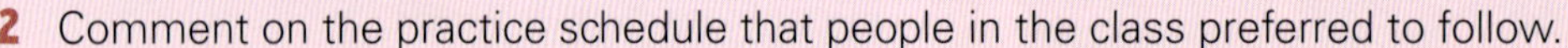

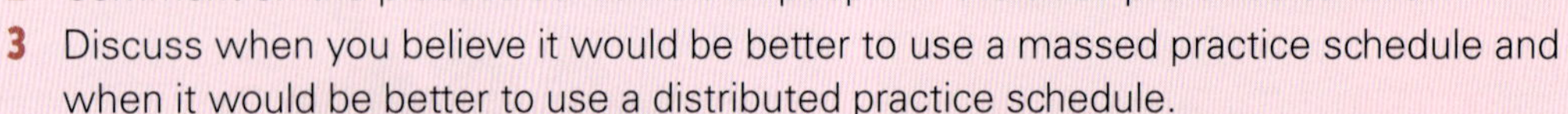

Discussion

1. Analyse the data. Which practice schedule led to better performance? Why do you think that was?
2. Comment on the practice schedule that people in the class preferred to follow.
3. Discuss when you believe it would be better to use a massed practice schedule and when it would be better to use a distributed practice schedule.

Worksheet 8.7

Variability of practice

Practice variability refers to how much skills change when you practise. You can practise more than one skill in a session (e.g. passing, catching and shooting). Practising more than one skill in a session improves skill acquisition. Blocked practice has less variability than random practice.

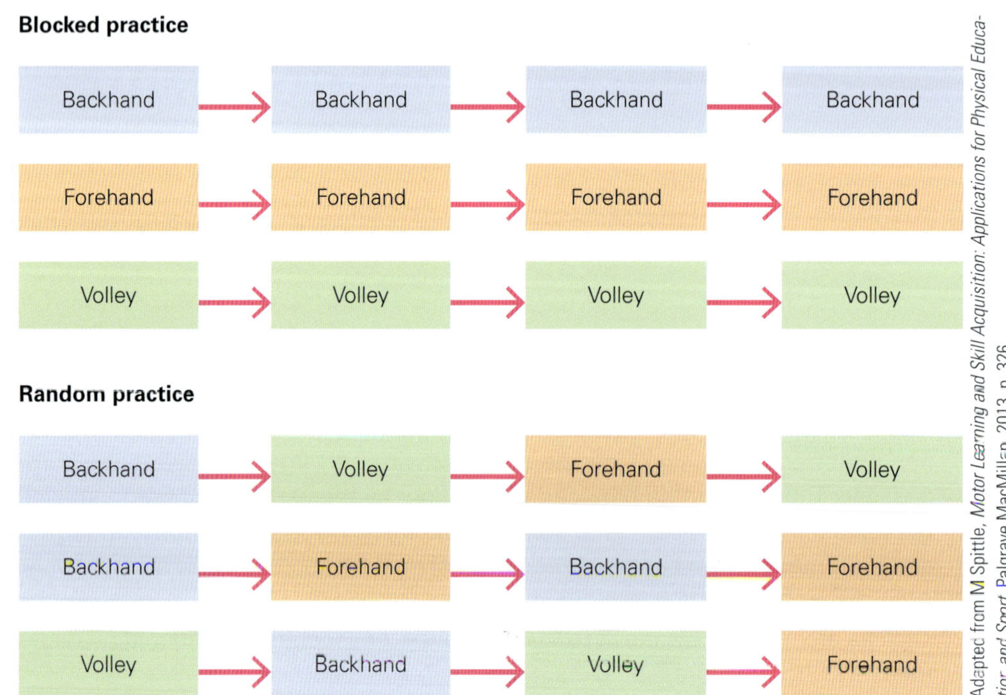

Figure 8.24 Random versus blocked practice for practising tennis

Adapted from M Spittle, *Motor Learning and Skill Acquisition: Applications for Physical Education and Sport*, Palgrave MacMillen, 2013, p. 326

9780170463102

- **Blocked practice** involves repetitively practising a skill for a period of time before practising another skill. For example, in basketball you might practise dribbling for 10 minutes, then passing for 10 minutes, then rebounding for 10 minutes.
- **Random practice** is alternating between skills so that you do not repeat the same skill twice in a row. For example, in basketball you might continually alternate between dribbling, passing and rebounding.

Getty Images/Trevor Williams

Figure 8.25 Blocked practice involves repetitively practising the same skill.

Blocked practice can be beneficial early in learning (for beginners or for younger players who have less experience). Random practice tends to produce better retention of skills, and skills that are more adaptable to match the performance situation. Blocked practice helps refine basic skills; random practice develops those skills further.

UP AND MOVING Practise

1. Pick a movement skill from one sport and design four drills or activities to practise the skill with a partner.
2. Evaluate how blocked or random the practice is for each drill.
3. Practise the skill using those drills and evaluate which of the drills was most effective for learning the skill.

IMPROVING TACTICAL SKILLS

To improve your tactical skills in games and sports, you can develop your knowledge of the game and your decision-making skills, use variability of practice, use feedback and questioning, practise tactical drills and adopt a game sense approach.

Improve your knowledge and decision-making

Developing knowledge and understanding of the game, including the rules of the game, the strategies and game plan, the playing conditions, your strengths and weaknesses and those of your opponent and your tactical options in specific game situations can help you choose the right tactics and make better decisions.

FACE TO FACE Effective tactics

Watch a game or sport and try to identify the tactics the players or teams are using. Discuss with a partner how successful or effective the tactics are.

Variability of practice

Variability of practice simulates more specifically how to use your skills in a game because you are constantly making decisions about how to use your skills. For example, netballers must choose who to pass the ball to, cricketers must decide what shot to play and tennis players must react to different shots from their opponent. Use practice activities to decide which movement skill to use rather than just repeating the same movement.

Feedback and questioning

As with developing movement skills, you need feedback on tactical skills. Too much extrinsic feedback (see page 337) can reduce how much you have to think for yourself. When helping others learn tactical skills, rather than directly telling them what to do, use questioning to help them think about their tactical skills. For example, asking questions such as 'Where is the best place to hit the ball?', 'What is the most effective way of defending in a particular instance?' or 'When should the ball be passed?' can all encourage a learner to think about their tactical skills.

Getty Images/Maurizio Borsari

Figure 8.26 Practising set-plays such as a corner kick in soccer can help improve team performance.

Tactical drills

You can practise using tactical drills to refine movement concepts and strategies so that when a specific situation occurs, you know what to do. These often occur when there is a set-play situation or a stoppage in the game (e.g. the centre bounce in Australian Rules Football, the lineout in rugby union, the centre-pass in netball). Tactical drills can also be simulations of game situations, such as having a cricket batting drill where players aim to score 30 runs in five overs to simulate the end of a limited overs game, or practising playing tie-breaks in tennis so that players can learn to adapt their tactical skills to the movement situation.

Game sense

This is an instructional approach to developing tactical and movement skills, in which you try to solve challenges presented in games rather than practising movement skills on their own. To create these challenges, elements of the games are modified. For example, creating a game where there are three attackers and two defenders can encourage attacking players to draw defenders to them when they have the ball and move into space when they don't have the ball. (Modifying games is explored on pages 332–6.)

9780170463102

FEEDBACK

When you practise movement skills you can use feedback (information about your performance) to help improve your performance. Feedback can come from internal sources (intrinsic) or external sources (extrinsic).

Intrinsic feedback

Intrinsic feedback is information received through your senses as you perform the skill. For example, when you hit a golf ball you get feedback such as seeing the ball travel through the air, feeling the movement of your arms as you swing and being aware of your grip on the club. With practice you will begin to better understand the movement skill and get better at using intrinsic feedback to correct and refine your movement. Elite performers in any sport become very aware of their own movements and how to detect and correct movement errors.

Extrinsic feedback

Extrinsic feedback is information about performance that comes from an external source. It supplements intrinsic feedback. For example, the coach telling you about your arm movement in hitting the golf shot adds to the intrinsic feedback you already had from the feeling of your arm moving. A GPS monitor can give extrinsic feedback about the intensity of your running, which can supplement your own feelings of effort.

Shutterstock.com/A_Lesik

Figure 8.27 Extrinsic feedback can be given to you by a coach or teacher.

TYPES OF FEEDBACK

There are two main types of intrinsic and extrinsic feedback that provide information you can use to improve your performance. These types are knowledge of results and knowledge of performance.

Knowledge of results	Information about the outcome of movement. For example, seeing your shot for goal going into the goal, or a coach telling you that you kicked the ball 30 metres.
Knowledge of performance	Information about the process of skill performance that led to the outcome. This is usually information about your movement technique. For example, your physical education teacher telling you that you need to follow through more on your throw, or a coach instructing you to keep your elbow bent when hitting a cricket shot.

Figure 8.28 Types of information you can use to improve your performance.

Knowledge of results feedback helps you understand whether you were successful at performing the skill. This means it is most useful when you can't work out for yourself whether you were successful. This could be because the skill is new or because you don't understand the goal of the skill. For example, you might think that you successfully passed the ball to a teammate but actually the pass was too high. Knowledge of results can also help motivate you by letting you know when you have been successful.

Knowledge of performance feedback should be provided when helping someone learn movement skills, because it is generally more useful for the learner. This is because you can usually work out the results of performance yourself, but it can be harder to work out what your movement technique was like. For example, you will know whether you scored a goal or not (knowledge of results) but will have more difficulty determining what you did that made you miss the goal or why the ball didn't go where you wanted it to (knowledge of performance).

MAIN ROLES OF FEEDBACK

Feedback provides information about your performance, motivates you to continue practising and reinforces specific movements and movement outcomes.

Information

Feedback provides you with information about your performance or the result of your performance. This information can help you to detect and correct errors in your movement skills so that you can analyse, adapt, and refine your movemnt skills.

Motivation

Getting feedback can motivate you to keep practising, especially if the feedback is positive. Getting information about your performance allows you to track your progress and improvement, which can motivate you to work on improving your movement skills.

Reinforcement

Feedback can provide reinforcement to encourage you to repeat the movement again. Positive feedback (such as the coach saying 'Well done') provided after correctly executing the movement tells you that you should do the same thing again.

FREQUENCY OF FEEDBACK

How often you give feedback can influence skill acquisition. For example, your coach could give you feedback every time you hit the ball in tennis, or every fifth time, or only when you make a mistake.

Giving too much feedback hinders learning; so does not giving enough feedback. Getting extrinsic feedback every time you perform a movement skill can stop you from thinking about the movement and using your own intrinsic feedback, and it can lead you to rely on the extrinsic feedback.

When helping someone learn a movement skill, you should give feedback (including positive reinforcement) at frequent intervals, so the learner has some additional information on the movement skill but also has an opportunity to work things out for themselves using their own intrinsic feedback. Early on in learning you may want to provide more feedback, as the beginner learner will have more difficulty correcting and detecting errors in their movement skill.

UP AND MOVING: Using feedback to improve performance

1 After your teacher has shown you how to perform a basic movement skill from a game or sport, work with a partner to provide feedback to one another while practising the skill.
2 After you have practised, reflect on the experience. Did the feedback help you learn the skill? If so, what role(s) did the feedback play?
3 Did you find that you were giving more performance or results feedback? Which did you and your partner find most beneficial?
4 How often did you give feedback? Did you want more or less feedback from your partner?

INVESTIGATION

HOW DOES FREQUENCY OF FEEDBACK AFFECT MY PERFORMANCE OF A MOVEMENT SKILL?

Purpose

To explore the effects of different frequencies of knowledge of results (KR) feedback on performing a movement skill.

Materials

- Example skill: **Bean bag throwing** (you also could complete this activity with another movement skill, such as a sofcrosse throw to a wall target, a frisbee throw to a hoop target or a bocce throw to a floor target). Throw the bean bag at a target using an underarm throw while blindfolded and record the score. Complete 50 throws. The target should be about 7 metres from the thrower.
- Equipment: blindfolds, bean bags, floor targets (e.g. archery targets stuck to floor), cones (to mark throwing distance), tape measures (to measure throwing distance)

Method

Your class will practise a movement skill using different feedback frequencies and compare them, in order to determine which feedback frequency led to better performance.

Procedure: Practise a movement skill when intrinsic feedback is removed (using blindfolds) and when different frequencies of KR feedback are provided. The class should be divided into four different feedback frequency conditions: No KR (no feedback), 25 per cent KR (feedback after every fourth throw), 50 per cent KR (feedback after every second throw) and 100 per cent KR (feedback after every throw). Each student should only complete one condition.

Complete the activity in small groups to collect the data (e.g. pairs or groups of four), with one person acting as the participant and other group members as the experimenters. Once you complete all practice trials and the transfer test, the group members can swap roles so that you all complete the exercise as a participant.

Collate the class scores to get a class average for each of the four feedback frequencies.

Discussion

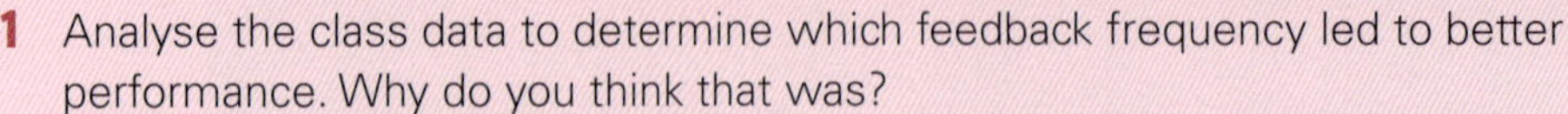

1 Analyse the class data to determine which feedback frequency led to better performance. Why do you think that was?
2 Which feedback frequency did people in the class prefer?
3 When do you think it would be better to give a lot of feedback and when do you think it would be better to give less?

Worksheet 8.8

TRANSFER OF LEARNING

Transfer of learning is the influence of previous experiences on the learning or performance of a new skill. This means that you can transfer and adapt skills and strategies from previous experience to new situations. For example, catching in cricket or basketball may transfer to catching in baseball, football or netball; serving in tennis may help in learning to serve in badminton, squash or volleyball; and skateboarding may adapt to learning to surf or wakeboard.

Transfer of movement skills

Movement skills in games and sports are often similar. For example, throwing is a movement skill that is used in a range of games and sports, such as baseball, softball, cricket and rounders. Sometimes the movement skills learnt in one game or sport can be transferred to other games or sports. This means that you do not always have to start from the beginning each time you learn a new skill. For example, you might adapt the striking action for hitting the ball in cricket when you first try to hit a golf ball.

Transfer of movement skills

Practise a series of movement skills from one type of game (e.g. invasion game or target game) and see if any of the movement skills transfer to another movement skill. For example, if you chose net/wall games, have some members of your class practise hitting a tennis forehand using a bat. Some practise a normal tennis forehand and the rest practise a down-ball strike.

After a while practising one skill, move on to practise the other skills. You could record your scores for each skill and see which skill you performed best.

1 Did any of the practice on one skill help you in performing one of the other skills?
2 Which skill seemed to transfer the most to the other skills?
3 Why do you think that was?
4 Report your findings back to the class.

9780170463102

Transfer of tactical skills

Different games have different tactics; however, there are often a number of similarities between games. This means that the strategies and tactics learnt when playing one game might adapt to performance in another game. For example, in invasion games (such as netball, football and hockey), moving into space to receive the ball is a common strategy used to gain an advantage. In net/wall games (such as tennis, badminton and volleyball), a strategy you could use is to place the ball away from your opponent. This transfer can help you transition between games, and encourages participation in a range of games and sports.

Alamy Stock Photo/Jonathan Larsen/Diadem Images

Alamy Stock Photo/tom carter

Figure 8.29 Moving into space to receive the ball is a tactical skill that can be transferred from rugby to hockey.

UP AND MOVING Transfer of tactical skills

1. Play a game of hockey, lacrosse or soccer and then play a game of netball or basketball.
2. Think about the tactics and strategies that you and your team are using when defending and attacking. Can you use the strategies you used in one game in another?
3. Are there similar strategies that could be used in another sport? Identify all the similar strategies used in the games and report on these to the class.

Worksheet 8.9

MODIFYING GAMES

Games can be modified to challenge the players. Game modification can make the game match your skill level and then allow you to progressively adapt and refine your skills by making further modifications to the game to increase difficulty as you prepare for the traditional sport. For example, a five-a-side game of soccer requires many of the skills of traditional soccer but is less complex, and will probably allow each player more opportunity to get the ball. Similarly, playing T-ball could help you prepare for playing baseball. Using a tag for tackling in football or rugby rather than full-contact tackling can reduce physical contact and make the game safer for your current skill level, as well as making the game faster.

Traditional sports are the official versions of the game or sport played according to the official rules.

Modified games are adapted from the traditional game and allow practice and development of skills in an environment that is appropriate for the current skill and developmental level.

Alamy Stock Photo/Dennis MacDonald

Dreamstime/Russ Ensley

Figure 8.30 Modified games have been adapted to help skill development.

INVESTIGATION HOW CAN I USE MODIFIED GAMES TO ADAPT AND REFINE MY SKILLS?

Modified games are games adapted from the traditional game that allow you to adapt and refine your skills in an environment that is appropriate for your current skill and developmental level. This can allow you to build your skills by gradually increasing difficulty as you prepare for the full game. For example, a modified game of soccer may still require many of the skills of the full game of soccer but is less complex, and will probably allow you more chances to get the ball. Aspects of the game that you can modify include the rules, number of players, space, equipment and tactical problems.

Purpose

To explore the modification of games on skill performance.

Materials

Below is an example of hypothetical games analysis data from a comparison of a modified game and a standard game of soccer played by skilled junior soccer players. The games comprised 10 minutes of play with the modified game played on a 30 × 25 metre field and the full-sided game on a 100 × 70 metre field. The modified game was 4 vs 4 players (including a goalkeeper for each team), and the full game was 11 vs 11 (including a goalkeeper for each team).

	Modified game (4 vs 4) 30 × 25m field	Full game (11 vs 11) 70 × 100m field
	Frequency	Frequency
Passes	122	77
Receive	102	56
Dribbles	16	12
Shots	36	7
Tackles	12	7
Headers	2	10
Interceptions	4	19
Turns with the ball	17	21

Method

Compare the data for the modified and full game to determine the effects of game modification on skill performance.

Discussion

1 Identify two game modifications made in the small-sided game.
2 What were the benefits of the modified game for practice of skills in soccer?
3 What were the disadvantages of the modified game for practice of skills in soccer?

UP AND MOVING — Exploring game modification

Explore how game modification can increase or decrease the complexity of a game by participating in the following activities for about five minutes each:

1 Start by playing a game of two-on-one keepings off in a small square (e.g. 10 × 10 metres) with a netball. The aim is for two players to keep the ball off one defender by running with or passing the ball. Players can move anywhere in the square and there is no contact allowed.
2 Play the same game, but this time, players are not allowed to run with the ball (netball rules).
3 Play the game so that the ball cannot be passed above head height.
4 Play the game again, but this time use a disc instead of a netball.
5 Play the game using soccer rules (dribbling with the ball and can't use your hands).

6. Now play the game as four-on-two keepings off in a larger square (e.g. 20 × 20 metres).
7. As a class, reflect on which versions of the game were most complex and which were the easiest. Explain why.

Aspects of games and sports to modify

Aspects of games and sports that can be modified include the rules, the number of players, the space, the equipment and the tactical elements.

Aspects of games and sports that can be modified

Rules

Knowing the rules of the game and understanding which are primary rules and which are secondary rules can help you modify games to make them appropriate for specific age groups or learners. When modifying games, don't modify the primary rules, otherwise the nature of the game changes. For example, changing the rules of a football game so that a player must not hold the ball for more than one second emphasises passing, but it is still a game of football.

Space

Manipulating space can simplify the game or make it more complex. Changing the size of the playing area, such as making it smaller or bigger or a different shape, can change the way the game is played. For example, making a volleyball or bat tennis court smaller encourages longer rallies because there is less space to hit the ball away from the other players.

Number of players

Reducing the number of players in games usually makes the game simpler; increasing the number of players usually makes the game more complex. For example, a two-on-two half-court basketball game is less complex than a five-on-five game.

Equipment

Equipment can be modified to challenge yourself and progressively increase the difficulty of the game and movement skills involved. Modifying equipment can also help match the current developmental levels of players so skills can be used appropriately in a game. For example, lowering net height in tennis or volleyball, or lowering the basket height in basketball or netball with younger players can allow them to use their movement skills effectively.

Figure 8.31 Aspects of games and sports that can be modified

9780170463102

UP AND MOVING Equipment and task modification

1. Practise catching a ball with a partner.
2. Make some task modifications that could influence the task difficulty. Practise the skill with each of the following:
   - small, medium and large ball
   - light, medium and heavy ball
   - round, oval and odd-shaped ball
   - small distance, medium distance and long distance between you and your partner
   - slow and predictable throw from your partner or a fast and unpredictable throw.
3. When you perform the skill, discuss with your partner which modifications make the skill simpler to perform and which make it more complex. Report your findings back to the class.

Tactical elements

Games and sports have tactical elements that create challenges for players to solve. These elements can be modified to make the game more or less complex. This allows tactical elements to match the skill level and needs of the players. For example, having one player who is a five-point player in Ultimate Disc encourages tactics of trying to create space for and getting the disc to that player for the attacking team and trying to deny the player space by the defending team. Using zones in invasion games such as football, where players are restricted in where they can move, encourages more players to be involved because players must move the ball from zone to zone.

Alamy Stock Photo/neal and molly jansen/

Figure 8.32 Lowering the height of the basketball basket can allow younger players to practise appropriate movement skills

UP AND MOVING Design your own game

Using your knowledge of how you can modify games, design your own game or a modified version of a sport that could be played by four people per team. You can do this individually or in groups of four. Decide on a name for the game and outline all the rules, the playing space, the equipment and the tactical elements in the game. Once you are happy with your game, teach it to a group of your peers and then play it. Compare your game with games that others have developed.

1. Reflect on the game – did it work? Was it fun?
2. Discuss if the game provided opportunities to develop movement skills and/or tactical skills.
3. List the three top elements of a good game.

Worksheet 8.10

1 Distinguish between massed practice, distributed practice, blocked practice and random practice.
2 Distinguish between knowledge of results and knowledge of performance feedback.
3 Describe what is meant by transfer of learning.
4 List four aspects of games that can be modified to challenge players and match skill level.

1 For each of the following movement skills, decide whether massed or distributed practice within a session would be beneficial:
   - hockey goal-shooting
   - sprinting
   - freestyle swimming
   - soccer passing
   - rugby goal-kicking.
2 Consider when it would be preferable to use blocked practice in improving movement skills.
3 Discuss the main roles of feedback in improving skills in games and sports.
4 Propose why equipment modification is important for improving skills of younger players in some sports.

1 Assess the transfer of tactical skills by completing the same practice drill using different equipment. For example, play a simple game of three-on-one keepings off played in a 15 × 15 metre square. The aim is for the three attackers to keep the ball off the defender. Modify the equipment and the rules of the game to match different sports: netball (throwing and catching), soccer (passing and trapping), hockey (passing and trapping), lacrosse (passing and catching) and football (handpassing and catching).
2 List the tactical similarities between the games.
3 Briefly explain how the games were different.
4 Discuss why some players who are good at one sport are often good at other sports.

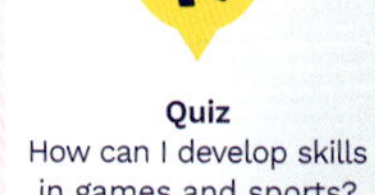

Quiz
How can I develop skills in games and sports?

HOW SHOULD WE PLAY AND PARTICIPATE IN GAMES AND SPORTS?

Games and sports involve more than just improving and using your movement and tactical skills. Games and sports involve competition and interaction with others. This means that how you participate is important. Participating in games and sports requires players to abide by certain requirements in order to ensure that everyone can play, but they should also participate in ways that make the activity fun for everyone. There are a variety of aspects that contribute to this, including ethics, fair play, codes of behaviour, sportspersonship, teamwork, motivating yourself and others, and leadership and collaboration.

ETHICAL BEHAVIOUR

Ethics is the study of ideas about what is good and bad behaviour. Ethical issues in games and sports include lack of sportspersonship by opponents, inappropriate behaviour by spectators, parents or guardians and coaches, cheating by opponents and the use of performance-enhancing or other drugs.

FACE TO FACE Fairness and ethics

Search online for organisations involved in promoting fairness and ethical behaviour in sport, such as the Australian Sports Anti-Doping Authority, sporting tribunals, the Anti-Discrimination Commission and the Court of Arbitration for Sport.

With a partner, discuss how these organisations influence your participation and those of others in games and sports.

FAIR PLAY

Fair play includes playing by the rules and also playing in the right spirit. It is a way of thinking and behaving that involves respect for others, friendship, team spirit, respect for the rules of the game, care for others, striving for excellence and experiencing enjoyment and fun in the game. Everyone involved in the game (players, coaches, teachers, parents, or guardians spectators and referees) has a responsibility to play and behave fairly.

Fairfax Syndication/Adam Hollingworth

Figure 8.33 Spectators also have a responsibility for fair play.

CODES OF BEHAVIOUR

Sporting organisations often provide codes of behaviour for participation. These codes identify key principles for participation to encourage fair play and fun, friendly and safe environments. For example, the Australian Sports Commission has general Codes of Behaviour for participation in sport by those performing different roles, such as coaches, officials, players and spectators.

The code of behaviour for players in junior sport lists the following guidelines:

- Play by the rules.
- Never argue with an official.
- Control your temper.
- Work equally hard for yourself and/or your team.
- Be a good sport.
- Treat all participants in your sport as you like to be treated.
- Cooperate with your coach, teammates and opponents.

- Participate for your own enjoyment and benefit.
- Respect the rights, dignity and worth of all participants regardless of their gender, ability, cultural background or religion.

Source: Adapted from http://www.ausport.gov.au/__data/assets/pdf_file/0007/115576/8._JnrCodesofBehaviourbrochure.pdf

SPORTSPERSONSHIP

Sportspersonship is about how people participate, behave and conduct themselves before, during and after the game. Being a good sportsperson involves a concern and respect for the rules, officials and opponents, as well as participating fully in the game or sport with a positive approach.

UP AND MOVING Regulating your own game and rules

Ultimate (sometimes called Ultimate Disc) is an invasion game played with a flying disc where teams score points by passing the disc to a player in an end zone. The interesting thing about Ultimate is that it is refereed by the players, not by an official. This occurs at all levels of the game, even at World Championships. This requires players to honour the rules of the game and participate fairly within the rules and with good sportspersonship in calling their own fouls.

1. To experience this, participate in a game of Ultimate as a class. Remember that you are all responsible for applying the rules.
2. When you play, reflect on how not having an umpire changes the nature of participation.
3. Report back to the class and discuss your experience:
   - Did you think your class would be able to regulate the game effectively before you started playing?
   - How did you resolve disputes in the game?
   - How could you use these skills in everyday life?

Newspix/George Salpigtidis

Figure 8.34 Positive self-talk can help motivate you and boost your confidence.

TEAMWORK

Games and sports provide an opportunity to cooperate with others and work towards specific team goals. For example, in netball, no matter how good any one player is, the ball must be moved up the field by several players before a goal can be scored.

MOTIVATING AND HAVING CONFIDENCE IN YOURSELF AND OTHERS

Motivation is the direction and intensity of your effort.

- Direction of effort is what you put your effort towards. If you are motivated, you will turn up and participate in all the activities.
- Intensity is how much effort you put in. If you are motivated, you will focus your energy on playing the game and learning the skills of the game.

Participating and performing in games and sports requires you to be motivated and to motivate others to participate.

WELLBEING CHECK IN

WHAT ARE MY VALUES?

This activity has been developed in collaboration with Dr David Bakker of MoodMission

Identify

Alright, here's a big question: What sort of person are you and who do you want to be? (Told you it was big…)

Understand

Research shows that people are happiest and most fulfilled when they are living by their values. Values are the things that you believe are important to characterise your actions. They aren't goals that you can complete; they're ongoing. There are no right and wrong values, and everyone values things differently. Picking your most important values can tell you what you already are doing to live a good life, and can help clarify how to live a more valued life.

Practise

1 Below is a short list of commonly held values. Read through them all and identify your top five values. These are the things that you believe are most important to living a good life:

- a acceptance
- b compassion
- c collaboration
- d creativity
- e determination
- f excitement
- g fairness
- h flexibility
- i grace
- j honesty
- k humour
- l independence
- m kindness
- n order
- o power
- p respect
- q responsibility
- r safety
- s self-awareness
- t skillfulness
- u trust

2 For your top five, write your definition next to each value. For example, if you picked 'trust' you might write 'to be reliable for myself and others, and try to always tell the truth'.

3 Out of your top five, identify one value that you don't think you're living by as successfully as you'd like.

4 What's something small that you could do to live more by this value? For example, if you want to improve your kindness, you could do one kind act for someone else this week. Note down your idea.

Reflect

If you had to come up with your top five personal values without the help of that list, do you think you would have come up with a similar top five? Did any of the values that you picked surprise you?

Confidence

Confidence is about believing that you can do something. This belief often influences what you do, so it is important to have confidence that you can play the game and participate. The thoughts you have while playing can influence your motivation and confidence.

Self-talk is the process of talking to yourself while performing. It is what you say and think to yourself. Elite players often talk to themselves, telling themselves to focus or concentrate or keep going. Self-talk can be positive ('Keep working hard') or negative ('I'm no good at passing'). Positive self-talk can improve your motivation and confidence to perform and participate, but negative self-talk undermines it. You can modify your own self-talk and also provide positive talk to your teammates by recognising when you are starting to talk negatively to yourself and replacing those thoughts with more positive ones. When you talk to other players, remember to be positive in what you say.

CHECK IN

FIGHTING MY BIASES

Identify

Brains take shortcuts all the time. Sometimes these shortcut thoughts are helpful, other times they lead to some pretty negative feelings.

This activity has been developed in collaboration with David Bakker of MoodMission

Understand

Thinking shortcuts (sometimes call heuristics, thinking traps, or cognitive biases) are used to save time. For example, If you need to make a decision about where to go for dinner, it can be easy to use a thinking shortcut rather than list out every single option and consider every single positive and negative factor. However, if we overuse thinking shortcuts, they can turn into thinking traps, which can lead us to incorrect or unhelpful conclusions.

Practise

Below are some of the most common thinking traps. Which ones do you think you use the most?

Black and white thinking, e.g. 'If it's not perfect, I have failed'

Over-generalising, e.g. 'nothing good ever happens'

Filtering out positives, e.g. 'that doesn't count, it's only going to get worse from here'

Weblink
Unhelpful thinking styles

Jumping to conclusions, e.g. 'They hate me because they looked at me funny'

Catastrophising, e.g. 'If I get a bad mark, I won't get into further study, then I won't get get a job, then I…'

Using should, e.g. 'it should be better'

Labelling, e.g. "I'm an idiot"

Emotional reasoning, e.g. using our emotions as evidence, "I feel embarrassed so I must be an idiot"

Reflection

What could you do to be more aware of your thinking traps? Is there a way of reminding yourself to think things through more slowly, or accounting for other types of evidence?

LEADERSHIP AND COLLABORATION

Leadership is a process of influencing others to move towards supporting a common goal. A leader knows where a group or team is going, provides direction and resources to get there and gives the team vision. Good leadership involves making decisions motivating others, giving feedback, acting ethically, ensuring everyone participates with a sense of fair play, working with others and helping people work as part of a team. Collaboration involves working with others on a shared goal and games and sports provide opportunities to collaborate with teammates on a shared goal. Taking on a leadership role and collaborating with others is important and can help you develop these skills in other areas of your life.

Worksheet 8.11

iStock.com/miodrag ignjatovic

Figure 8.35 Games and sports provide opportunities for leadership.

1 Define ethical behaviour.
2 Explain if sportpersonship is different to ethical behaviour.

1 How can participating in games and sports develop your teamwork skills?

1 List three things you might do to improve your confidence and motivation in games and sports.
2 With a partner, develop a code of behaviour for participation in a specific sport. The code should include how players should behave and five principles that players should abide by in that sport.

Quiz
How should we play and participate in games and sports?

CHAPTER 8 REVIEW

1 A sport is a game but with three additional elements. List these three additional elements.

2 Provide an example of a primary rule in a game or sport you are familiar with.

3 Explain the primary strategy of each of the four categories of games: invasion, net/wall, striking/fielding and target.

4 Clearly state the difference between fundamental movement skills and specialised movement skills.

5 Distinguish between locomotor, manipulative and stability movement skills.

6 State how gross and fine movement skills are different.

7 Provide an example of a discrete, a serial and a continuous movement skill. For each skill, explain why you have classified it that way.

8 Distinguish between an open and a closed movement skill.

9 Explain what tactics, strategies and game plans are.

10 Performance assessment can be subjective or objective. What is the difference between these?

11 List four pieces of information that you could code in a basic games analysis.

12 Provide an example of both a massed and a distributed practice schedule for practising basketball free throws if you had five hours a week to practise.

13 Distinguish between blocked and random practice.

14 Distinguish between intrinsic feedback and extrinsic feedback.

15 Explain when you would provide knowledge of results and knowledge of performance to someone you were helping learn a movement skill.

16 Discuss how you can use transfer of learning to help others participate and perform in games and sports.

17 State two differences between modified and traditional games.

18 Consider how reducing the number of players in a game influences the game.

19 Describe what you understand 'fair play' to mean.

20 Consider how participating in games and sports can develop your leadership and collaboration skills.

PHYSICAL ACTIVITY PLANS

Shutterstock.com/GoodStudio

IN THIS CHAPTER

You will explore how to design, implement and evaluate individualised physical activity programs. Training principles and methods are outlined and several lifelong physical activities that originated from other cultures and that develop the mind, body and spirit connection are examined.

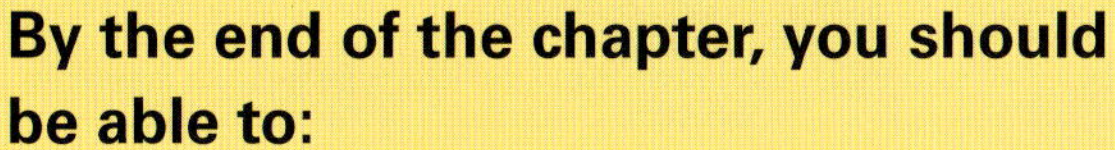

By the end of the chapter, you should be able to:

- design a personalised plan to improve or maintain your physical activity levels and fitness
- implement a personal physical activity and fitness plan
- evaluate a personal physical activity and fitness plan
- participate in a range of physical activity options to design, implement and evaluate participation strategies that promote health and social outcomes.

video
What is a personal activity and fitness plan and how can I develop one?

WHERE DO YOU START WHEN DESIGNING PERSONAL PHYSICAL ACTIVITY PLANS?

Quiz
Pre-chapter

Before you start, take the pre-chapter quiz to find out how much you already know.

Shutterstock.com/Halfpoint

iStock.com/RyanJLane

Shutterstock.com/pikselstock

Getty Images/FG Trade

Figure 9.1 How are you including lifelong physical activities in your day-to-day living?

This section explores how to develop a personal physical activity plan to improve your health- and skill-related components of fitness and health and wellbeing. A basic goal of a personal physical activity plan is to include a variety of movement opportunities across the following domains:

- incidental physical activity (e.g. walking from one classroom to another when the bell rings)
- household chores and work in the garden
- occupational physical activity (the physical demands of your job or activity during school time)
- active commuting (e.g. walking, skating or riding to work or school)
- leisure-time physical activity, such as lifestyle physical activity and structured sport or exercise for fitness.

Worksheet 9.1

9780170463102

NEEDS ANALYSIS

Set your goals

What do you hope to achieve from your personal physical activity plan? Examples include enhanced feelings of energy and wellbeing, and/or improved specific fitness components.

Current physical activity level

You need to assess your current physical activity habits and sedentary behaviours to determine your baseline physical activity level. You could use a survey (such as the CLASS Instrument or an online survey) or complete a physical activity journal. Alternatively, you could wear a device such as a pedometer or a smart watch with a pedometer function for two to three days and record your daily steps. Record your totals each day and calculate the average by adding the daily totals together and dividing this figure by the number of days monitored. (There is more detail on using a pedometer on page 382.)

Establish your priorities

Identify which energy systems, fitness components and muscle groups you want to develop. If you are designing a plan for everyday life rather than for a specific sport, you would apply the recommendations from the Australian 24-hour Movement Guidelines for Children and Young People (5–17 Years).

Sometimes personal trainers help clients establish their priorities for health and fitness. If you play a specific sport, your coach or physical education teacher may be able to help you determine which fitness components (health-related and/or skill-related) are important for your sport, and possibly even for the position(s) you play – this is known as specificity. Often coaches complete game analysis to determine what areas are important to work on. Alternatively, you may just have a particular area you want to improve in.

Getty Images/Klaus Vedfelt

Figure 9.2 Setting aside time for rest and relaxation is important for your wellbeing.

Dreamstime.com/Darkworx

Figure 9.3 Pedometers are an inexpensive option to assess how active you are.

Alamy Stock Photo/Tetra Images

Figure 9.4 The most effective personal physical activity plans are tailored to your specific individual needs.

Shutterstock.com/brizmaker

Figure 9.5 The 1.6 kilometre run test is an easy option to complete, either during class or for homework, to assess your aerobic power.

normative data
data that show what is usual in a defined population (e.g. for Australian women aged 15 to 18)

Fitness testing

Based on the priorities identified, select a fitness test that could be used to assess your relevant health-related and skill-related components of fitness. Ensure you select a recognised standardised test that will allow you to compare your results to **normative data**. Then you can determine your strengths and the areas where you would like to improve. There are many fitness test descriptions and normative data available online.

Tailor to your interests

Write down your interests (e.g. walking, social netball, yoga). You are much more likely to stick to your program if it is tailored to things you enjoy, places you like to go and the people you like to spend time with.

Weblink
As a starting point to creating a personal physical activity and fitness plan, you should complete a basic needs analysis. Be sure to consider all of the factors listed in your needs analysis.
This Physical Activity Readiness Questionnaire will assist you.

Worksheets 9.2 and 9.3

Getty Images/Mark Edward Atkinson

Figure 9.6 If you enjoy the activity, you're more likely to want to keep doing it.

Available time and resources

You need to create a program that works in with the other commitments in your life, such as school, work, social events, family expectations and daily chores. The program needs to be based on resources you have access to; for example, the equipment you have at home (e.g. bicycles) or the facilities you have access to within your local community (such as parks, facilities and trails).

Appropriate training methods

Are the training methods appropriate for you? Can you access the appropriate facilities? For example, do you need to be able to access a weights room? What training methods will be used and what is the justification for their selection? Training methods and the application of basic principles of training,

Alamy Stock Photo/Elizabeth Given

Figure 9.7 Local skate parks are an accessible option for many youth.

Dreamstime.com/Dmitriy Melnikov

Figure 9.8 Resistance training should form part of everyone's physical activity plan.

such as **overload** and specificity, are discussed in detail later in this chapter when we look at implementing your personalised program.

overload in fitness training, overload is doing more than usual

Physical condition

Are you injury-free? Are the activities selected appropriate for your level of physical activity and fitness? If you have not been active for a while, consider visiting your doctor before starting a program. It is common practice to complete a physical activity readiness questionnaire.

WELLBEING CHECK IN

FINDING YOUR MOTIVATIONS

This activity has been developed in collaboration with Dr David Bakker of MoodMission

Identify

Why do we do what we do? The answer is more complicated than it seems.

Understand

There are two types of motivation: extrinsic and intrinsic. Extrinsic motivators are reasons for doing something that come from outside, like money or status. Intrinsic motivators come from within. These include feelings of pleasure, accomplishment, autonomy and belonging. Lots of research has found that extrinsic motivations can wane over time and don't often help our mental health. On the other hand, if we do things because we are intrinsically motivated to, our motivation will stay strong and we will feel happier and more fulfilled.

Practise

1 Starting and maintaining physical activity can sometimes be a struggle. But we still dig deep and attempt it. Think of a regular physical activity you could add to your weekly schedule. List the pros and cons for doing it:
   a PROS: e.g. keeps me healthy
   b CONS: e.g. lack of time means I need to give up something else
2 List some of the extrinsic reasons why you might want to participate in physical activity. Remember, these are rewards that come from outside, such as winning a game, approval from others or other longer-term goals.
3 List some of the intrinsic reasons why you want to participate in physical activity. Remember, these are things we feel from within, such as feelings of achievement, pride or connection to others.
4 If you ever feel unmotivated or get hung up on the extrinsic reasons for doing something, take a moment to look over your list of intrinsic reasons and remind yourself of the rewards waiting for you.

Reflect

Are there things you do that you do mainly for extrinsic reasons, such as jobs that earn you money or assignments that you do to get good marks? Do you think you also get intrinsic rewards out of doing any of these things? Would it help to focus on these intrinsic rewards in future? How can you use this focus on intrinsic motivation to help you maintain your physical activity plans?

BECOMING INDEPENDENTLY PHYSICALLY ACTIVE AND FIT

School students are generally required to participate in a range of physical activity opportunities (e.g. physical education classes, sport education and other active organised activities). Some students often actively commute to and from school several days a week.

Students should progressively assume more responsibility for their own activity, fitness and wellness. Once you finish compulsory physical education and sport or school altogether, you will need to organise regular physical activity opportunities yourself. If you do not participate in regular physical activity, many of the associated health benefits gained while you are active at school will quickly disappear. This is known as **reversibility**, or detraining. Your physical education experience at school should help you develop the habit of activity and exercise, and so encourage lifetime fitness. This requires you to go from a level of dependence to independence. The illustration below shows how.

reversibility a rapid return to pre-training levels

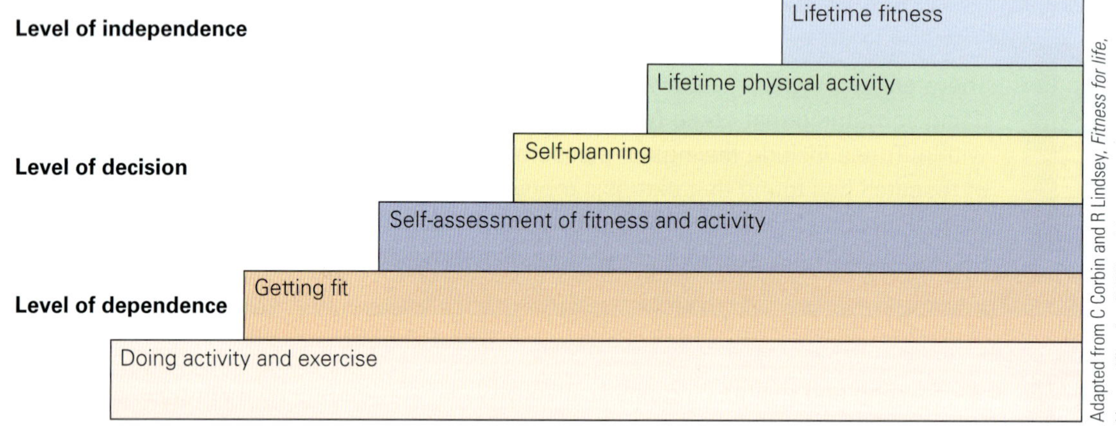

Figure 9.9 Stairway to lifetime fitness

1 Identify what your priorities are for your own physical activity.

1 Consider and analyse what physical activity facilities you have access to in your local area that you could incorporate in your individual physical activity plan.

1 Refer to Figure 9.9 and identify where you think you would be in relation to the levels shown. Explain where you think you would remain once you finish participating in compulsory physical education at school. Justify whether you believe you will be regularly active once you finish secondary school.

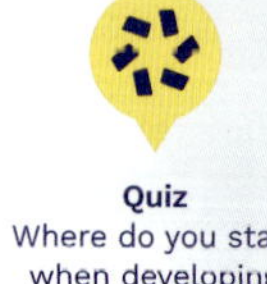

Quiz
Where do you start when developing personal physical activity plans?

WHAT ARE THE ESSENTIAL ELEMENTS TO IMPLEMENTING A PERSONAL PHYSICAL ACTIVITY PLAN?

IMPLEMENTING TRAINING PRINCIPLES

There are 10 key principles to consider in any physical activity program, particularly for developing and maintaining fitness levels. These include specificity, intensity, duration, frequency, progressive overload, detraining (reversibility), maintenance, individuality, diminishing returns and variety.

1. Specificity

To keep up a physical activity program, the program must be tailored specifically to the needs and interests of the person as mentioned earlier in this chapter. If you play a sport, the program can be specifically tailored to the demands of the sport or more specifically to the position you play. You can focus on the energy demands, fitness components and muscles used. For example, an AFL player who plays an 'on ball' position would have very different demands on their body than a full back player. Training should be tailored to work the relevant energy systems, fitness components and muscles used, and based on the skills you need to perform.

Getty Images/Jupiterimages

Figure 9.10 A basketball player needs to train for explosive power so they can jump high to shoot and rebound.

2. Intensity

Intensity refers to how hard your heart, lungs and muscles work during physical activity. A specific intensity is required to train particular energy systems, fitness components or muscle groups for particular activities. Most health-related fitness benefits associated with being physically active occur when you are working at least at a moderate intensity. Moderate-intensity physical activity usually consists of sustained rhythmic movements. These are characteristics of working at a moderate intensity:

- able to carry out a conversation comfortably
- working at 50 to 70 per cent of your maximum heart rate
- expending three to six times the energy that you would while at rest.

Moderate-intensity physical activities include brisk walking, cycling, raking leaves, mopping the floor, sweeping, lifting weights, doing aerobics, golfing and paddling in a pool.

Methods of determining intensity include measuring your heart rate, breathing rate and ability to talk comfortably (the talk test). There are also many apps available to help measure heart rate. A common way to determine the intensity of training is to refer to target training zones during aerobic activity. Working at different intensities has different effects on the body.

FAST FACTS

There is a common saying that goes, 'No pain, no gain'. But while exercise can be difficult, it should never hurt. If you exercise at an intensity beyond discomfort and actually feel pain, you are significantly increasing your chance of becoming injured.

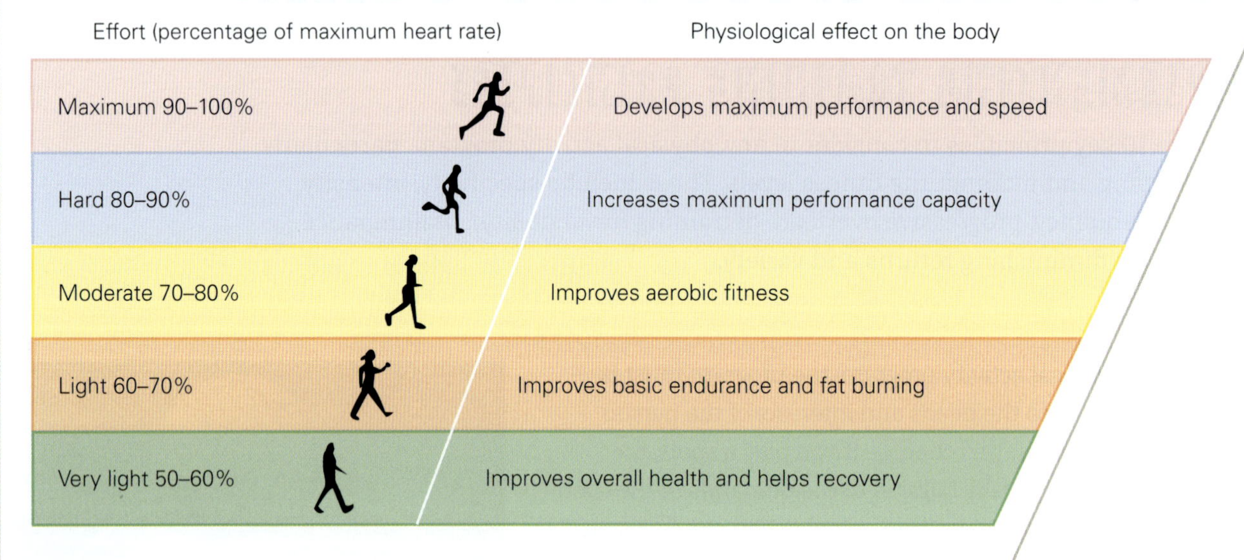

Figure 9.11 Training at different intensities affects the body differently.

FACE TO FACE Calculating maximum heart rate

There are at least seven recognised equations to calculate maximum heart rate.

1 Collaborate and search online for recommended training zones. Based on these, calculate the heart rate zone you will need to be within to achieve an improvement in your aerobic fitness.

2 Discuss what heart rate (bpm) you would need to work at to achieve improvements in basic aerobic endurance.

FAST FACT

Moderate-intensity training and physical activity is far more attractive to most people than vigorous-intensity physical activity. Studies consistently show that vigorous intensity can be a deterrent to many people being active, which is why most physical activity guidelines for adults encourage 'at least moderate-intensity' physical activity, rather than long durations of vigorous-intensity physical activity.

CASE STUDY ⇨ PHYSICAL ACTIVITY FOR MENTAL HEALTH

Identify

Cycling or walking to and from work are among the best exercises for our mental health, according to a paper published in the journal *Mental Health and Physical Activity*. Along with active transport, any leisure activity we enjoy, whether it's playing football, hitting the gym or going for a stroll, also reap mental health rewards. At the other end of the scale, physical activity at work and housework have the least positive impact.

Understand

For the analysis, lead author Dr Megan Teychenne, from Deakin University's Institute for Physical Activity and Nutrition, and her colleagues examined the existing research on the effect of different types of activity … They found no difference between aerobic or muscle-strengthening activities, however, those who do both have the 'lowest likelihood of depressive symptoms'. Being active in nature also boosts mood more than indoor activity.

The level of intensity, the duration and the dose also made less of a difference than just doing something.

'We know for physical health benefits that people should be undertaking 150 minutes of moderate physical activity or 75 minutes of vigorous. But for mental health benefits the dose of physical activity is likely not to be so important,' Teychenne says. 'We know that even low doses of physical activity – going for a walk twice a week – has been shown to be linked to a reduced risk of depression.' This is significant given that the majority of the population don't meet the physical activity guidelines and that people who experience mental illness are even less likely to meet them …

The activity doesn't have to be of a high-intensity to be effective, she adds. In fact, while the evidence around high-intensity interval training (HIIT) and mental health is limited, some research suggests it is associated with more negative affective states during participation.

'It does come down to your preference,' Teychenne says. 'If someone enjoys HIIT, working at a vigorous intensity, and many people do, then they're more likely to gain that mental health benefit.

'If you're being forced or you really don't enjoy vigorous-intensity activity or your body isn't able to cope with that then you're less likely to enjoy it and you're less likely to see the mental health benefits. It's not a one-size-fits-all.' …

It is the choice and enjoyment, or lack there-of that may explain why work or domestic-related activities seem to have little benefit from a wellbeing perspective.

'If you're vacuuming the floors at home, for example, you might not be getting enjoyment from that activity because you're feeling forced to do it so you may not see the mental health benefits,' Teychenne explains.

Source: 'The exercise that best supports your mental health' by Sarah Berry, *Sydney Morning Herald*, 20 November 2019.

Discuss

1 Discuss why you think the researchers found that physical activity accumulated during work or housework had the least positive impact on mental health.
2 Outline the key to gaining mental health benefits from being active.
3 Explain why doing HIIT is not necessarily going to result in an increase in psychological outcome for everyone.

3. Duration

Duration refers to how long an activity lasts. All of the following are examples of duration:

- how long a training program goes for (e.g. 12 weeks pre-season)
- the minimum time needed to train to achieve a health or fitness benefit (e.g. at least six weeks)
- how long an actual training session would need to last to achieve a health or fitness benefit (at least 20 minutes to achieve **aerobic adaptations**)
- how long a bout of exercise is within a training session (e.g. a 10-minute bout of vigorous-intensity running).

aerobic adaptation a physiological adaptation to repeated bouts of exercise as a result of aerobic training; e.g. increased blood volume is a cardiovascular adaptation to aerobic training

4. Progressive overload

Ultimately, the aim of a training program should be to achieve long-term health and fitness benefits that can be maintained. The human body is very adaptable and tries to constantly adjust to whatever activities are performed. The principle of overload is the most basic. When you introduce a new activity requiring a higher intensity or different fitness component or muscle group, a stress response occurs. In response to this stress, the body adapts and accustoms itself to the new demands, and then the adaptation **plateaus**. To gain further improvements, the body needs to be progressively overloaded with additional stress. This might be achieved by running two extra minutes, using the next size dumbbell up or doing an extra session per week. For an adaptation to occur, the body needs to rest and recover. The body can also be overloaded by increasing the number of **repetitions** or **sets** of an activity that are completed, by increasing the intensity of an activity or by reducing the rest interval. When overloading it is important to ensure only one of these variables be increased at a time.

plateau to reach a level or period where no change is observed

repetitions (reps) the action of repeating something; e.g. 8–12 reps of bicep curls

set several exercises intended to be done in a series; e.g. bicep curls 3 sets × 8–12 reps

Scout Kozakiewicz

Figure 9.12 To overload, either the weight, repetitions or sets of repetitions can be increased.

iStock.com/G&J Fey

Figure 9.13 Resistance bands help to maintain strength and muscle mass without overdoing it.

9780170463102

5. Detraining (reversibility)

Detraining means a rapid return to pre-training levels. This is usually caused by either stopping a training method or stopping training altogether. Training might be stopped because of illness, injury, boredom or increased commitments at work or school. Ultimately, there is a reversal in the training benefits achieved. The fastest reversal tends to occur to the **VO_2max**, where there can be a decrease of 18 per cent in just three months.

VO_2max the maximum amount of oxygen that can be taken up, transported and utilised per minute by the body

6. Frequency

Frequency in this context refers to how often an individual is active. If a person wants to improve a fitness component, they probably need to train that component three times a week. It is also important to allow adequate rest between sessions to allow adaptations to occur. Often, people who do **resistance training** train every second day. Those who do resistance training every day generally alternate the muscle groups they work; this is known as a split routine. For example, you might work the upper body (chest, shoulders and arms) one day and back, legs and abdominal muscles the next. This gives your body the opportunity to recover and maximises adaptations.

Figure 9.14 Frequency of physical activity sessions can be mapped on a calendar.

resistance training to move against a force provided by a person's own body weight, machine or weight such as barbell, dumbbell, kettlebell or medicine ball

7. Maintenance

You can maintain fitness gains by training twice a week. Some people might play sport several times a week, so it is important to maintain fitness but still allow enough time to recover. It is important to continue training twice a week to avoid detraining.

8. Individuality

Each person responds and adapts differently to exercise and different training methods and workloads. These differences are due to individual factors, including:

- initial fitness levels
- interests (likes and dislikes)
- hormones
- genetic factors
- motivation
- nutritional requirements
- recovery
- injury.

9. Diminishing returns

Have you ever wondered if there is an absolute limit to your fitness potential? Each person's potential for improvement in each fitness component is genetically predetermined. It has been jokingly said that if you want to be an Olympic athlete you need to pick your parents carefully. Of course this isn't possible, but the statement illustrates that environmental influences and training can only take you as far as your genetic potential will allow. A person who is new to training has a much greater potential to improve compared with an elite athlete who trains daily and is near their maximum potential. As you get closer to your ultimate potential (ceiling), the rate at which you can improve slows down significantly. Figure 9.15 illustrates how time and performance level impact improvements.

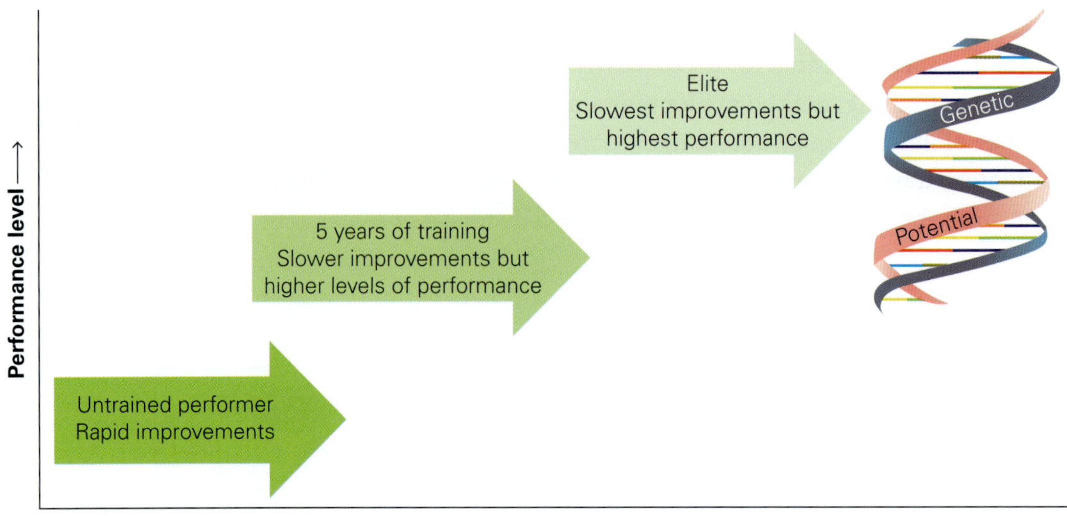

Figure 9.15 Diminishing returns based on time, performance level and genetic potential.

10. Variety

As the saying goes, 'Variety is the spice of life'. It is also the key to not becoming bored with a fitness training and physical activity program. It is important to change things up at times by alternating different aspects of your physical activity plan, such as:

- where you work out
- who you are active with
- the types of activities you perform, by using cross-training (e.g. for aerobic training you could run, swim, ride a bike or exercise bike, use a treadmill or rowing machine or elliptical, or do an aerobics or boxing class)
- duration, intensity and rest intervals.

Elite athletes need variety but they also need consistent routines. They must be careful that variety does not compromise the specificity of their training. For example, all elite athletes will obtain health and fitness benefits from cross-training. But if elite swimmers want to be race-ready, they need to do more swimming than any other form of cross-training to ensure they are race-fit.

Worksheet 9.4

TRAINING METHODS

The training methods you select will be based on the following factors:

- goals
- needs and interests
- energy demands, muscle groups used, actions and movements performed and fitness components important to the physical activities or sports you participate in
- available resources (such as time, facilities available at home, school or in your local community, transport, opportunities such as training sessions).

There are a variety of training methods.

Continuous training

Continuous training is low-slow-distance training and is very useful for developing aerobic fitness. It involves work periods conducted in the aerobic training zone of 70 to 85 per cent of your maximum heart rate with no rest during the session. Sessions are usually a minimum of 20 minutes to achieve aerobic benefits. Typical activities use whole body or large muscle groups; examples of this type of training are aerobics, swimming and running.

Fartlek training

This is a variation of continuous training that combines continuous activity with short bursts of high-intensity effort at regular stages during the training session. Many team sports requiring a solid aerobic base use Fartlek training because it can closely replicate the demands on your energy systems during a game, which often involve short, sharp high-intensity bursts with lower-intensity walking or jogging rest periods throughout (this maintains specificity). Most people new to training often start with continuous training to develop a base level of fitness before moving to Fartlek.

Worksheets 9.5 and 9.6

Interval training

This involves periods of work followed by periods of rest. Interval training can be classified as either long, intermediate or short depending on the intensity, duration and recommended rest intervals. Interval training is most commonly undertaken with running; it can, however, be made more specific by using swimming or cycling to 'train' specific muscle groups in the actions that are going to be repeated in the performance or competition. Generally the higher the intensity and the longer the rest, the more the interval training will work the anaerobic energy systems (ATP-PC and the anaerobic glycolysis). Interval training known as long interval training has a work to rest ratio of longer or the same periods of work to develop the aerobic system. Interval training is also good for general health and training for team sports and specific sports, such as sprinting or swimming. High-Intensity Interval Training (known as HIIT) has become popular in recent years. HIIT involves short periods of high-intensity work followed by periods of lower intensity. After completing a thorough warm-up, an example HIIT session might include: 30-second maximum effort on a stationary bike followed by four minutes rest, repeat four to six times.

Speed training

Speed training involves training the nervous and muscular systems to work more efficiently together and, in a sense, brings about the most efficient stride frequency and length. Activities might include sprinting with a band connected to a tyre or parachute.

Weights/resistance training

Resistance training is important for everyone for general health, not just for improving sporting performance. There are three types of weight/resistance training.

Isokinetic training

Involves the resistance being adjusted to allow you to exert the maximal force that you can throughout the entire range of movement. This can only be achieved through the use of machines such as Cybex or Nautilus, which might be found in a gym.

Isotonic training

The most commonly used resistance training, this involves the use of 'free weights' (dumbbells or barbells) or a pin-loaded machine. Pin-loaded machines only allow a restricted range of movements. Free weights allow you to mimic actions used in real game situations. Resistance bands can also be used.

Isometric training

Involves a contraction in the muscles against the resistance (weight) at a particular joint angle for a period of time with no change in muscle length. Gains are only made at the angle performed, so this type of training is most useful for activities such as skiing, gymnastics or weight lifting.

Figure 9.16 Types of resistance training

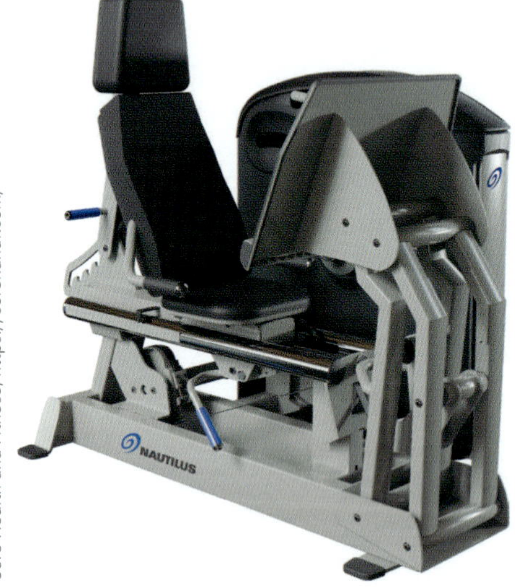

Core Health and Fitness, https://corehandf.com/

Figure 9.17 Isokinetic training can only be achieved through use of machines like this at the gym.

Shutterstock.com/TORWAISTUDIO

Figure 9.18 Isometric training can be achieved with exercises such as the plank, where core muscles are tensed for an extended period of time without changing position.

iStock.com/kali9

Figure 9.19 Isotonic training

Alamy Stock Photo/Cultura Creative (RF)

Figure 9.20 Which type of training is this?

Table 9.1 outlines the differences among the three methods of resistance training: strength training, power training and muscular endurance training.

Strength training	Power training	Muscular endurance training
% of 1 RM†: 80–95	% of 1 RM†: 30–50	% of 1 RM†: 40–60
Sets: 2–5	Sets: 3–5	Sets: 2–4
Reps: 2–4	Reps: 3–5	Reps: 15–20
Speed: slow/medium	Speed: fast	Speed: medium/fast
Rest: 3–4 mins	Rest: 3–4 mins	Rest: 1–2 mins

*Note that the sets and repetitions are just a guide to give an idea of the difference between the methods and will vary with each individual.
†1 RM = the maximum amount of weight one can lift in a single repetition for a given exercise

Table 9.1 Weight training for muscular strength, power or endurance*

Circuit training

Circuit training involves performing a number of activities at various 'stations' in sequence during a workout. These can be tailored to specific components of fitness, systems, muscle actions and distances required for a particular sport or position you play. Even work and rest intervals can be worked out to mimic the demands of the sport played during competition. There are three types of circuit training: individual load, fixed load and fixed time:

- Individual load involves taking into account the strengths and weakness and different loads you can manage at each station.
- Fixed load and **fixed time** circuits involve performing a set number of repetitions or work against a resistance for a set time, completing as many repetitions as you can. Activities might include using your own body weight (push-ups or skipping) or weights.

Plyometrics training

Plyometrics training involves stretching or lengthening a muscle (eccentric contraction) and then a rapid shortening (concentric contraction) to bring about an explosive action. This is known as the stretch-shortening cycle. Example activities include depth jumping, clap push-ups, bounding or leaping and throwing a medicine ball. Activities usually only involve the person using their own body weight, or jumping from a platform to the ground and back up onto a platform or over obstacles. Athletes who need to be explosive (basketball and netball players, volleyball, soccer and hockey players, and high- and long-jumpers) use this form of training.

Getty Images/grandriver

Figure 9.21 Depth jumping is one of the most common forms of plyometrics training.

Flexibility training

Flexibility is critical in allowing maximal ranges of motion to be achieved. There are three major classifications of stretching (see Figure 9.21).

How to select a training program

Just as we have a cardiovascular and respiratory system, we have three energy systems that work together to provide the energy required for all movements. This includes all activities, from writing notes in class and walking to your locker, all the way to competing in a triathlon or running on a hockey field.

Figure 9.22 Three classifications of stretching

Dynamic stretching
Involves either slow controlled movements or ballistic explosive movements through a full range of movements.

Proprioceptive neuromuscular facilitation (PNF) stretching
Involves taking a joint to its maximum range and then contracting isometrically (no movement of muscles) against a resistance provided by a partner or a band. This is repeated to increase the range until no further increases in range of motion can be made. Flexibility training is also important for everyone's general health and mobility.

Static stretching
Involves gradually stretching a joint to its maximum range and then holding this position for 10 to 20 seconds. This is often used at the end of training sessions to help the individual relax.

Worksheet 9.7

The ATP-PC system provides energy for short, explosive movements lasting up to 10 seconds. If we work at high intensities for longer than 10 seconds, the anaerobic glycolysis system takes over as the main provider of energy. Both these systems are known as 'anaerobic' because they essentially do not need oxygen delivered to working muscles. They both provide energy quickly, but they quickly fatigue, and that's when your body switches to the aerobic energy system.

The aerobic energy system relies on delivery of oxygen to working muscles. It is used when you work sub-maximally (when you can still chat with someone working beside you). Typically, the aerobic energy system is used for activities requiring a more sustained or long-term effort.

Table 9.2 shows the different training methods and how you would select which physical activities to include in a personalised program. First identify which training methods are linked to the various health- and skill-related fitness components.

Table 9.2 Training methods used to improve health- and skill-related fitness components

Fitness component training method	Aerobic power	Local muscular endurance	Muscular strength	Anaerobic capacity	Muscular power	Agility	Flexibility	Speed or speed endurance
Energy system	**Aerobic energy**		**ATP-PC energy**					**Anaerobic glycolysis**
Continuous	✓	✓						
Fartlek	✓	✓						
Long interval	✓	✓						
Intermediate interval		✓		✓				✓
Short interval				✓				✓
Speed				✓				✓
Weights/ resistance		✓	✓		✓			

Fitness component training method	Aerobic power	Local muscular endurance	Muscular strength	Anaerobic capacity	Muscular power	Agility	Flexibility	Speed or speed endurance
Energy system	**Aerobic energy**		**ATP-PC energy**					**Anaerobic glycolysis**
Circuit		✓	✓	✓	✓	✓		✓
Plyometrics				✓	✓	✓		✓
Flexibility							✓	
High Intensity Interval Training				✓	✓			✓
Swiss ball and Pilates		✓	✓				✓	

MAINTAINING YOUR PHYSICAL ACTIVITY AND FITNESS TRAINING

By far the most challenging aspect of a personalised plan is to develop something that is sustainable and that can become a habitual part of your life. Engaging in regular physical activity requires highly developed self-management skills, including the ability to:

Worksheet 9.8

- record your goals
- commit yourself by putting together weekly training schedules and training session outlines
- enlist social support and be active with someone
- use reminder systems so you don't forget to be active
- reward yourself at different intervals.

WELLBEING CHECK IN

FIGHTING GUILT

This activity has been developed in collaboration with Dr David Bakker of MoodMission

Identify

Most people feel great after a workout. But do you ever feel bad because you haven't done something?

Understand

All emotions motivate us to do something. Being scared motivates us to avoid or run away, while being angry motivates us to attack. Guilt, however, is a tricky one. Sometimes it can motivate us to try harder next time. For example, if we feel guilty about not sharing something, we will probably try to share it. But guilt can sometimes make us feel sad and ashamed. For example, if you miss a workout and think 'I should've tried harder, I'm so lazy', your guilt might leave you wanting to curl up in a ball and eat ice cream.

Practise

1 Guilt can often come from 'should' thoughts. What are some thoughts that use the word 'should' that have made you feel guilty before?
For example: 'I should have tried harder.'

2 What are some situations when you might think these thoughts?
For example: Missing football training, not studying for a test

3 What are some other ways you could think that might make you feel less unhelpful guilt?
For example: 'I'll make a better plan next time'; 'What happened happened; I can try differently next time'

Reflect

How can you remind yourself to have more helpful thoughts the next time you're thinking 'should's? Do you think there are some situations when it's okay to think in 'should's?

CASE STUDY  ANALYSING A TRAINING SCHEDULE

Identify

Below is an example of a weekly training schedule for 16-year-old Madison. Their personal goal is to meet the Australian 24-hour Movement Guidelines for Children and Young People (5–17 Years) for a 16 year old and improve their aerobic capacity.

Understand

Table 9.3 Example weekly training schedule

	6–9 a.m. (before school)	9 a.m.–3 p.m. (school hours)	3–10 p.m. (after school)
Monday	Walk to school; 10 min		High-intensity interval training class; 60 min
Tuesday		Physical education class; 100 min	
Wednesday	Walk to school; 10 min		After-school interschool sport (volleyball); 90 min
Thursday	Boxing stations; 20 min		Walk the dog; 45 min
Friday		Physical education class; 50 min	Home-based fitness circuit; 45 min

	6–9 a.m. (before school)	9 a.m.–3 p.m. (school hours)	3–10 p.m. (after school)
Saturday	Push-ups × 30 Sit-ups, 3 sets × 25 reps	Bike ride; 90 min	
Sunday			Walk the dog; 70 min

Discuss

1. Did Madison meet the Australian 24-hour Movement Guidelines for Children and Young People (5–17 Years)? Justify your answer (ignore the screen-time component for the purpose of this exercise).
2. Identify which sessions would most likely enhance each of the health-related components of fitness.
3. Create a blank table using the same format and complete a weekly training schedule for yourself.

Worksheet 9.9

Table 9.4 provides an example of a training session based in a gymnasium. The components include a warm-up, weight (resistance) training exercises and a cool down. The program alternates which muscle group is being worked.

Table 9.4 Example template for training session (gym session)

Date: 20/06/2020	Location: YMCA gym	Time: 9 a.m.–10.30 a.m.
Warm-up: Walk on treadmill: 5 min Exercise bike: 15 min		Rowing machine: 5 min Stretching: 5 min
Weight training exercises	**Repetitions**	**Sets**
Lat pull-down	12–15	2–3
Leg press	15–20	2
Crunches	20–25	3
Bench press	5–10	2–3
Leg curls	12–15	2–3
Bicep curls	5–10	2–3
Lunges	10 each side	2
Triceps extensions	12–15	2–3

Getty Images/Chijo Takeda

Figure 9.23 Research consistently shows that dog owners walk more than non-dog owners.

Date: 20/06/2020	Location: YMCA gym	Time: 9 a.m.–10.30 a.m.
Sit ups	20–25	1–2
Bent over row	10–15	2–3
Cool down: Exercise bike (slow): 5 min Stretching: 10 min		

Safety tips for resistance training

- Older people must have a medical clearance from their doctor before starting a training program.
- Perform an adequate warm-up and stay warm during your workout.
- Avoid holding your breath while lifting.
- Avoid dangerous exercises.
- Progress slowly.
- Do not arch your back during any exercise.
- Always keep weights close to your body – even when you lift something.
- Never do a full squat.
- Always check that collars of weights (on dumbbells and barbells) have been tightened.
- Do not pause between repetitions.
- Do not drop weights or let them bang together.
- Wear gloves to maximise grip, and clean equipment after use.

FAST FACT

'Women will become bulky from resistance training' is an exercise myth. Women have less testosterone than men and do not bulk up from resistance training to the same extent as men. The higher percentage of body fat that women have prevents the muscle definition that's possible in men.

iStock.com/zamrznutitonovi

iStock.com/Zinkevych

Figure 9.24 Free weights (dumbbells and barbells) are an excellent option for home-based exercise.

9780170463102

CASE STUDY  HEALTH AND WELLBEING AT COMMUNITY CENTRES

Identify

The following table outlines the weekly offering across a range of local community centres across the Brisbane area.

Understand

Monday	Tuesday	Wednesday	Thursday	Friday	Saturday
F45 Cardio 8–9 a.m.	RPM/cycle 6–7 a.m.	Strength and conditioning 6–7 a.m.	Boxing 6–7 a.m.	Yoga in the park 7–8 a.m.	Run/walk clubs 8–9 a.m.
Mums and bubs fitness 10–11.15 a.m.	Women's self defence 1–2 p.m.	Hatha Yoga 11–12 p.m.	Core fit class 7–8 a.m.	Run/walk clubs 11–12 noon	Disc golf 10–11 a.m.
Tai Chi 1–2 p.m.	Teen fit 4.30–5.30 p.m.	RPM/cycle 6–7 p.m.	Lightweight hand weights for older adults 10–11 a.m.	Metafit 4–5 p.m.	RPM/cycle 12–1 p.m.
Yoga in the park 3–4.30 p.m.	Bollywood Dance/Zumba fitness 7–8 p.m.	Hip Hop dance 7–8 p.m.	F45 Cardio 6–7 p.m.	Aqua Tone 7–8 p.m.	F45 resistance training 12–1 p.m.
Aqua Tone 7–8 p.m.	Adults Jazz Dance 7–8 p.m.	Yoga 7–8 p.m.	Strength and conditioning 7–8 p.m.	Tai Chi 7–8 p.m.	Family boot camp 1–2 p.m.

Discuss

1 Select five different sessions from the weekly offerings and identify three health-related fitness components that each session could potentially improve.
2 Identify three sessions that might be of interest to you, and research whether any of these are offered within your local community.
3 Describe two benefits other than physical benefits of attending a group fitness or wellbeing session.

Fairfax Syndication/Jacky Ghossein

Figure 9.25 Exercising at the park

USING NON-SPECIALISED EQUIPMENT

You don't need specialised gym equipment such as treadmills, barbells or machines to improve your fitness. There is a range of non-specialised equipment at home or at the local playground that can be used to develop your health- and skill-related fitness. A pair of dumbbells can cost between $20 and $120. However, as an inexpensive alternative, you can fill a pair of 2-litre milk or cordial bottles with sand or water. You could use a variety of household items to make up your own fitness circuit.

INVESTIGATION

HOW I CAN USE RECYCLED MATERIALS TO KEEP FIT?

Purpose

The purpose of this practical activity is to come up with a range of ideas for activities that you could engage in at home to improve your health and fitness.

Materials

Either your school will provide or ask you to bring one or several of the items shown in Table 9.5. The items have been classified as household items, recycled items or inexpensive sporting items. The table shows just some examples; use your imagination to come up with more examples.

Table 9.5 Examples of non-specialised fitness and play equipment

Household items	Recycled items	Inexpensive sporting items
blankets	tyres	Frisbees
tarps	boxes	tennis balls
brooms	milk crates	pool noodles
broom handles	chaff sacks	skipping ropes
buckets	rope	play balls
small steps	milk/cordial bottles	hacky sacks
hay bales	sand	hoops

9780170463102

Method

1 In pairs or small groups, list as many physical activities as possible that could be incorporated into either a fitness circuit or backyard game. You need to come up with at least two activities for the fitness circuit and two games.

2 Each group or pair should present their ideas to the rest of the class. You will need to write down all your ideas and have a few extra ideas in case another group comes up with the same activities.

3 Create a table that outlines which health- and skill-related components of fitness are developed for each of your activities within your circuit.

Discussion

1 Discuss three advantages and three limitations to the ideas you came up with and suggest several modifications that could made to improve your ideas.

INVESTIGATION

HOW CAN I GET FIT AT THE LOCAL PARK?

Purpose

The purpose of this practical activity is to use your creativity and design, to implement and evaluate a fitness circuit you could complete at the local park.

Materials

You will need to access a local park with playground equipment and/or exercise equipment and, if possible, natural and man-made features like seating, trees, logs, hills, grass, etc.

Method

1 As an in-class or homework activity, go to a local park that has a range of playground equipment such as monkey bars, slides, other climbing equipment, swings or seating.

2 Design a fitness circuit. It could consist of either of the following formats:

» Fixed time at each station (e.g. one minute on the monkey bars doing chin-ups, swinging across the ladder or hanging) OR

» Fixed load – a continuous circuit where a set number of repetitions is performed at each station before moving on to the next station (e.g. complete 20 step-ups on the park bench seating).

Your circuit needs to incorporate the playground equipment and other natural and built features at the park as much as possible. Your circuit could include the following activities:

» crunches
» hanging from a bar
» step-ups
» push-ups

» swinging across monkey bars
» sliding down slide
» chin-ups
» sprints
» swinging on swing
» squats
» sit-ups
» hill runs
» lunges
» jumping over lines
» log rolls down hill
» dips
» stretches
» building sandcastles.

3 When you design your circuit, make sure you consider the following points:
   » Write down the name of the activity to be completed at each station and take a photo of the equipment or location to be used. If possible, include a photo of someone performing the activity safely. You may need to do some research online to look up the key tips for completing that exercise safely.
   » Ensure you build in a warm-up before completing your circuit.
   » Outline the time (fixed time format) or repetitions (fixed load format) you would complete at each station of your circuit.
   » Arrange your circuit to alternate the muscle groups and to allow adequate rest (e.g. avoid doing two upper-body exercises in a row, such as push-ups followed by chin-ups).
   » You could make a poster of your circuit that shows the order of activities photographically, the location of each station and the key safety tips.
4 Now try out you circuit.

Weblink
Safe exercise techniques

Figure 9.26 Household items are all you need to design your own fitness circuit at home.

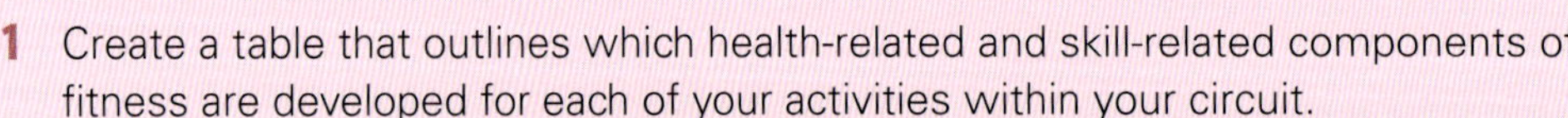

Discussion

1 Create a table that outlines which health-related and skill-related components of fitness are developed for each of your activities within your circuit.
2 Explain how you could modify the circuit if you had someone who wasn't very fit to ensure they could still safely participate at their own level of fitness.
3 Describe the benefits other than improved fitness to completing a fitness circuit at home or at your local park.
4 Evaluate whether you think your circuit addressed at least five different components of fitness. Discuss the aspects of your circuit that could be modified to make you even more likely to engage in it again.

Worksheets 9.10 and 9.11

9780170463102

EVALUATING A PERSONAL PHYSICAL PLAN

This section considers how to evaluate personal physical activity and fitness levels. There are four basic steps:

1. Setting goals
2. Establishing your baseline physical activity and fitness levels
3. Identifying strengths and areas for improvement
4. Monitoring progress.

Step 1: Setting goals

Your goals need to be **S**pecific, **M**easurable, **A**cceptable, **R**ealistic, **T**ime phased, **E**xciting and **R**ecorded. Goals might relate to meeting the Australian 24-hour Movement Guidelines for Children and Young People (5–17 Years) or to improving particular fitness components for a sport you play.

Once you set your goals, you will be able to monitor your progress towards reaching those goals.

Step 2: Establishing your baseline physical activity and fitness levels

The only way to determine whether your physical activity and fitness levels have improved, maintained or gone backwards is to monitor your progress towards your set goals. Important self-management skills that help to monitor progress include goal-setting, self-monitoring, rewarding and reminding.

Before starting a new physical activity or fitness program, you need to establish a starting point. This is the baseline level of your physical activity and fitness. Data can be collected before, during and after programs on any of the following characteristics:

- basal heart rate, which is generally measured when the person is at rest and relaxed and hasn't exercised or exerted themselves recently
- blood pressure, which is the pressure exerted by circulating blood upon the walls of blood vessels. It usually refers to the arterial pressure of the systemic circulation. During each heartbeat, blood pressure varies between a maximum (systolic) and a minimum (diastolic) pressure.
- hours of sleep per night
- results of tests relating to the various health- and skill-related fitness components (see Tables 9.6 and 9.7)
- number of steps taken per day
- time spent in light-, moderate- or vigorous-intensity physical activity
- types of physical activities and sedentary behaviour engaged in each day
- training sessions
- whether you meet the Australian 24-hour Movement Guidelines for Children and Young People (5–17 Years).

When you have assessed one or several of these factors, you can analyse the results by comparing them to the guidelines.

Table 9.6 Tests used to assess health-related components of fitness

Fitness component	Assessment test
Aerobic capacity	⇨ VO_2max treadmill/ergometer test ⇨ Cooper's 12-min run ⇨ Harvard step test ⇨ 20-m shuttle run/beep test ⇨ Critical swim speed test ⇨ 1.6-km jog test ⇨ Yo-Yo intermittent recovery test
Anaerobic capacity	⇨ Phosphate recovery test ⇨ Running-based anaerobic sprint test (RAST) ⇨ Wingate anaerobic cycle test (30 sec)
Body composition	⇨ Body Mass Index (BMI) ⇨ Waist circumference ⇨ Skinfold sums ⇨ Body fat percentage
Muscular strength	⇨ Handgrip dynamometer ⇨ 1 RM bench press; 1 RM leg press ⇨ Seven-level abdominal test
Muscular endurance	⇨ Timed sit ups, timed push ups ⇨ Pull-ups and modified pull-ups ⇨ Flexed arm hang ⇨ Partial curl-ups
Flexibility	⇨ Modified sit and reach ⇨ Shoulder and wrist elevation test ⇨ Trunk and neck extension ⇨ Ankle extension test ⇨ Shoulder rotation test ⇨ Ankle (dorsi) flexion test

Table 9.7 Tests used to assess skill-related components of fitness

Fitness component	Assessment test
Balance	⇨ Stork balance stand test ⇨ Standing balance test
Reaction time	⇨ Ruler-drop reaction test ⇨ Online reaction tests
Coordination	⇨ Alternate-hand wall toss test ⇨ Soft-drink can test
Agility	⇨ Illinois agility test ⇨ Semo agility test
Speed	⇨ Sprint standing start tests: 15 m, 40 m and 50 m ⇨ Running: 40-m sprint test

Muscular power	Standing long jump Vertical jump test Basketball throw

Step 3: Identifying strengths and areas for improvement

Once you have collected early data on a range of key variables, you can select appropriate activities for the areas you want to improve or focus on. Your health and physical education teacher, personal trainer or coach should be able to assist you.

Step 4: Monitoring progress

There are many ways to monitor your progress towards set goals. The most effective strategy is to consistently use the same technique to measure activity level or fitness. For example, it is not useful to conduct a baseline (pre-program) measurement of physical activity using only a self-report recall survey and then after 12 weeks to use a pedometer. These two measures cannot be compared to determine improvement because they measure different factors! It is okay to use a combination of measures, but to determine improvement you must repeat the test using identical measures or fitness tests. Measures used to monitor progress include diaries and logs, surveys, pedometers, accelerometers, heart-rate monitors, fitness tests and global positioning systems (GPS).

Based on the monitoring of your progress, you may need to adjust what you are doing in your personal physical activity plan and fitness program to achieve your set goals. You may need to change your program to make it more effective by adjusting the intensity, duration, frequency or type of activities and exercises being performed. You may decide to adjust your set goals once you achieve them or to make them more achievable or more challenging. Ultimately, the main goal should be to become as active and fit as you can be. Even doing something small regularly is better than doing nothing, or doing something only once a week!

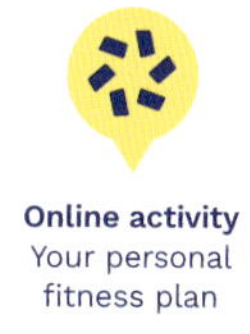

Online activity
Your personal fitness plan

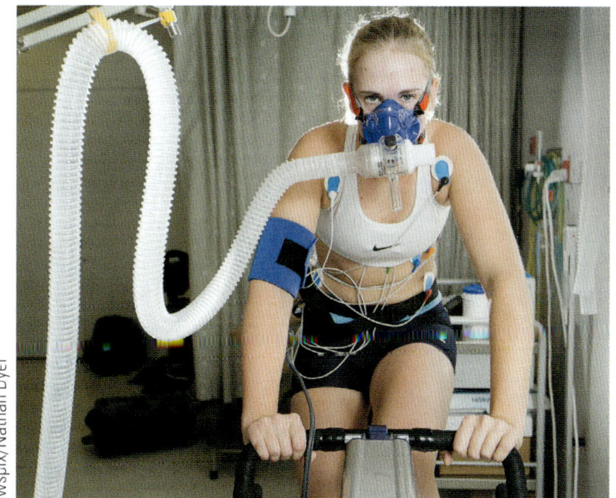

Newspix/Nathan Dyer

Figure 9.27 How is the principle of specificity being applied in this VO_2max test?

Getty Images/Jordan Siemens

Figure 9.28 Digital devices can provide immediate feedback about physical activity intensity, location, speed, heart rate, energy expenditure, steps and total distance covered.

CASE STUDY  PEDOMETER FUNCTION ON SMART WATCH OR WEARABLE TECHNOLOGY

Identify

Most digital smart watches (e.g. Garmin or Apple watch) these days contain a pedometer function, which is relatively accurate for recording the number of steps the wearer takes.

Understand

A pedometer function is a really powerful tool for raising awareness of your activity level, as it can provide you with immediate feedback. It is recommended that adults accumulate at least 10 000 steps per day, and young people 15 000 steps per day to achieve health benefits such as maintenance of healthy weight. Pedometer step counts can be a source of motivation for people, as they can set a daily goal and monitor their progress towards achieving that goal. Pedometers can also be used to estimate energy expenditure and total distance covered. The more steps you complete per minute, the higher the intensity you are working at. Wearable technologies calculate most of the data you need once you have put your age, height and weight into the device.

iStock.com/pixelfit

Figure 9.29 Wearable technologies such as a smart watch generally include a pedometer function, which allows people to monitor the number of steps they take daily.

Discuss

1 Explain why pedometers are a useful tool for measuring and maintaining physical activity.
2 Wear a pedometer for a day. Are you hitting the recommended step count of 15 000 steps a day?
3 Consider and state ways you could increase your daily step count.

1 Outline six safety tips for participating in resistance training and describe why this form of training should be included in everyone's regular physical activity.
2 Create a simple graph demonstrating your understanding of diminishing returns.

1 Explain the difference between fixed time and fixed load (repetitions) when designing your circuit training program. Provide an example to demonstrate the difference in the two approaches.
2 Describe how you might compare to an Olympic athlete if you started a new training program, in terms of your potential to improve.

9780170463102

3 Use the internet to research and describe four different examples of plyometric exercises.
   a Identify an example action for each of three different sports that would require explosive power.
   b Explain how a baseball player could use plyometrics training to improve their throwing power.

1 Refer to Table 9.2, which outlines the training methods used to improve various health-related and skill-related fitness components. Consider your typical week and analyse which of these training methods feature in your weekly routines for sports training or fitness regime. Identify which fitness components are being developed/maintained and which do not get much attention in your current physical activity plan.
2 Explain why training as hard as an Olympic athlete would not guarantee that you would attain the same level as fitness as them.
3 Use the web to complete the following table by outlining the factors each measure can be used to assess.
   a Search for some examples of fitness test protocols and norms to compare your results to.
   b Discuss which of the measures in the table are accessible to you to use.

Measure	Format (e.g. electronic)	What does this measure assess?
Diaries and logs	Paper, online or apps	
Surveys	Paper, online or apps	
Pedometers	Digital	
Accelerometers	Digital recording personal devices	
Heart rate monitors	Digital recording personal devices	
Fitness tests	Apps	
Global positioning systems (GPS)	Digital recording personal devices	

Scaffold
Assessment tests

Quiz
What are the essential elements to implementing a personal physical activity plan?

WHAT ARE SOME EXAMPLES OF LIFELONG PHYSICAL ACTIVITIES THAT PROMOTE HEALTH AND SOCIAL OUTCOMES?

In this section, we will take a look at a range of physical activities that promote health and social outcomes. The multicultural nature of Australian society has led to the introduction of a large variety of physical activities, outdoor recreational pursuits and sports. People migrating from all over the world have brought their culture, including their physical activities, with them. Sometimes people take up physical activities their families engaged in over past generations because they were exposed to these either through role modelling, observing friends or family or being encouraged to participate by their family or friends.

Some cultural groups provide wonderful social support and friendship through their participation in sport or physical activity. Over decades, a huge diversity of physical activities have been imported to Australia from other nations and adopted by Australians. The Irish brought Gaelic football and curling, the Germans brought Pilates and the Italians brought Bocce. Our physical activity options are as diverse and rich as our culture. The sports Australians play not only influence their health physically; they can also shape their identity, their values, their social circles and rituals and strengthen their sense of belonging and social connectedness. In fact, research has shown that participating in community support has a range of social benefits, such as reduced risk-taking behaviour and depression, and increased social support.

Alamy Stock Photo/Jeff Gilbert

Getty Images/Suhaimi Abdullah

Figure 9.30 Zumba and sepak takraw are two activities introduced to Australia by other nations.

9780170463102

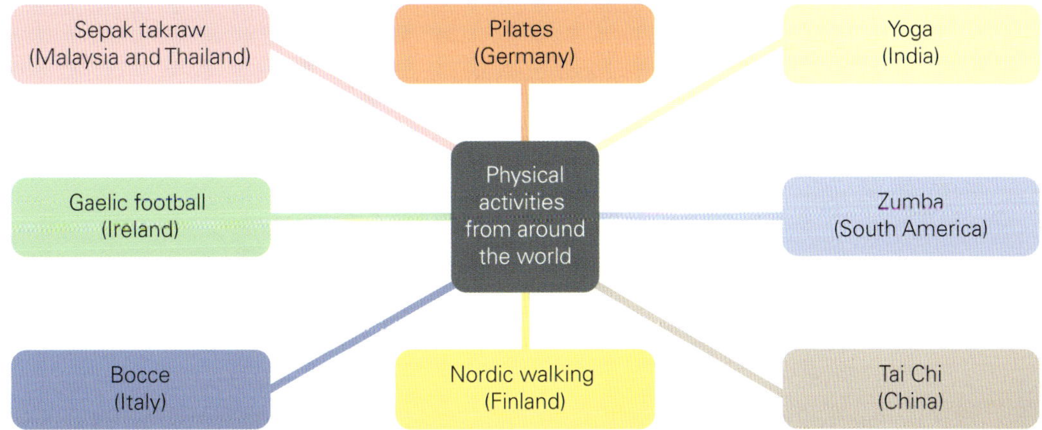

Getty Images/Brent Winebrenner

Getty Images/Pam McLean

Figure 9.31 Australians have embraced physical activities from cultures around the world.

CASE STUDY → GAELIC FOOTBALL

Identify

Gaelic football and Hurling have been played across Australia and New Zealand for many decades by Irish immigrants, visa workers and backpackers keen to maintain a connection with the national sport of their youth.

Understand

In 2001, Gaelic football met the stringent criteria of the Australian Sports Commission to become an officially accredited sport in Australia. To safeguard the continued development of Gaelic games, a strong focus is currently being place on youth development with the introduction of 'Go Games' specifically targeted at children under 12. Perth are leading the charge with the establishment of their Gaelic Games Junior Academy. Similar to other field sports, Gaelic games are generally played over the winter months. However, in South Australia, Tasmania and New Zealand, they have switched to a summer season, allowing them to attract local-born AFL and Rugby League players keen to maintain match fitness over their off-season.

Source: Gaelic Football & Hurling Association of Australasia

Discuss

1. Locate some clips online of Gaelic football and review them.
2. Identify three major components of fitness necessary to play, and explain how each component would be important to playing Gaelic football.
3. Explain three benefits associated with participating in physical activity with others.

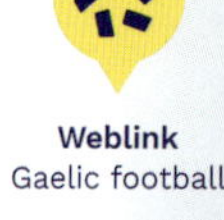
Weblink
Gaelic football

1. Outline five physical activities from around the world and identify which countries they originated from.

1. Survey your class to examine a range of cultural physical activities your grandparents engaged in. Determine whether these were also participated in by subsequent generations (e.g. your parents, uncles, you, your siblings or cousins).

1. Research online one physical activity played by Australians that originated in another country. Identify the rules and skills needed to participate.

Quiz
Pre-chapter

HOW CAN PARTICIPATION IN PHYSICAL ACTIVITIES IMPROVE THE MIND BODY SPIRIT CONNECTION?

Pilates, Tai Chi and yoga are three popular lifelong physical activities designed to improve the mind, body and spirit connection.

PILATES

Pilates was created by Joseph Pilates during the 1920s in Germany. It consists of more than 500 stretching and strengthening activities designed to condition the mind, body and spirit. The exercises are performed either on the floor or by using machines. Pilates is based on a strong belief that a healthy person needs to develop a strong mind in order to develop a strong body. The exercises used in Pilates are therefore designed to improve both the mind and body. Pilates is based on six key principles:

9780170463102

- **Centring** – refers to the centre or core of the body, where all movements in Pilates begin.
- **Concentrating** – focusing the mind to control the body.
- **Control** – every exercise is performed from the centre in a controlled manner.
- **Precision** – the way each exercise is performed, as slow controlled smooth movements with the emphasis being more on 'how' than 'how many'.
- **Breath** – all exercises are performed to a rhythm of breathing to achieve optimal circulation of oxygenated blood to the body.
- **Flow** – graceful and flowing succession of one exercise to the next.

Adapted from Muscolino J, and Cipriani S (2004), *Pilates and the 'Powerhouse' Journal of Bodywork and Movement Therapies*, Volume 8., 2004, p.16

Dreamstime.com/Nomadsoul1

Figure 9.32 Pilates is often practised on a machine like this.

TAI CHI

Tai Chi originated in China during the 16th century and was derived from martial art folk traditions. Tai Chi is a set of systematic callisthenic exercises with an emphasis on exercising the mind and body. It cultivates and stimulates the internal energy (Chi circulation) within the body, keeping you fit, relaxed, energised and rejuvenated. Tai Chi involves smooth, flowing, continuous movements going from one posture to the next. Tai Chi is considered a moderate-intensity physical activity and has many health benefits:

- improved cardiorespiratory (heart and lungs) function
- improved musculoskeletal (muscles and joints) function
- increased concentration and awareness
- decreased stress levels, including reduced blood pressure
- enhanced feelings of peace, relaxation and wellbeing
- improved sleep quality.

YOGA

Yoga is believed to have originated in ancient India around 2500 BCE. Yoga practice consists of breath work to connect the mind and the body. The ultimate aim of yoga is to connect thoughts and feelings to movement. There are many different types of yoga, ranging from slower paced programs (which place more emphasis on meditation) through to faster flowing movement sequences.

© Randy Glasbergen/glasbergen.com

Figure 9.33 The race to relaxation

UP AND MOVING Do a class

Participate in either a Pilates, Tai Chi or yoga class (e.g. Ananda, Ashtanga, Bikram, Hatha, Iyengar, Jivamukti, Kripalu, Kundalini, Power Yoga, Sivananda or Svaroopa). Afterwards, answer the following questions.

1 Describe what types of activities were completed.
2 Explain how you felt before, during and after the session.
3 Identify the nearest two locations to your home where you could participate in this activity.

Check out the Yulunga Traditional Indigenous Games resource online

1 Search for a game for students in a similar year level to you using the filter and select a game category (e.g. skipping games, tag games).

Weblink
Yulunga Traditional Indigenous games

1 Select a game you would like to participate in and outline the background, language, a short description, players needed, playing area, duration, equipment and game play and basic rules. (e.g. 'ii-ye' played by the Wogadj people of central Australia).
2 Participate in the game if possible and describe what you enjoyed about learning a new physical activity.

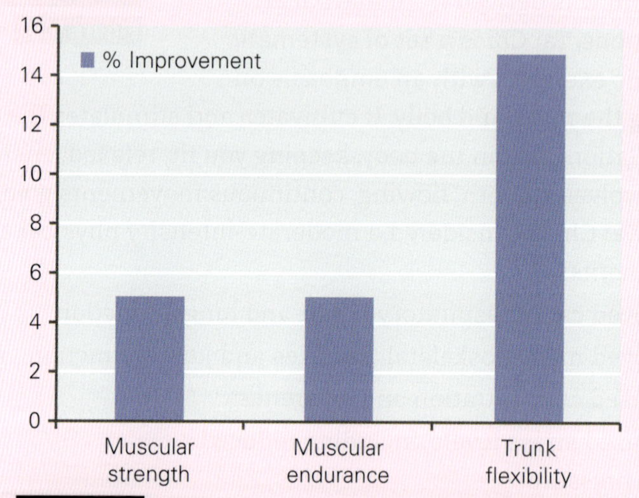

Figure 9.34 Improvements to health-related fitness components from yoga training

1 Describe the social benefits associated with participation in physical activities.
2 Explain why you think some of the Traditional Indigenous Games are still part of Australian play culture.

Quiz
How can participation in lifelong physical activities improve the mind–body–spirit connection?

9780170463102

CHAPTER 9 REVIEW

1 Identify a variety of movement opportunities you need to consider when planning your personal physical activity plan (e.g. maximising incidental physical activity, active commuting).

2 List 10 key training principles that need to be considered when designing and implementing a physical activity program.

3 Describe the principle of progressive overload using resistance (weight) training as an example.

4 Describe how plyometrics training works and provide some example activities.

5 Compare and contrast muscular strength, power and endurance as types of resistance training.

6 Evaluate the effectiveness of Fartlek training compared to short interval training for an endurance athlete.

7 Evaluate the advantages and limitations for using household items that could be used to design your own fitness circuit at home.

8 Outline three methods you could use to monitor your progress during your personal physical activity program.

9 Identify at least one fitness test for each health- and skill-related component of fitness.

10 Describe three physical activities that can improve health and social outcomes.

FIT TO DANCE

IN THIS CHAPTER

You will explore dance as an athletic pursuit by investigating how it can enhance the components of fitness and by examining the fitness levels of elite dancers. You will also anlayse the health-related benefits of dance for the general population and develop fun, simple and innovative ways to use dance to enhance personal and community fitness.

Shutterstock.com/Nicetoseeya

By the end of the chapter, you should be able to:

- investigate the fitness components within dance
- analyse the fitness requirements of an elite dancer with that of other elite athletes
- examine how dance can assist in developing individual and community health-related fitness
- use existing knowledge of fitness principles to create an innovative cardiovascular training dance by adapting an existing social dance to develop a moderate-intensity training dance
- develop an aerobics-style workout that could be used in your own personal physical activity program or at a before- or after-school program to promote physical activity opportunities for all students

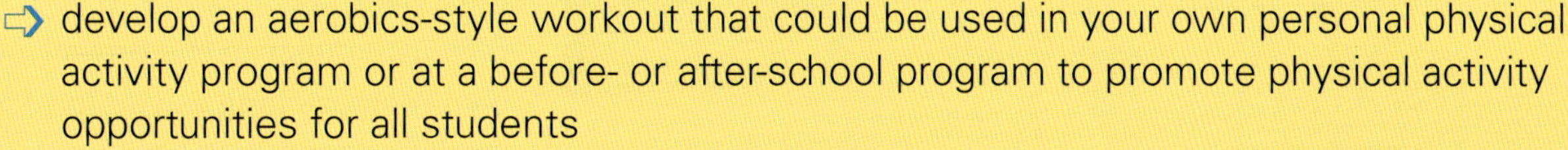

- investigate and implement safe dance practices
- examine the knowledge and skills of partners and group members when redesigning and performing an existing social dance
- apply decision-making strategies to work cooperatively and constructively with others when making decisions about movements, tempo, music and placement
- evaluate dance routines based on set criteria
- investigate the health-related benefits of social and cultural practices of physical activities from Asian countries, such as Tai Chi
- investigate the term 'mindfulness' and how it can be applied to health and wellbeing
- design and implement a safe mindful stretching routine to music to improve flexibility.

IS MINDFULNESS HELPFUL IN FLEXIBILITY TRAINING? 406

CAN WE ENCOURAGE COMMUNITY FITNESS THROUGH 'MILD'? 413

HOW CAN DANCING KEEP US FIT?

Quiz
Pre-chapter

Before you start, take the pre-chapter quiz to find out how much you already know.

People dance for many different reasons. The desire to get up and dance can be instigated by personal, cultural and social situations. But whatever the reason for dancing, there is no ignoring the fact that it gets people moving and that movement is good for us.

Regular physical activity is important for health, but sometimes finding the time and right place to exercise can be difficult. For many people, the lack of time or a convenient place to exercise can be a problem. Statements such as, 'I haven't got time to go to the gym', or 'I wish the swimming pool was closer' highlight some reasons why people don't exercise on a regular basis. So how can we find a way to encourage people to exercise without taking up too much time or having to go out? Maybe there is a solution!

Shutterstock.com/Aila Images

Figure 10.1 Dance can easily be done in the privacy of your own living room, making it an easy way to participate in physical activity.

One of the best reasons to dance is the fitness benefits it offers. Dance can be a cheap and time-efficient way to achieve a fitter and healthier lifestyle. You can dance to your favourite music in the privacy of your own room, as long as you have a safe working space. You don't have to go anywhere and you don't need any special equipment, except perhaps a pair of runners.

CASE STUDY ⇨ VIRTUAL DANCING!

Identify

Music is pounding. Neon lights are flashing all around me. I duck, dodge and side-step, my hands lash out in every direction. But this isn't some new martial arts fitness craze or a high-energy spin class. It's a VR game called Beat Saber that I'm playing on the Oculus Quest headset. I'm working out in VR – and it's fun.

9780170463102

Understand

Launched in May 2018, Beat Saber has quickly become one of the most popular VR games … The premise of Beat Saber is simple. You have two 'sabers', which it feels like you're really holding thanks to the motion controllers that come with VR headsets. You use these sabers to rhythmically slice boxes that fly toward you, in sequences of varying heights, widths and sequential complexities – think of it as like one of those dance machines you see at seafront arcades, but from the future …

'For many, Beat Saber has become a daily fitness routine,' Michaela Dvorak, Head of Marketing at VR studio Beat Games, the company behind Beat Saber, tells TechRadar. 'People are sharing with us their pictures before and after they started playing Beat Saber and results are insane – many people have lost more than 20–30 pounds since the game released.' …

'We began to fully realise the fitness potential of Beat Saber soon after the release when players started to post pictures with their sweaty faces and comments like these: "Hey Beat Saber team, your game tricked me into working out! How dare you!"' Dvorak tells us.

'People who were never too much into working out found out they can create a healthy daily routine with Beat Saber,' Dvorak says. It's easy to see why VR fitness games appeal to those who may be put off by gyms or fitness classes – you can participate from the comfort of your own home, go at your own pace and there's no judgement. For these same reasons, Dvorak explains that Beat Saber has also become popular among people who are slowly trying to overcome injuries or health problems …

One of the most obvious considerations when it comes to VR fitness is staying safe. VR is a new medium for many people and, because of that, there are different dangers to bear in mind – the main one being your senses are immersed in the virtual environment and not the real one.

'We want to make sure people will have enough space around them when playing,' Dvorak tells us. 'But also, they need to approach playing as a form of high physical activity and act like they are working out – stretch muscles, drink enough water, rest regularly. It is very important to keep this in mind.'

… you need to make sure you have enough space, remove obstacles, be careful when you take a headset off and ensure you've stretched beforehand so you don't pull a muscle or hurt yourself when you have lots of Beat Saber boxes whizzing at your face.

Discuss

1 Why might Beat Saber and other games like it encourage rhythmical movement and fitness at the same time?
2 What was the original aim of the game and what unintended outcomes were noticed by the team that developed the product?
3 Are digital tools such as virtual reality games a fun, safe and engaging way to encourage more people to move? Justify your answer and use examples from the text to support your ideas.
4 Critique the safety practices encouraged by the game's developer to maximise participant safety. Are they adequate?
5 Propose ideas for a Virtual Reality Dance Fitness concept.

How can dance improve fitness?

Fitness can be defined as the ability of the body to adequately perform a desired task. However, the level and type of fitness depends on the task you are asking your body to perform. For example, a football player's task is different from that of a swimmer or a gymnast, yet they all have similarities and differences in the way they require their bodies to move. Generally, overall fitness can be expressed as the ability of the heart, lungs, working muscles and joints to perform adequately and consistently over a period of time to meet the needs of the task.

Fairfax Syndication/Katherine Griffiths

Figure 10.2 What fitness components are required for dance?

Overall fitness is made up of different areas, called core components, which include:

- aerobic capacity
- anaerobic capacity
- muscular strength
- muscular endurance
- flexibility
- body composition.

When looking at the fitness requirements of a sport, these components need to be considered in more detail.

Table 10.1 Components of fitness

Health-related fitness	Skill-related fitness
Aerobic capacity	Balance
Anaerobic capacity	Reaction time
Body composition	Coordination
Muscular strength	Agility
Muscular endurance	Speed
Flexibility	Muscular power

When training, the demands of the sport or activity need to be considered, and the specific components of fitness required for that sport or activity need to be trained. This is the principle of specificity and all athletes, including dancers, need to be aware of the specific fitness requirements of their discipline.

FACE TO FACE Fitness requirements

1. What do you think are the specific fitness requirements for dance?
2. Identify the fitness requirements of an elite-level ballet dancer.
3. Analyse why each component of fitness is necessary. Provide an example where possible.
4. Identify the fitness requirements of an elite 100-metre track sprinter. Analyse why each component of fitness is necessary. Provide an example where possible.
5. Compare and contrast the similarities and differences between the elite ballet dancer and the 100-metre track sprinter and discuss your overall findings with the rest of your class.

DANCERS AS ATHLETES

Traditionally dance has been viewed as an artistic and aesthetic endeavour. Dancers as athletes need to have the required fitness levels to be able to successfully complete a performance; however, dance is so much more than just fitness. Dancing is about combining skill, technique and expressive interpretation to deliver a pleasing performance.

Newspix/Katrina Tepper

Figure 10.3 Australian ballet dancers Ben Davis and Olivia Bell rehearse *Dyad 1929*, watched by choreographer Damien Welch, at Australian Ballet headquarters in Melbourne.

Fitness

There have been a relatively limited number of studies examining the fitness levels of dancers, and it is uncertain whether existing tests are valid and reliable for assessing stage dance performance. There are many forms of dance, and the specific fitness requirements of each different style or genre make it difficult to generalise on the limited data available.

When performing, dancers generally move rhythmically and sporadically. Most performances require a dancer to come onto the stage and give a short, usually high-intensity effort followed by rest breaks. Studies have compared classical ballet with sports such as tennis and soccer, where explosive bursts of exercise are followed by moments requiring precision and skill. It has been suggested that during a choreographed ballet solo, a dancer can have a mean lactic acid blood level as high as a top-class football, squash or hockey player during a match.

A highlight of any ballet performance is the skill and technique exhibited during the dynamic jumps, leaps and turns performed. Muscular power and endurance are necessary elements for dancers – a large reserve of power is required for explosive jumps and high elevation that last for a few seconds; muscular endurance is needed to maintain a relatively high power output for up to a minute.

Weblink
Visit the Science 2.0 website to find out who's more fit: ballet dancers or swimmers.

Flexibility and joint mobility are very important to ballet dancers, as well as lean body mass.

Endurance

Studies show that dancers, in general, require less cardiorespiratory endurance than comparable athletes because of the intermittent nature of their performances. However, elite ballet dancers do require many years of consistent training to become highly skilled at their craft, and as a result they become refined technicians. Very few sports require such precision of movement, or such long hours of training to achieve it.

Success in many sports depends on the fitness level of the athlete. But ballet dancing requires so much more than just fitness. Skill, precision, musicality and expression are all important in developing the dancer, both as an artist and an athlete.

Fairfax Syndication/Dallas Kilponen

Figure 10.4 It takes long hours of practice to get to this high level of skill and precision.

CASE STUDY: INTERVIEW WITH BEN DAVIS, SOLOIST WITH THE AUSTRALIAN BALLET COMPANY

Identify

Ben Davis is a former soloist with the Australian Ballet. He danced at an elite level with the company for more than 10 years. Watch the interview with Ben and check out his profile below.

Video
Watch the interview with Ben Davis

Understand

Born in Melbourne in 1982, Ben used to tag along to his sister's jazz ballet classes and copy the routines from the back of the studio before he started attending his own classes at age seven. He trained with Leeanne Rutherford of Ballet Theatre Australia and joined the Australian Ballet in 2005. Ben was promoted to coryphée in 2009 and to soloist for the 2011 season.

After retiring from the Australian Ballet in 2018 Ben has worked as an instructor at Ballet Theatre Australia.

The Australian Ballet

Fairfax Syndication/Simon Schluter

Figure 10.5 Ben Davis and Amy Harris rehearsing *Cinderella* for the Australian Ballet

Discuss

1 After watching the interview with Ben Davis, reflect on your list of core fitness components for dancers. As a class, discuss anything else you would add to your list.

Worksheets 10.1 and 10.2

INVESTIGATION: FITNESS SPECIALIST

Purpose

To explore the role of a fitness specialist by analysing requirements for a sport of your choice.

Method

1 You have been employed by your favourite sport association to investigate the fitness requirements of their elite athletes. Your task is to undertake research on your chosen sport, using the discussion questions below to guide your research.
2 Present your findings on the sport to the class.

Discussion

1. Identify the fitness components required by an elite athlete for your favourite sport or activity.
2. Examine each component and why each component is necessary.
3. Reflect on the interrelation between components and the extent to which these need to be trained.
4. Propose how each fitness component can be trained.
5. Make overall recommendations on training specificity for this sport or activity.
6. Present and report your findings to the class.
7. Compare and contrast your ideas with the rest of your class.

1. Discuss the term 'specificity' as it relates to fitness training.

1. How can dance be a convenient and effective form of exercise?
2. As well as providing a fitness effect, why is dance also considered an artistic and aesthetic endeavour?

1. Examine how the fitness training of a professional hip hop dancer might differ from that of an elite ballet dancer.

FAST FACTS

A 2003 study in the New England Journal of Medicine by researchers at the Albert Einstein College of Medicine discovered that dance can decidedly improve brain health. The study investigated the effect leisure activities had on the risk of dementia in the elderly. The researchers looked at the effects of 11 different types of physical activity, including cycling, golf, swimming and tennis, but found that only one of the activities studied—dance—lowered participants' risk of dementia. According to the researchers, dancing involves both a mental effort and social interaction and that this type of stimulation helped reduce the risk of dementia.

Dancing and the Brain, by Scott Edwards, Harvard Medical School, 2015.

IS DANCING A GOOD OPTION FOR HEALTH-RELATED FITNESS?

Not everyone can be an elite athlete; in fact, only a very small percentage of the population will ever compete, perform or play at the top level of their sport. So, for the majority of people, sport and physical activities are hobbies and ways to keep fit and healthy.

When most people talk about fitness for the general population, the focus is on the different types of fitness levels required for a health benefit. Most people don't need the cardiorespiratory endurance of a marathon runner to have a healthy cardiorespiratory system.

Shutterstock.com/bbernard

Figure 10.6 Dance is often used as a low- to moderate-intensity exercise.

Dancing, when performed over an extended period of time, can provide great opportunities for your body to move and therefore helps to improve cardiorespiratory functioning.

When most people get up to dance, it is usually casual movement that requires low- to moderate-intensity effort, not the high-intensity demands of professional dancers.

This means that people can usually enjoy the movement of the dance and have a conversation at the same time. This in itself can be used as a training tool because both low- and moderate-intensity level movement are beneficial to health.

See Chapter 9, pages 361–2 for an explanation of training intensities.

Worksheet 10.3

CAN SOCIAL DANCE GIVE US A FITNESS BENEFIT?

Video
Watch *The Nutbush* dance video

Social line dances are great for getting everyone on the dance floor! For the following activities *The Nutbush* will be used as the base for creating a fitness dance. However, your teacher may use another line dance to complete the same activities.

The Nutbush is an example of a social line dance that became famous during the 1970s. It was performed to the song 'Nutbush City Limits' by Ike and Tina Turner.

Alamy Stock Photo/Directphoto Collection

Figure 10.7 *The Nutbush* is a social line dance from the seventies that is still popular today at parties.

The great thing about *The Nutbush* as a dance is that it has four base moves that keep repeating throughout the song, allowing it to be learnt quickly for mass participation. This is one of the reasons it has become so popular at social gatherings and parties.

The other great thing about *The Nutbush* is that it generally goes for about three to four minutes, making it an aerobic-based activity that people can participate in while talking and enjoying the company of others.

UP AND MOVING The Nutbush

1. As a class, perform *The Nutbush*. Chat with a partner during the dance if you wish; dancing should be fun!
2. Record your resting heart rate before you begin and then again after you have completed the dance.
3. As you perform the dance, perform the 'talk test' fitness test.
4. Do you think *The Nutbush* would be classed as moderate or vigorous exercise, given the results of the 'talk test' fitness test and your heart rate test?

CREATING A TRAINING VERSION OF *THE NUTBUSH*

Moderate-intensity activity can be a great training tool for cardiorespiratory endurance. Most people who dance in a social setting benefit from low- to moderate-intensity activity, whether they are highly energenic on the dance floor or just performing a social line dance. Depending on the level of effort put into the dance, *The Nutbush* can generally be classified as a moderate-intensity physical activity.

Now that you have completed the original version of *The Nutbush*, it is time to create your own dance!

MILD moderate intensity line dance

With a partner you will create your own **m**oderate-**i**ntensity **l**ine **d**ance. The acronym for this is **MILD** – an appropriate term for a moderate-intensity activity! At this point, your new dance will be called *The MILD*, unless you and your partner come up with a new name for your dance.

To be able to change the dance, the original needs to be broken into its various parts.

The Nutbush consists of 32 counts of movement that are continually repeated from the beginning of the song to the end. These 32 counts are broken down into four groups of eight counts. Each eight-count phrase has its own specific movement sequence. Can you work out what these are?

Fairfax Syndication/Elesa Kurtz

Figure 10.8 What moves could you use for your moderate-intensity line dance?

UP AND MOVING The MILD

1. With your partner, perform the first eight counts of *The Nutbush* to establish the first eight-count movement sequence.
2. Once you have done this for the first eight counts, continue doing the dance for the second, third and fourth group of eight counts using three different movement sequences.
3. Document each movement sequence.

Worksheet 10.4

THE NUTBUSH DANCE

Not only is *The Nutbush* a great dance for moderate-intensity exercise, it also highlights some other important exercise principles. When you were listing the movement sequences for each group of eight counts, you will have noticed that each movement sequence is repeated on both the right and left side. This is very important for developing even amounts of muscular strength and endurance and joint mobility. Imagine if you just did everything on one side of your body! You would end up very lopsided and probably with muscular and joint problems. A good exercise or training routine for any sport requires a balanced approach.

THE MILD

Now that you have broken down *The Nutbush* dance to establish that it has four movement sequences that repeat, and that the movements are balanced on both sides of the body, work with your partner to create your own MILD.

Safe dance practices

As with any sporting activity, participating in dance requires you to be proactive about safety. It is essential to consider safe dance practices before undertaking any dance-related activity.

Consider and apply the following when dancing and creating your MILD:

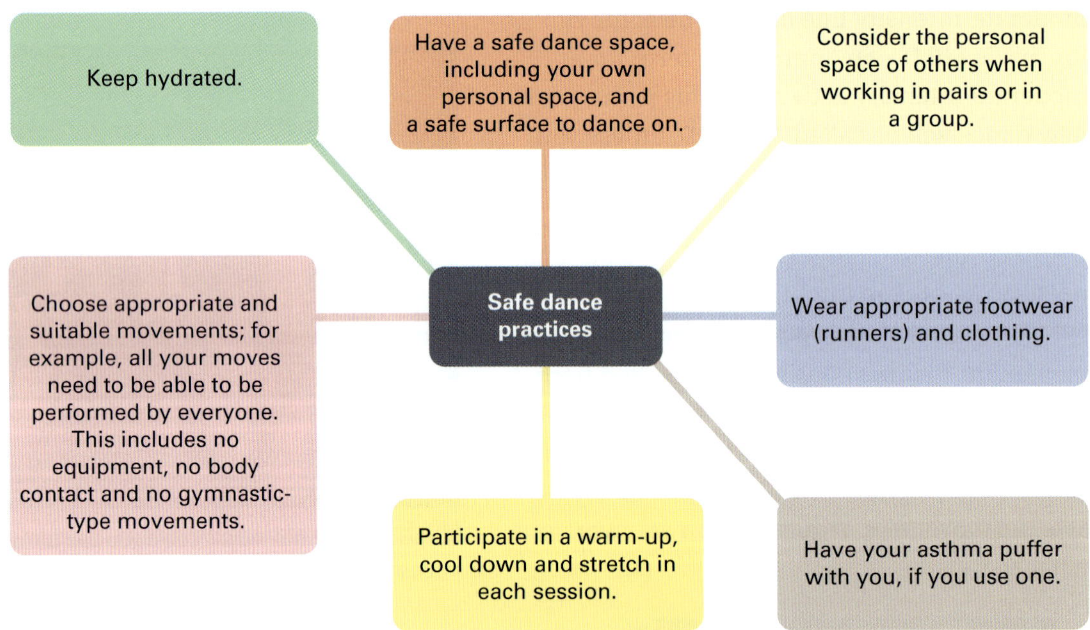

Figure 10.9 Safe dance practices

Safety: no breakdancing or gymnastic moves

Students must be aware that safe dance practices are essential. Breakdancing and acrobatic or gymnastic styles of movement require specialised instruction and matting, so are not to be included in your MILD without a qualified supervisor. The following movements are **not** to be performed in your dance:

- rotation of the torso over the back, head and neck
- body support using the back, head or neck
- gymnastic rolls, cartwheels and handstands, etc.

9780170463102

FACE TO FACE Think, pair, share

1 Working cooperatively with your partner, think, pair and share your ideas to come up with four new movement sequences to replace the movement sequences in the original dance.

2 When creating your new MILD, consider that your dance needs to meet the following criteria. It must:

- consist of four movement sequences with an equal number of counts. *The Nutbush* uses eight counts for each movement sequence; however, you can use 16 or 32 counts as long as the counts are the same for each movement
- consist of movement on the right and left side of the body and opposing muscle groups
- continue the moderate intensity level of the dance.

3 As *The Nutbush* dance turns in an anticlockwise direction to rotate, you can choose to add this to your MILD or keep your dance facing the same direction.

MOVEMENT IDEAS

Worksheet 10.5

FACE TO FACE Brainstorm

Now that you have the sequences, you and your partner need to brainstorm possible options for creating new moves for your MILD.

1 How about giving your new routine a hip-hop flavour? Check out some moves that could turn your MILD into a hip-hop training routine.

Video
Watch the hip-hop moves video to get some ideas.

2 Instead of hip-hop, your MILD could use any of the following movements:

- aerobic-based steps, such as heel digs, easy walks or flick kicks
- social dance–based steps like the grapevine, double shuffle or step touch
- sport-based movements such as a basketball bounce and set shot
- any safe and great moves that you can create!

If you are unsure of what some of these moves are, look at the video clips on Nelson MindTap.

3 Once you have made a decision about your changes, get up and have a go to ensure that your new dance can be performed successfully. Practise your new dance to the music of 'Nutbush City Limits' by Ike and Tina Turner, *or* to a song that you choose with your teacher so that everyone in your class can practise at the same time.

Video
Watch the video to get some other movement ideas.

4 Keep brainstorming with your partner to make the necessary changes until you are satisfied with the movement sequences of your MILD, then record these in your workbook.

Shutterstock.com/Pressmaster

Figure 10.10 What other styles of dance could you use moves from?

iStock.com/gradyreese

Figure 10.11 Practice makes perfect!

NOW FOR A CHANGE OF MUSIC

For many people, music is the main motivation for dancing. Have you ever heard a favourite song and unconsciously started to move to the beat? Although 'Nutbush City Limits' is an iconic song, and one that remains popular, sometimes when we hear a song too often it lessens its impact.

Now that you have a feel for the tempo of the song, a different choice of music could be substituted for the class to practise the routines you have learnt. You can even perform these at home as part of your personal exercise program.

A change of music requires a few prerequisites:

- **Tempo**: 132 to 138 beats per minute (BPM), similar to *The Nutbush*. Any faster and the dance could become high intensity; any slower and it could become low intensity.
- **Time**: Music is to last for at least three to four minutes.
- **Lyrics**: Lyrics and context of the song should be appropriate for school.

FACE TO FACE Music

As a class, brainstorm some music suggestions for your MILD.

Tempo

The tempo of a song is found by counting the number of beats for one minute. For example, if you were to count the beats in *The Nutbush* for one minute, you would count 132, which means that the tempo of the song is 132 beats per minute (BPM). To find another song with the same or close to the same tempo as *The Nutbush* you would need to counts the beats for one minute of that song to establish its BPM.

UP AND MOVING Good to go?

Once you have a few options, you could vote as a class on the best choice, or perhaps have a few different songs to use. You can try these to see if they are motivating and enjoyable to dance to.

9780170463102

WELLBEING CHECK IN

MUSIC TO LIFT

This activity has been developed in collaboration with Dr David Bakker of MoodMission

Identify

There are so many reasons why people create and listen to music. But most of them have something to do with feelings.

Understand

Research shows that music can create powerful emotional effects. Some songs might make us feel sad, while others can make us feel angry, excited or joyful. Some scientists believe that music and spoken language are closely related. Before our species started talking, we may even have been communicating using something that sounded a bit musical.

Practise

1. Let's make some playlists to change the way we feel.
2. If we're feeling sad and want to feel happy again, it can be useful to listen to upbeat songs. Take a moment now to make a playlist of 4–5 songs that make you feel happy.
3. If we're feeling angry and want to work that anger out of our system, it can be useful to listen to angry or high-energy songs. Take a moment now to make a playlist of 4–5 songs that could help when you're angry.
4. If we're feeling on edge or anxious, it can be useful to listen to calming music. Take a moment now to make a playlist of 4–5 songs that make you feel relaxed.
5. The next time you want to change how you feel, try listening to one of your playlists.

Reflect

Are there times when you feel sad and just want to listen to lots of sad music? Does it make you feel any better, or just even sadder? This can be different for different people, so it's important to know how sad music affects you.

Sometimes when you are dancing or exercising, once the song finishes you have to wait for the next song to start and hope that it has the same tempo as the last one.

Another option when using dance as a form of exercise is to use continuous music that has been sequenced and pitch-controlled to a steady BPM. This means that all the songs run into each other without a beginning or end, and continue at the same tempo for the entire course of the music.

To make this activity as easy as possible, your teacher will provide a copy of a whole soundtrack of appropriate, continuous and sequenced music that is ready to use.

Fairfax Syndication/Elesa Kurtz

Figure 10.12 Adding arm movements can increase the intensity of the dance.

You can also look at some other continuous playlists and tracks at Music and Motion or other websites.

When you have finished creating your MILD you can rename your dance if you wish.

UP AND MOVING

Arm movements

Now that you and your partner have developed the footwork for your MILD, you can extend your moderate-intensity line dance by adding in specific arm movements for each of the moves. Adding arm movements can improve your dance by creating an opportunity to exercise the muscles of the upper body and arms and is, of course, another element for artistic expression.

To focus on the exercise and fitness opportunity that your MILD can provide, you need to ensure that the arm movements consist of movement on the right and left side of the body and opposing muscle groups.

When choosing the arm movements, consider how they may help to improve your overall fitness and muscular endurance and increase your heart rate.

Worksheet 10.6

1 Work with your partner to develop arm movements to go with your MILD.
2 Experiment with these until you find movements that flow naturally with the footwork, then practise your new improved routine to the music. Record your arm movements in your workbook.

Film it!

It might also be fun to record your dance so you can review it and check that your MILD meets the set criteria. Filming your dance can also help you and your partner to self-assess and make any necessary changes to the MILD. As this is a school activity, only school-based equipment can be used.

As a class you now have a number of new moderate-intensity line dances that can be used as a way to improve cardiorespiratory fitness. Now you can share ways to exercise. By learning another MILD you can add another exercise routine to your own personalised program.

Teach it

1 With your partner, teach your MILD to two other students and then learn their dance.
2 As a group of four, reflect on the differences between both routines.

CONSIDERING THE WELLBEING OF OTHERS

Dancing and the way people express themselves through dance is very personal. Your dance style is completely unique to you. Therefore, when working in groups it is important to remember that each individual has their own dance style, which cannot be judged. However, feedback can be provided to your peers about the set criteria for any given task. When reflecting on a MILD, your reflection should only be about the following criteria:

- Did the MILD contain four new movement sequences consisting of equal counts?
- Did the MILD contain four new movement sequences consisting of movement on the right and left side of the body?
- Did the MILD contain four new movement sequences continuing the moderate-intensity level of the dance?

FACE TO FACE Reflection and feedback

Look at the other class MILDs and then write your reflections in your workbook. Remember to base your feedback on the criteria for the movement sequences, not on individual dance style.

CLASS INTERVAL TRAINING SESSION

If you used a series of dances such as *The Nutbush* in a row, a moderate-intensity training routine could be created to enhance cardiorespiratory endurance levels; it would be like an aerobics routine. Developing a moderate-intensity dance routine that lasts for more than 30 minutes can provide a training effect. This means that such a routine can improve participants' cardiorespiratory fitness levels.

A training effect can also be achieved with a form of progressive interval training. Interval training is fun because it allows you to work aerobically for a period of time, then have a quick break before a change of activity.

UP AND MOVING Class interval training session

1. To create a class interval training session, you and your partner will lead the class in two repetitions of your MILD. Performing your MILD for two repetitions allows the class to learn your dance and provides an aerobic training effect. To lead the class, you and your partner will need to be in a position where everyone can see you.
2. Another pair will then take over to perform and lead the class in two repetitions of their MILD, until all dances have been performed. By the time all the class have presented their MILD, you will have participated in at least 30 minutes of moderate-intensity activity and had a great cardiorespiratory work out!
3. If you and your partner are enjoying the opportunity to create a dance-based MILD, you could extend the dance by adding more steps. This can be done by adding in new movement sequences and ensuring the dance is balanced on both the right and left sides of the body.
4. With your partner, you have created your own moderate intensity dance-based training routine.

This routine could be adapted to include higher intensity or higher impact movements to create a greater cardio-respiratory effort and training effect by turning your MILD into a high-intensity line dance (HILD).

To do this, choose two movements from your MILD to change into higher intensity movements.

For example, a simple knee lift could be performed with a hop to increase the effort required, thus increasing the intensity and workload on the cardiovascular system and specific muscle groups. Ensure you check the safety of your new moves with your teacher before including them.

1. Reflect on the difference between low-, moderate- and high-intensity movement and how these can impact fitness.
2. Examine the importance of exercising on both sides of the body.
3. Propose how adding arm movements to the MILD improves the fitness benefits of the dance.

1. Investigate the possible fitness and social benefits of social dancing.
2. Reflect on why music choice is so important.

1. Investigate how the MILD could be adapted into a high-intensity line dance.

Worksheet 10.7

IS MINDFULNESS HELPFUL IN FLEXIBILITY TRAINING?

As well as cardiorespiratory fitness benefits, dance can provide opportunities to improve flexibility. Flexibility is the ability of the joints and muscles to adequately complete the range of motion necessary to perform a required task.

The range of joint motion for ballet dancers is far greater than the range of joint motion required for many other sports. For example, an extreme range of motion in the hip is required to be able to perform movements such as splits, split leaps and arabesques.

Flexibility training depends on the requirements of each sport or activity. However, general flexibility is a very important component of overall fitness for everyone. Having a good range of motion and muscular flexibility allows you to perform everyday tasks efficiently and without injury or soreness.

iStock.com/Stigur Már Karlsson/Heimsmyndir

Figure 10.13 Flexibility is a key fitness component that dance and other movement exercises can improve.

WELLBEING CHECK IN

CHILD'S YOGA POSE

MoodMission

This activity has been developed in collaboration with Dr David Bakker of MoodMission

Identify

What if there was a way of doing more with less?

Understand

Yoga can improve mental health in three different ways. Firstly, it's a type of exercise, and we know that exercise can release feel-good chemicals throughout the body. Secondly, it can be quite relaxing because it's slow and quiet. And thirdly, it places a lot of emphasis on mindfulness and being present. When doing a yoga pose, it's important to pay attention to how you feel in your body. Yoga encourages a non-judgemental approach to our bodies as we curiously explore how they feel when we put them in certain positions.

Practise

1. Kneel on your hands and knees, so your hands are in line with your shoulders and knees in line with your hips. Lift your head so your neck and spine are straight.
2. Move your hands forward so they are slightly in front of your head and bring your big toes together. Press into your palms while moving your knees to the edge of your mat, one leg at a time (if you're not using a mat, your knees should be slightly wider than your hips).

Shutterstock.com/Dmitry Rukhlenko

Figure 10.14 Child's pose

3. Inhale, look ahead of you, and begin to exhale as you bend slowly to sit on your heels. Keep the side of your torso long and your arms stretched out straight in front of you with your palms flat on the mat.
4. Inhale, reach forward and exhale as you sink back onto your heels, bringing your head and chest to the mat with relaxed chest, shoulders and elbows.
5. Hold this position for 30 to 60 seconds. You can slowly rock your hips from side to side if you wish.
6. Repeat the above steps again to move in and out of the pose one more time.

Reflect

This was just one yoga pose. How do you think you might feel if you did a whole class? Some people might worry about being embarrassed in a class. Are there ways you could overcome this embarrassment to try out more yoga?

Alamy Stock Photo/Richard Ellis

Figure 10.15 Tai Chi

Worksheet 10.8

TAI CHI

The importance of general flexibility has been recognised by many cultures for centuries.

Many Asian cultures acknowledge the importance of harmony within the mind and body, and have developed practices through martial arts and controlled movement routines to enhance the mind–body connection. One such practice is the traditional Chinese art of Tai Chi.

Tai Chi has a number of different forms but has generally evolved into a movement practice that promotes harmony through mental alertness and suppleness of the body. This is achieved through the slow, controlled and coordinated performance of specific movements and sequences that reflect the natural movements of the body.

There is an old Chinese saying:

Whoever practises Tai Chi regularly will in time gain the suppleness of a child, the strength of a lion and the peace of mind of a sage.

CASE STUDY  TAI CHI IS GROWING IN POPULARITY WITH MILLENNIALS

Identify

Tai Chi is an ancient martial art developed in China that's often referred to as a 'moving meditation'. … Tai Chi's slow, graceful movements are accompanied by deep circular breathing. Though Tai Chi is practiced slowly for health benefits – stress relief, improved balance and flexibility – it can be sped up and used as a fighting form in very advanced classes.

Understand

Chinese physicians prescribe Tai Chi as a gymnastic form of medicine to complement other traditional treatments such as acupuncture and herbs, according to Tai Chi master Terry Dunn, who helped popularize it in the West. The movements are working with what is called 'qi' or life force, a type of 'flow' that, according to Tai Chi practitioners, everyone has …

Though studies show that most of those who practice Tai Chi are 50 years old and up, several instructors report a renewed interest among younger folks looking for an antidote to stress. York does Tai Chi three times a week, which he says helps him be more patient and go with the flow …

It also helps him be more patient in his job as classroom aide for special-needs children. 'A couple of times I have guided them through some breathing exercises, and within minutes, they're more focused. I'm more relaxed, they're more relaxed, and we're able to move on to the lessons,' York said.

Dunn, who has been teaching Tai Chi in Los Angeles for more than three decades, says that in the past year and a half, he's gotten an uptick in calls from young men in the tech industry … Google headquarters has been offering Tai Chi to its employees for the past couple of years.

Master David Chang, owner of the Wushu Central Martial Arts Academy in San Jose, says he has several students in their 20s and 30s.

'It used to be all senior citizens. It was uncommon for anyone younger than 30 learning Tai Chi. I've seen it's shifted quite a bit.' Chang, 41, says those students are looking for both a good source of exercise and something to relieve stress …

'Tai Chi has a stigma of being for old people because it's slow. Tai Chi is great for young people because it helps you to develop that slowness, which can be very beneficial in the world when things are stressful,' York said. He says that although it's practiced slowly, it's not easy. 'The slowness of Tai Chi is deceptive. It's more difficult to remain slow and connected to the breath. Each movement uses almost every muscle. So when we're standing and we're in a form and we're centered low – the legs are engaged, the torso, the arms – everything's engaged but not stressed like it would be in a workout in the gym. My legs have become much stronger.'

Taniela Irizarry, 37, says it helps with her injuries from years of beating up her body as a professional dancer ...

Most health research into the benefits of Tai Chi focus on people over 50. But the earlier you start in life, the more you reap the rewards, says Dr Michael Irwin, director of UCLA's Mindful Awareness Research Center. He says the benefits are cumulative. 'A younger person that's experiencing stress, there's a way for them to target that stress so that it does not accumulate over time – increasing the likelihood of them developing disease.'

Irwin has carried out more than a dozen peer-reviewed studies evaluating the ability of Tai Chi, yoga and mindfulness to improve health outcomes. Irwin's earliest tai chi study in 2003 found an increase in the number of disease-fighting 'T cells' that fight off shingles in patients who practiced Tai Chi. The 15-week study showed no increase in those who did not participate. In a later study, Irwin found that practicing Tai Chi gave the same immunity boost as a vaccine developed for shingles. In 2015, a study found that Tai Chi, over several weeks, reduced cellular inflammatory responses in patients suffering from insomnia, Irwin says.

'As we age, we find you can be more at risk for an infectious disease, or you can be more at risk for an inflammatory disorder,' he said. 'Tai Chi importantly impacts both elements of the immune system. It improves our ability to fight off infectious disease. It also decreases our inflammation.'

Irwin is now studying whether Tai Chi can be as effective as cognitive behavioral therapy to treat depression and anxiety.

Discuss

1. Examine the various health benefits of Tai Chi.
2. Why are millennials in particular picking up Tai Chi more?
3. Why is it recommended to start a practice such as Tai Chi earlier in life?
4. Read the article at the Nelson MindTap link discussing the benefits of Tai Chi in aged care. Compare the benefits of practising Tai Chi for the aged to the benefits for millennials.
5. Use your own research skills to find two more articles that highlight the benefits of Tai Chi. Attempt to uncover at least one article that has been written by a medical practitioner, health-care professional or that has been based on research findings.

Weblink Benefits of Tai Chi in aged care

Worksheet 10.9

THE IMPORTANCE OF POST-ACTIVITY STRETCHING

Flexibility training is important to ensure optimal joint mobility and muscular performance for any activity, whether at an elite level or just for performing everyday tasks. Flexible joints allow freedom of movement, so joints need to be kept in good working order.

One of the most important reasons to stretch is to return muscles to a state of equilibrium after

Getty Images/Tim Platt

Figure 10.16 Young and old people benefit from practices like Tai Chi.

activity. While you are performing any sport or physical activity, the joints and muscle groups involved are in constant states of contraction through flexion and extension, and they need to be released back to their resting state.

Stretching should be performed after the core muscle temperature has been raised, so any pre-sport or activity stretching should only occur after you have warmed up, (usually after the first 5–10 minutes of the initial warm up). To prevent soft-tissue injury and any post-activity soreness, the working muscle groups should be stretched at the end of each activity session.

Getty Images/Sarah Casillas

Figure 10.17 Pre- and post-activity stretching is important to prevent injury.

Therefore, after completing any physical activity, such as your MILD or class interval training session, the joints and muscle groups involved should be stretched through their full range of motion.

One way to achieve safe stretching is to participate in static stretching of the muscle groups involved in the activity that has been undertaken. Static stretching requires you to stretch each muscle group by taking the joint to the full range of motion and elongating the muscles involved. Once the muscle is stretched to the point of resistance but not pain, the stretch should be held for approximately 10 to 20 seconds and then released. This should be repeated three times.

Creating a mindful stretch routine

In addition to improving cardiorespiratory fitness, dance can also be used to create flexibility routines, promoting general flexibility and helping to prevent soft-tissue injury and post-activity soreness.

You now have an opportunity to create a routine that you can use after you have performed any exercise or sport training session. You can change the stretches to work the muscle groups that you have targeted in your training session.

Exploring stretching and mindfulness

Mindfulness is a term that has its origins in Buddhist philosophy and practice. The term relates to a state of alert presence where attention is focused on what is being experienced or performed in the present moment. Mental alertness is required when stretching to ensure we stretch correctly. If we are distracted or talking during stretching, then we lose awareness of the body and can inadvertently over or under stretch the muscle. As with Tai Chi, in stretching the best results come from focusing both the mind and body to create harmony, ensuring that we are fully aware of what is happening to our body. By being fully present or mindful when we stretch, we can ensure we stretch correctly and don't overextend the muscle or joint capacity.

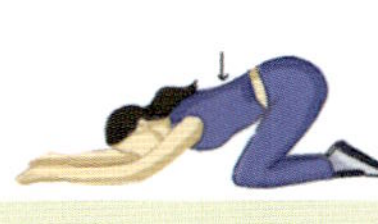

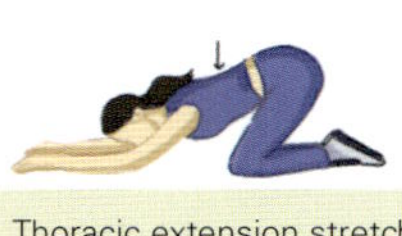
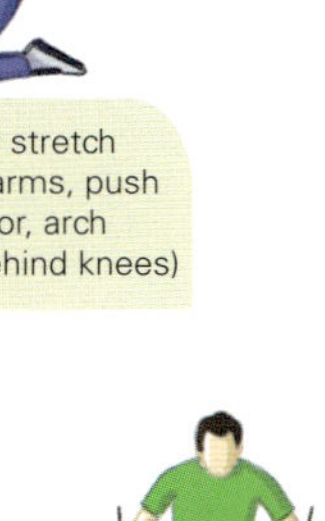

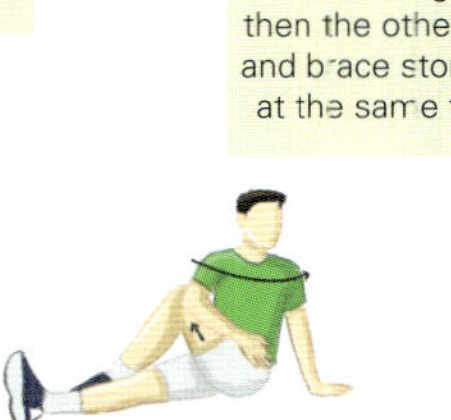

Figure 10.18 Stretching exercises

Mindful stretch routine

WB Worksheet 10.9

Your teacher will allocate you and your partner one muscle group to design a mindful static stretch for. (Refer to the stretch diagrams provided in Figure 10.18.) You will also develop a script around the delivery of your stretch, describing the mindful performance of the movement involved.

1 With your partner, begin by creating a slow, controlled and coordinated static stretch for each opposing muscle in the group using the safe stretch technique. For example, a quadriceps stretch would be followed by a hamstring stretch. Each stretch must meet the following safe stretch criteria:
   - performed as static and not bounced
   - performed in a slow and controlled manner
   - lengthened to the point of resistance, never pain
   - held for 10 to 20 seconds
   - repeated three times
   - followed by slow rhythmical movement to proceed into the opposing muscle stretch
   - performed on a stable, safe and shock-absorbing surface (not concrete!).

   Always remember that safe technique is very important, and that you should never force a stretch. It is important to perform stretching on a safe surface. You need to have a mat to protect your back and knees from the ground.

2 Once you have developed your opposing muscle stretch and had your teacher check it, you can now develop a mindfulness script describing how to perform the stretch safely with presence and alert focus so you become more aware of your own body and how it is feeling.

   With your partner, research how to teach a mindful stretch. Things to consider in your mindful stretch script include:
   - description of how to safely perform the stretch
   - use of the breath – inhaling and exhaling
   - kinaesthetic sensation through your senses such feeling, seeing, hearing
   - relaxation of the muscle group.

3 Teach another pair your stretch using your mindful approach.

Weblink
Example mindfulness stretching script

4 As a class, explore some slow, rhythmic background music that you like. You could play free meditation music to develop a relaxing, mindful stretch session to perform after exercise.

   The use of slow, rhythmic movement allows the stretches to be performed in a more coordinated, controlled and sustained manner. It also enables the movement between stretches to flow rhythmically to create a dance-like feeling throughout the performance of all muscle group stretches.

Weblink
Meditation music for stretching
Peaceful workout music
Slow music playlist

5 Try your stretch to the music that the class has chosen and practise so that you and your partner can lead the class in your stretch and mindfulness script when it is your turn.

6 Once each pair has taken their turn to present and lead their stretch, you should have completed at least one stretch for each of the major muscle groups. You now have a mindful stretching routine (MSR) that you can use in the future.

1 How would you define flexibility?
2 Why is Tai Chi more than just a form of stretching?

1 What is an adequate level of flexibility for the general population?

1 Create a mindful stretch app from the class creations.

CAN WE ENCOURAGE COMMUNITY FITNESS THROUGH 'MILD'?

DANCE FOR FITNESS SESSION

Worksheet 10.10

You have created your own MILD, learnt one that has been created by another pair and then linked these to create a class interval training session. These MILDs could now be shared with others who might not have had the experience of using dance to improve their cardiorespiratory fitness.

Using the class interval training session, you could set up a dance fitness session for students in other classes, demonstrating how easily dance can be used to facilitate physical activity opportunities and fitness. You can even use your mindful stretch routine as the cool down and post-activity stretch.

Using what you have already developed as a class, you could try running a before-school, lunchtime or after-school dance for fitness session. This would provide a great physical activity opportunity within the school environment.

As a class, along with your teacher, plan your class dance for fitness session. You will need to decide when and where the session will occur and choose the format of the event. You can then develop the necessary flyers and posters to advertise and promote your session.

Newspix/Erin Byrne

Figure 10.19 You need to create appealing and bright flyers to attract people to your class.

CASE STUDY  DANCE MOVEMENT THERAPY

Identify

Why do we stop dancing when we grow up? … It is surprising to me that dance/movement therapy (DMT) is not more popular within the fields of psychology and psychotherapy globally …

Understand

Dance/movement therapy goes beyond simply dancing. DMT uses dance and movement to promote insight, integration and wellbeing, as well as to diminish undesirable symptoms in various clinical populations … Just like with more traditional psychotherapies, DMT can be applied in a wide range of ways. It may involve talking, different types of music or no music at all. It can be done in groups, with individuals or with couples. Therapists sometimes dance with their clients and at other times observe.

A group therapy session may involve a warm-up and check-in as to where we are at emotionally, mentally and physically. It may be followed by the development of a theme, which emerges spontaneously or has been prepared by a therapist (for example, working with difficult emotions). It ends with grounding (reconnecting with our bodies and our selves in the present moment) and closure (for example, a gesture, a sound, a word). All of this is done with our bodies in motion or stillness, but some verbal sharing, journaling, drawing and other elements may be added …

One of the main reasons people dance is to modify their emotional state; typically, they strive to feel more joy and happiness and to reduce stress and anxiety … A recent systematic review of research on dance/movement therapy specifically found it to be effective in the treatment of adults with depression.

Dance typically involves learning sequences of steps and movements in space, in coordination with music. In other words, it requires substantial physical and cognitive engagement and, as such, it should improve not only muscle tone, strength, balance and coordination, but also memory, attention and visuospatial processing.

When comparing relatively long-term dance interventions (of six and 18 months) to conventional fitness training, several studies have found improvements in attention and verbal memory and neuroplasticity in healthy older adults. Researchers also found improvements in memory and cognitive function for older adults with mild cognitive impairment after a 40-week dance program.

In addition, a recent meta-analysis of seven randomized controlled trials comparing the effects of dance therapy to non-dance interventions in Parkinson's disease found that dance was especially beneficial for executive function, the processes that help us plan, organize and regulate our actions …

Overall, these studies are compatible with the idea of using dance and DMT in various neurological and psychiatric disorders – such as Parkinson's disease, Alzheimer's disease and mood disorders – as well as in the general population.

'From depression to Parkinson's disease: The healing power of dance' by Adrianna Mendrek, *The Conversation*, 5 December 2019, https://theconversation.com/from-depression-to-parkinsons-disease-the-healing-power-of-dance-123748

Weblink
Find out more about research on DMT and watch the video introducing DMT.

Discuss

1 The author gives their opinion on what they think is one of the main reasons people dance. Do you agree or disagree? Elaborate on your answer.
2 Describe and reflect on dance/movement therapy?
3 What are the three main parts to a DMT session and what do they include?
4 Who might benefit from DMT? Explain how the participant may benefit.
5 In the first video clip in the article, the speaker talks about the opportunity that DMT provides to awaken us. Can you elaborate on what this statement implies?

FACE TO FACE The benefits of dance

Worksheet 10.11

1 As a class, discuss and propose the health benefits of dance for people of your own age group and decide as a class which of these benefits would be appropriate to promote as a part of your dance for fitness session.
2 Decide what needs to be included on the poster and flyers. You will need to include the main health benefits and the information your participants will need to be able to attend your session, including the following details:
   - time
   - place
   - what to wear
   - what to bring (e.g. water bottle and asthma puffer, if necessary).
3 Create posters and flyers to advertise your session and then display them in the most appropriate places around the school to generate maximum attendance.
4 With your teacher's guidance, consider other publicity opportunities to promote your dance for fitness session within the school.

Supervision and safety

For your dance for fitness session to run safely, you will need to organise school permission and staff supervision for your session. Your teacher will assist you with this.

It is important to be proactive about the safety of your participants, so consider and apply the following when conducting your dance for fitness session:

- have a safe dance space, including your own personal space, and a safe surface to dance on
- consider the personal space of others
- wear appropriate footwear and clothing
- bring asthma puffers if required
- participate in a warm-up, cool down and stretch
- ensure you are given appropriate and suitable movements
- keep hydrated.

Rehearsal

You will need to practise your class interval training session to ensure that you can teach this in the dance for fitness session. The participants will be following along for two repetitions of each MILD led by each pair.

As a group, consider the following issues:

- you may want to change the order of the MILDs
- it would be wonderful for each pair to lead their MILD, but if you get stage fright you could consider another role to fulfil on the day
- make sure your continuous music is organised.

UP AND MOVING Cool down and mindful stretch routine

Complete your dance for fitness session with a warm-down and stretch using your MSR.

Be sure your participants know how to perform each stretch following the mindful instructions:

- performed as static and not bounced
- performed in a slow and controlled manner
- lengthened to the point of mild discomfort, not pain
- held for 10 to 20 seconds
- repeated three times
- followed by slow rhythmical movement to proceed into the opposing muscle stretch
- performed on a stable, safe and shock-absorbing surface (not concrete!).

Once you have considered any issues, you are ready for your dance for fitness session.

Remember, your aim is to provide your participants with a fun, safe, physical activity and fitness opportunity. Do this in conjunction with your teacher's supervision. Enjoy yourself and have fun!

1 Describe the methods that were chosen to advertise your dance for fitness session. Name one advantage and one disadvantage of each of your forms of advertisement.

1 Is it important to offer school or community-based dance sessions?
2 Do you think the time and place selected for the dance for fitness session was effective in attracting students to attend? How could you improve accessibility in future sessions?
3 Evaluate the effectiveness of your session.

1 Explore the possibility of developing ongoing dance for fitness sessions within the school.

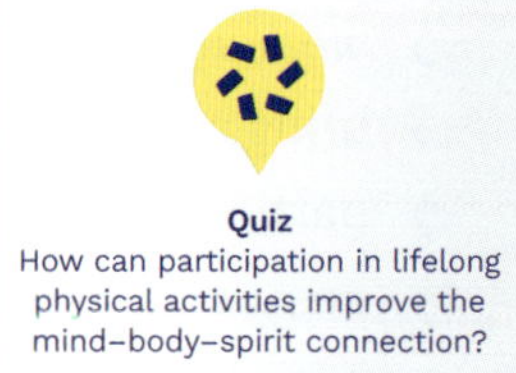

Quiz
How can participation in lifelong physical activities improve the mind–body–spirit connection?

9780170463102

CHAPTER 10 REVIEW

1 Identify and explain how dance can address the components of fitness.
2 Discuss how dance can be used to provide a low-, medium- and high-intensity fitness effect.
3 Compare/contrast the fitness requirements of an elite soccer player with those of an elite gymnast.
4 Explain what is meant by a training version of a line dance.
5 Explain how dance can assist in developing individual and community health-related fitness.
6 Explain the acronym MILD.
7 List three examples of safe dance practices.
8 How did you evaluate the MILD routines based on set criteria?
9 What are the important points in performing a safe static stretch?
10 Define mindfulness and why is it important when we stretch.

INDEX

A

B

C

9780170463102

D

E

O

P

T

U

Z

9780170463102